Agents and Multi-agent Systems in Construction

Also available from Taylor & Francis

Understanding IT in Construction
Ming Sun, Rob Howard   Pb: 0–415–23190–6

Financing Construction
Russell Kenley   Pb: 0–415–23207–4

Innovation in Construction – An International
Review of Public Policies
Edited by: Andre Manseau, George Seaden   Hb: 0–415–25478–7

Cost Modelling
Edited by: M. Skitmore, V. Marston   Hb: 0–415–19230–1

Information and ordering details

For price availability and ordering visit our website **www.tandf.co.uk**

Alternatively our books are available from all good bookshops.

Agents and Multi-agent Systems in Construction

Edited by
C. J. Anumba, O. O. Ugwu
and Z. Ren

Taylor & Francis Group

LONDON AND NEW YORK

First published 2005
by Taylor & Francis
2 Park Square, Milton Park, Abingdon, Oxon OX14 4RN

Simultaneously published in the USA and Canada
by Taylor & Francis
270 Madison Ave, New York, NY 10016

Taylor & Francis is an imprint of the Taylor & Francis Group

© 2005 Taylor & Francis

Typeset in Sabon by
Newgen Imaging Systems (P) Ltd, Chennai, India
Printed and bound in Great Britain by
MPG Books Ltd, Bodmin

All rights reserved. No part of this book may be reprinted or reproduced or utilised in any form or by any electronic, mechanical, or other means, now known or hereafter invented, including photocopying and recording, or in any information storage or retrieval system, without permission in writing from the publishers.

British Library Cataloguing in Publication Data
A catalogue record for this book is available
from the British Library

Library of Congress Cataloging in Publication Data
Anumba, C. J. (Chimay J.)
    Agents and multi-agent systems in construction / C. J. Anumba,
O. O. Ugwu and Z. Ren.
      p. cm.
    Includes bibliographical references and index.
    1. Construction industry – Data processing.
  2. Construction industry – Communication systems.
  3. Intelligent agents (Computer software).
  I. Ugwu, O. O. II. Ren, Z. III. Title.
TH215.A58 2005
690′.0285′63–dc22   2004018905

ISBN 0–415–35904–X (hbk)

Contents

List of figures	vii
List of tables	xi
List of boxes	xiii
List of contributors	xv
Acknowledgements	xix

1 Introduction   1
  C. J. ANUMBA, O. O. UGWU AND Z. REN

2 Intelligent agents: fundamentals   6
  T. ARCISZEWSKI, Z. SKOLICKI AND K. DE JONG

3 Multi-agent systems in construction: an overview   31
  Z. REN, C. J. ANUMBA AND O. O. UGWU

4 Negotiation theories   87
  Z. REN, C. J. ANUMBA AND O. O. UGWU

5 Agent-support for collaborative design   102
  O. O. UGWU, C. J. ANUMBA AND A. THORPE

6 MASCOT: a multi-agent system for construction claims negotiation   162
  Z. REN AND C. J. ANUMBA

7 Specification and procurement of construction products using agents   186
  E. O. OBONYO, C. J. ANUMBA AND A. THORPE

8	Agent-based virtual marketplace for AEC-bidding M. SCHNELLENBACH-HELD, H. DENK AND A. ALBERT	211
9	Agent-based information search and retrieval M. SUN	233
10	Agents for standards processing H. KILICCOTE AND J. H. GARRETT, JR	249
11	Multi-agent-based procurement in the construction material supply chain C. E. UDEAJA AND J. H. M. TAH	272
12	Conclusions O. O. UGWU, C. J. ANUMBA AND Z. REN	310
	Index	321

Figures

2.1	Evolution towards decreased human involvement	8
2.2	Evolution towards the micro-level and increased use of fields	8
2.3	Increased complexity then simplification	9
2.4	Evolution towards increased dynamism and controllability	10
2.5	The average value of all attributes measured for every agent	20
2.6	A Q–Q plot against a normal distribution	21
2.7	Average attributes' values	21
2.8	Swarm agents vs. other agents	23
3.1	Nwana's requirements for agenthood	34
3.2	An agent architecture	35
3.3	The classification of learning methods for MAS	68
4.1	A general learning mechanism	97
4.2	The dual responsive model	98
5.1	Conceptual framework of the general agent management reference model	108
5.2	Process model of information logistics and data flow in steel fabrication	112
5.3	A framework for MAS development in collaborative design	113
5.4	Consumption and production of resources	118
5.5	Costing agent workflow for the task compute cost	119
5.6	ADLIB main Use Case diagram	120
5.7	Scenario interaction diagram for the Use Case: design frame structure	125
5.8	(a) Activity diagram for the Use Case: negotiate design changes. (b) Logical packages of the MAS based on information requirement. (c) Component packages showing the physical partitioning of the MAS. (d) Component packages showing examples of agent resources	126

5.9	Class diagram – structural engineer's viewpoint	129
5.10	Class diagram – steel fabricator's viewpoint	130
5.11	Class diagram – other collaborative design views	131
5.12	(a) Class diagram – object relationships in the ADLIB domain. (b) Class diagram – cost estimation ontology for steel frame structures: fabrication and installation	132
5.13	Joint/collaborative decision-making process in design space	134
5.14	Knowledge-level ontology for agent-based collaborative design	135
5.15	Symbolic representation of collaborative design processes	136
5.16	ADLIB ontology – concrete level domain concepts	136
5.17	Gradient decent algorithm	143
5.18	Flow chart of the ADLIB demo	150
5.19	Sample agent user interfaces	151
5.20	Simulated collaborative MAS design space	151
5.21	(a) Initial design proposals by three specialist task agents. (b) A typical traffic volume in communication between agents. (c) Agents converge on a design solution after automated negotiation	152
5.22	Typical email automated message in a negotiation deadlock	155
6.1	Tripartite negotiation	169
6.2	Bayesian updating mechanism	171
6.3	(a) An example of the contractor's negotiation zone. (b) An example of the agreement zone between agents	172
6.4	The MASCOT process model represented using the IDEF0 negotiation. (a) Negotiate claims. (b) Evaluate offer	174
6.5	(a) The contractor-agent's utility function. (b) The engineer-agent's utility function	176
6.6	The negotiation process	181
6.7	The input and negotiation display windows of MASCOT	182
7.1	The agent road map	188
7.2	The APRON prototype system	192
7.3	Philips Lighting Web site	193
7.4	Web-based search form	197
7.5	AutoCAD-based search form	199
7.6	The spiral model	200
8.1	Overview of different standards	215
8.2	Classification of agents on SiReAM	217

8.3	Architecture of SiReAM	219
8.4	Using data from the LVML to create facts and tasks	221
8.5	Extensions of a generic ZEUS agent's architecture	223
8.6	Architecture of a security agent	224
8.7	Sequence diagram for registration and establishing contact	225
8.8	Co-ordination concepts	227
8.9	Submission of variants and negotiating with subcontractors	229
8.10	Class diagram pkiPerformative	230
9.1	Collaborative construction information network	242
10.1	The overall architecture of SPF	252
10.2	Federation architecture	253
10.3	Agent community	253
10.4	A generic skeleton for SPF Agents	256
10.5	AND-OR graphs in ASPEN	262
11.1	Four-main use of the term supply chain	275
11.2	The supply chain structure	276
11.3	A typical material procurement process	277
11.4	Pictorial specification of alumaco-door task decomposition	284
11.5	SC role model	285
11.6	SC collaboration diagram	286
11.7	Transition diagram of a typical negotiation	288
11.8	Negotiation growth and decay functions of buyer/supplier	288
11.9	Supply and value chains	290
11.10	Multi-agent system ontological framework	291
11.11	ZEUS ontology – door fact description	292
11.12	Multi-agent system framework	294
11.13	Interaction diagram using the information discovery	295
11.14	Human actor federated system framework	296
11.15	MAS-construction material SCP architecture	297
11.16	A screenshot of the ontology realisation	300
11.17	A screenshot of task decomposition for a goal to assembly doors	301
11.18	A screenshot of supplier/buyer negotiation activities	302
11.19	A screenshot of frame supplier 1/buyer accepted negotiation graph	303
11.20	A screenshot of frame supplier 2/buyer failed negotiation graph	303

Tables

2.1	IAs vs. SAs	24
3.1	Agents in historical perspective	35
3.2	Agent qualities	38
3.3	Categories for agent negotiation	50
5.1	Requirements of agent-support in collaborative design and project management	106
5.2	Agents in ADLIB and their responsibilities	118
5.3	Components of the ADLIB ontology	138
5.4	Performative acts in ADLIB domain	141
5.5	Summary of agent transaction report: trail of conversation/negotiation between SDA and other task agents	154
6.1	The contractor's prior knowledge of the engineer's reservation value	176
6.2	The conditional probabilities of the engineer's offer given the contractor's hypothesis	176
6.3	The contractor agent's belief of the probability after the 2nd iteration	179
6.4	The negotiation process	181
7.1	Characteristics of agent technology phases	189
7.2	A snapshot of the lighting bulb information source	194
7.3	Results from questionnaire	202
8.1	PKI-message-types	231
11.1	Low-level role description/decomposition table	302

Boxes

5.1a	Informal specification of the ADLIB prototype system	120
5.1b	Structural Design Agent (SDA) roles	121
5.1c	Costing and Constructability Agent (CCA) roles	121
5.1d	Building Services Agent (BSA) roles	122
5.1e	Safety Advisor Agent (SAA) roles	123
5.2	Messaging syntax for ADLIB agents	139
5.3	Knowledge fragment for agent registration	139
5.4	Knowledge fragment for publishing the design proposal to all participating agents that are registered with the Interface Agent	140
5.5	Knowledge fragment for negotiation protocol	141
5.6a	Code fragments for CCA to handle exceptional cases during negotiation	144
5.6b	Package level resource definition for integrating CCA and WWW infrastructure	145
7.1	Extracting column titles	196
7.2	Extracting product attribute values	196
7.3	The output XML file	197
7.4	Agent dialogue in the procurement module	198

Contributors

A. Albert obtained a Diploma in Civil Engineering from the University of Kaiserslautern in 1994. Between 1994 and 1996, he worked in the Technical Department of Philipp Holzmann AG. He then held teaching and Research Assistant posts at Darmstadt University of Technology from 1996 to 2001, obtaining a PhD in 2001. He worked for Krebs und Kiefer, Beratende Ingenieure für das Bauwesen between 2001 and 2004 and has recently joined Bochum University of Applied Sciences, Germany, as a Professor of Concrete Structures.

C. J. Anumba is Professor of Construction Engineering and Informatics and founding Director of the Centre for Innovative Construction Engineering at Loughborough University, UK. His research work cuts across several fields including intelligent systems, advanced engineering informatics and knowledge management, and has received research funding with a total value of over £15 million from a variety of sources. He has over 350 publications in these fields. His industrial experience spans over 20 years and he is a Chartered Civil/Structural Engineer, holding Fellowships of the Institution of Civil Engineers, the Institution of Structural Engineers and the Chartered Institute of Building.

T. Arciszewski is Professor and Chair in the Civil, Environmental and Infrastructure Engineering Department in the Information Technology and Engineering School at George Mason University. He is currently involved in research on evolutionary design and on infrastructure security. Dr Arciszewski has published more than one hundred and twenty research and technical articles in various journals, books and conference proceedings. He is also an inventor, with patents in structural engineering, obtained in three countries (Canada, Poland, USA). More details are available at http://mason.gmu.edu/%7Etarcisze/index.htm

K. De Jong is Professor of Computer Science and the Associate Director of the Krasnow Institute at George Mason University. Dr De Jong's research interests include evolutionary computation, adaptive systems and machine learning. He is an active member of the evolutionary computation research community with a large number of papers, PhD students and presentations in this area. Further details are available at http://www.cs.gmu.edu/~eclab

H. Denk studied Civil Engineering at Darmstadt University of Technology (DUT) from 1992 to 1998 and undertook postgraduate study in Civil Engineering (leading to a Masters Degree) at Worcester Polytechnic Institute, Massachusetts, USA. Following this, he worked in the Technical Department of Dyckerhoff. He then worked as a Teaching and Research Assistant at DUT and obtained his PhD in 2003. He returned to industry in 2003, joining Ruffert & Partner, Ingenieurgesellschaft mbH, Beratende Ingenieure BDB, VSVI, Germany.

J. H. Garrett, Jr is a Professor of Civil and Environmental Engineering at Carnegie Mellon University, USA. His research and teaching interests are oriented toward CAE system development: ICT for construction and facility management; mobile hardware/software systems for field applications; and representations and processing strategies to support the usage of engineering standards.

H. Kiliccote received his PhD in 1997 from Carnegie Mellon University where he was a faculty member for three years. In 2000, he joined Atoga Systems which was later acquired by Arris International. His principal expertise is in the area of modelling distributed information systems. His past experience includes agent-based distributed frameworks, distributed databases, heterogeneous systems, computer security and representation and reasoning with inconsistent knowledge.

E. O. Obonyo has a strong background in Construction Informatics, Quantity Surveying, Building Economics and Architectural Technology. She recently defended her thesis for the award of an Engineering Doctorate (EngD) degree from Loughborough University, UK and is presently employed by the Balfour Beatty Group as an Innovation Engineer.

Z. Ren obtained his first degree from Tsinghua University in 1989. He then worked for ten years as a design engineer, international contractor and consultant in China, Sri Lanka and Thailand. He completed his masters degree at the Asian Institute of Technology (AIT), Thailand, and was awarded a PhD for research work at Loughborough University, UK. Currently, he is a Research Associate at Loughborough University, working on major European projects.

M. Schnellenbach-Held studied Civil Engineering at Ruhr University Bochum from 1982 to 1988 and obtained a PhD from the same University in 1991. She then worked in industry as a Structural Engineer and Construction Site Manager at Philipp Holzmann AG, Dusseldorf, Germany. She joined Darmstadt University of Technology as a Professor of Structural Engineering, from 1997 to 2004, and currently holds the same position at the University of Duisburg-Essen, Germany.

Z. Skolicki is a student at George Mason University working towards a PhD degree in Computer Science. He graduated from Jagiellonian University in Cracow, Poland, where he received an MSc degree also in Computer Science. Zbigniew's research interest is broadly in the field of Artificial Intelligence. His master's thesis was on graph grammars and qualitative reasoning; currently he is working on evolutionary computations (island models). He would like to understand theoretical and practical limits of computer 'intelligence', but also the methods of building powerful computational systems.

M. Sun is Professor of Construction and Property Informatics at the University of the West of England, Bristol. He has 15 years research experience in the field of construction IT. He has led several major UK research projects on applying IT to achieve integrated and more efficient design and construction processes.

J. H. M. Tah is Professor of Construction Information Technology at the University of Salford. His main research interests are in the application of advanced IT and artificial intelligence techniques to the provision of intelligent decision support within distributed computing environments for designing and managing sustainable construction projects and enterprises.

A. Thorpe is Professor of Construction Information Technology and Head of the Department of Civil and Building Engineering at Loughborough University. His research interests include the applications of advanced IT and communications technologies to construction organisations and the construction process. He has published widely in these fields.

C. E. Udeaja obtained a B. Eng. degree in Civil Engineering. He worked briefly as a Site Engineer and Design Engineer before undertaking a masters degree at Imperial College, London and postgraduate research in advanced IT in construction. On completion of his PhD, he joined the Architectural Informatics Group at the University of Newcastle-upon-Tyne, UK. Dr Udeaja's expertise is in the application of advanced IT and artificial intelligence techniques to the provision of decision support within a collaborative construction project environment.

O. O. Ugwu obtained a B. Eng. (Hons) degree in Civil Engineering from the University of Nigeria, an MSc in Construction Management from Strathclyde University, Scotland and a PhD from South Bank University, London. He did his post-doctoral research at Loughborough University, UK, and joined the Department of Civil Engineering, The University of Hong Kong as a Research Assistant Professor in 2002. His research interests cover theoretical foundations and applications of artificial intelligence, information and communication technology; sustainable construction, and project management. He worked on electrical power, roads and housing projects in industry prior to his academic career. He has published over 60 papers and reports in his research areas, and is listed in the 8th Edition of Marquis Who's Who in Science and Engineering (2005–6).

Acknowledgements

We are grateful to all our contributors who have spent a considerable amount of time putting together their chapters. We acknowledge the contribution of the various agencies and organisations that funded the research projects and initiatives on which this book is based. Mrs Colette Bujdoso and Mrs Sara Cowin (Professor Anumba's Personal Assistants) played a major role in collating the chapters. We are also indebted to our families, whose continued love and support make ventures of this nature both possible and worthwhile.

Screenshots reprinted with permission from Microsoft Corporation.

Professor Chimay J. Anumba
Dr Onuegbu O. Ugwu
Dr Zhaomin Ren
(Editors)

March 2005

Chapter 1

Introduction

C. J. Anumba, O. O. Ugwu and Z. Ren

1.1 Background

Collaborative working in construction remains a major precursor to achieving improved productivity resulting in direct business benefits to construction organisations. It is therefore imperative that researchers continue to explore state-of-the-art paradigms, techniques and technologies that would enhance efficient delivery of constructed products as measured using various metrics such as: quality, cost (life-cycle cost), safety and sustainability. The challenge is how to continuously improve the activities associated with product delivery including design, procurement (i.e. tender appraisal and construction material purchasing), specification, construction, project management and knowledge management across construction organisations.

Multi-agent systems (MAS) are a fast developing information technology (IT), where a number of intelligent agents (IA), representing real world entities, co-operate or compete to reach the desired objectives of their owners. The increasing interest in MAS is because of its ability to provide robustness and efficiency, to allow inter-operation of existing legacy systems and to solve problems in which data, expertise or control are distributed.

The general goal of MAS is to create systems that interconnect separately developed agents, thus enabling the ensemble to function beyond the capabilities of any singular agent in the set-up (Nwana and Ndumu, 1999). This is because the limited knowledge, computing resources and perspectives limit the capability of a singular agent. Thus, if a problem domain is particularly complex, large or dynamic (such as in the construction industry), then the only way it can be reasonably addressed is to develop a number of functionally specific and modular components (agents) that specialise in solving a particular problem. This decomposition allows each agent to use its best knowledge for solving the particular problem. Thus, when interdependent problems arise, the agents need to co-ordinate or collaborate with one another to ensure that interdependencies are properly managed from different perspectives.

Fragmentation is one of the major problems in the construction industry. Different project participants, often geographically distributed, need to co-operate and collaborate to perform various construction activities. This makes the need for effective information and communication technologies acute. Although various ITs such as expert systems, databases and the Internet have greatly enhanced the productivity of the industry, none of them can solve the fragmentation problem of the industry. The additional problems posed by the use of heterogeneous software tools are well known and need to be overcome by the adoption of new approaches such as the use of IA. IA consist of self-contained knowledge-based (k-b) systems that are able to tackle specialist problems and which can interact with one another (and/or with humans) within a collaborative framework (Anumba et al., 2002).

Construction problems also involve open and dynamic problems where the structure of the system itself is capable of changing dynamically. The characteristics of such a system are that its components are not known in advance, can change over time and can consist of highly heterogeneous agents implemented by different people, at different times, with different software tools and techniques. These capabilities require that agents be able to inter-operate and co-ordinate with each other, and to learn from one another and the environment. As a result, MAS have considerable potential to address some of the fragmentation problems of the industry.

1.2 Structure of the book

This book consists of chapters that describe current developments and future directions in the theory and application of IA and MAS in the Architecture, Engineering and Construction (AEC) sector. The book reflects an effort from an international network of construction IT researchers (and related disciplines such as computing), investigating different aspects of agent theory and applications.

The chapters contributed cover different perspectives and application areas. The reported research projects represent significant efforts to harness emerging technologies such as IA and MAS for improved business processes in the AEC sector. Although the chapters in the book can be read in any order, they can be broadly grouped into two areas: (a) agent theoretical foundations in Chapters 2–4; and (b) construction applications – Chapters 5–11. However, the authors also discuss relevant theoretical underpinnings in the respective application areas. Chapter 12 summarises the book and draws a number of conclusions.

Chapter 2 focuses on some perspectives and overview of IA in civil engineering. It explores the evolutionary dimensions of systems in engineering design and describes the metamorphosis from first-generation (k-b) systems to current systems that are underpinned by evolutionary computation.

It also takes a quantum leap into the future evolution of agent-based systems and discusses potential applications of evolutionary computing techniques to improve agent behaviours in problem solving. The chapter differentiates between IA and the emerging sapient agents, and identifies additional attributes for agent classifications. Using these attributes, it proposes a definition of an IA and a taxonomy based on: interaction range, interaction depth, learning and knowledge, structure and quantity. This contributes new dimensions to the evolving classification of agents within the research community.

Chapter 3 discusses theoretical foundations and other fundamental aspects of agent research and development. These include agent taxonomy, co-ordination, negotiation and learning, amongst others. Negotiation theories are discussed in detail in Chapter 4. This recognises the fact that negotiation as well as other aspects such as co-ordination and interaction are pivotal in a MAS environment. The two predominant models and paradigms in mainstream negotiation studies – economic and game-theoretic models are described with a clear articulation of the most appropriate situation for their use.

Chapter 5 presents the first MAS application, which addresses agent-based collaborative design. It includes methodological issues in MAS development, computational modelling of the identified business processes, design knowledge modelling and representation issues that underpin fully automated peer-to-peer negotiation as well as other reactive and deliberative behaviours and characteristics of agents in a MAS design space. The collaborative design of a steel-framed light industrial building is used to illustrate the potential of the resulting prototype system.

The application of agent technologies to construction claims negotiation is the subject of Chapter 6. The chapter presents all the main aspects of this domain including modelling the practical claim negotiation problem, analysing negotiators' roles in an agent-based system, designing the negotiation protocols and developing agent negotiation strategies. A particular agent-learning approach is also integrated into the negotiation mechanism. The chapter also provides a general methodology for the development of MAS negotiation mechanism in construction. This is particularly important because very little has been published on this aspect. A prototype system, MASCOT, is presented which encapsulates the methodology and illustrates the potential of agent-based construction claims negotiation. Examples of the working of the system are also included.

Chapter 7 describes the use of agents to specify and procure construction products. The difficulties associated with aggregating product information from heterogeneous sources are highlighted and the approach adopted to overcome these described. A prototype system, APRON, which facilitates product specification and procurement is presented and its functionality illustrated using an example.

Chapter 8 presents the application of agents in bidding, during which the agents facilitate procurement solutions through the sourcing and purchase of building materials, equipment, and supplies in a virtual marketplace and in compliance with the German standards and regulations. The resulting prototype application demonstrates extensions of the functionality provided by agent development platforms to address construction-specific issues, such as the use of Public Key Infrastructure (PKI) to offer security to agents participating in the tendering process. The chapter also discusses various software models for virtual AEC bidding.

Information search and retrieval is the focus of Chapter 9, which discusses the application of agents to provide a key generic support service that cuts across various application domains. Information search, retrieval and delivery to users using IA provide a solution to the problems of information overload as more context-specific and user-oriented information can be delivered. This chapter describes how this can be done.

Chapter 10 explores the application of agents to standards processing. It demonstrates agent application to standards evaluation and validation, which cut across various domains in engineering design. It starts with a review of related work on standards processing and presents a distributed framework for agent-based standards processing. The proposed MAS tackles scalability and other problems, with each agent using a different representation and reasoning method rather than a single, very general method.

Chapter 11 discusses another application of agent-based systems in procuring construction materials. It gives a detailed analysis of construction supply chain operations including several existing problems. The chapter then identifies how MAS can address these perennial problems and contribute to streamlining the business processes in materials procurement and delivery. It also demonstrates another extension of the contract-net protocol (Smith, 1980), in the context of agent application for construction materials procurement. The chapter focuses on materials procurement as part of the product delivery process at micro-level operations.

In deploying new technology in any sector, it is important to address the technical implementation issues (technology-related factors) as well as non-technical issues (soft or human factors). This view of organisations as socio-technical systems has been the subject of considerable research in mainstream computing (specifically software) engineering. It is also an area that demands focused research by researchers in the AEC sector, because both technical and social factors affect the success or failure of an IT project (as measured by implementation, adoption and depth of usage to enhance productivity through process innovation). In the light of this, Chapter 12 discusses some of the issues raised in various chapters of the book. It highlights the socio-technical dimensions and draws some conclusions on agent applications in construction.

We hope that after reading this book, readers would have a deeper understanding of theoretical and practical dimensions that underpin MAS (and other aspects of intelligent systems) research in construction. We also hope that this would generate new interest in the MAS paradigm, as well as other state-of-the-art IT and computing techniques.

References

Anumba, C. J., Ugwu, O. O., Newnham, L. and Thorpe, A. (2002) 'Collaborative design of portal frame structures using intelligent agents', *Automation in Construction*, 11(1): 89–103.

Nwana, H. and Ndumu, D. (1999) 'A perspective on software agents research', *The Knowledge Engineering Review*, 14(2): 125–42.

Smith, R. G. (1980) 'The contract net protocol: high level communication and control in a distributed problem solver', *IEEE Transactions on Computers*, C-29(12): 1104–13.

Chapter 2

Intelligent agents
Fundamentals

T. Arciszewski, Z. Skolicki and K. De Jong

2.1 Introduction

The progress of our civilisation is mostly driven by the changes in science. First, in the era of agrarian societies, the progress was driven by the advances in agriculture. Next, in the industrial societies, the progress was a result of improvements in manufacturing. Today, we are living in the post-industrial societies and their evolution results mostly from the advances in information and knowledge processing. For this reason, our times are often called the era of Information Technology Revolution, as a way of describing the complex synergistic process of changes occurring in the society and within engineering. These changes are driven by new sciences, technologies and tools, all related to information and knowledge processing.

In Civil Engineering, the Information Technology Revolution is reflected in the emergence of knowledge-based (k-b) systems during the last 20 years. Such systems are intended for various decision-making purposes including design, planning, maintenance of infrastructure systems, etc. Recently, there is a growing interest in a new category of such systems called 'Intelligent Agents' (IAs). IAs constitute the most advanced form of (k-b) systems and many researchers and practitioners expect them to become the dominant form of (k-b) systems, which will soon change civil engineering practice entirely. At the moment, the notion of IAs is still not fully understood, both within Computer Science and Civil Engineering communities. Therefore, various definitions of IAs are introduced and discussed in Section 2.3. Our definition is thoroughly explained and justified in the same section. However, it is given here in order to provide sufficient context for the introduction. It is as follows:

> An intelligent agent is an autonomous system situated within an environment. It senses its environment, maintains some knowledge and learns upon obtaining new data and, finally, it acts in pursuit of its own agenda to achieve its goals, possibly influencing the environment.
> (Skolicki and Arciszewski, 2003a)

The process of changes in Civil Engineering within the domain of (k-b) systems can be understood as a process of evolution. As such, it can be considered in the context of Directed Evolution (Clarke, 2000). This term reflects an emerging Engineering Science that deals with the analysis of the evolution of engineering systems over periods of time. The basic premise of Directed Evolution is that the evolution of engineering systems is driven by objective patterns of evolution. Eight such patterns have been identified as a result of studies of the evolution of many engineering systems developed in various countries over long time periods. In the case of (k-b) systems, four patterns seem to be particularly relevant. Therefore, they will be used to explain why IAs are emerging and why this process is so significant. These relevant patterns of evolution and implications on the evolution of (k-b) systems are discussed later.

2.1.1 Evolution towards decreased human involvement

When the evolution of an engineering system is considered over a period of time, it can be easily observed that the required human involvement is systematically decreasing. For example, when a car is considered, every year improvements are made to reduce human involvement in the operations of the car's engine, brakes or steering. Similarly, each new computer system design attempts to simplify and reduce human involvement.

In the case of (k-b) systems, this evolution pattern means that with time their use will require less and less human input. The last 20 years of evolution of (k-b) systems have already confirmed the validity of this evolution pattern. When the first (k-b) systems were developed they were written in symbolic languages (LISP or Prolog), their user interfaces were poor or non-existent, and their use required tremendous effort. Gradually, various shells for building (k-b) systems with excellent user interfaces have been developed, and their introduction significantly reduced the amount of effort required to use such systems. Obviously, the natural trend is in the direction of autonomous systems. That may mean the reduction of human involvement in the case of the majority of (k-b) systems and the emergence of highly autonomous systems. This second group may become gradually dominant. The systems in this group would require only very limited human involvement to perform their functions and that involvement most likely will be on the level of meta-rules representing abstract knowledge. Ultimately, the pattern of Evolution towards decreased human involvement means in this case that IAs are the natural goal of evolution of (k-b) systems and that their emergence should be simply expected. The gradual shift from systems totally controlled by humans to systems interacting with humans and ultimately to autonomous systems is illustrated in an abstract way in Figure 2.1.

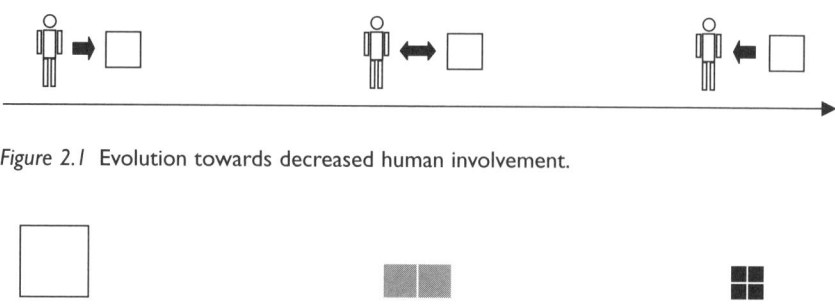

Figure 2.1 Evolution towards decreased human involvement.

Figure 2.2 Evolution towards the micro-level and increased use of fields.

2.1.2 Evolution towards the micro-level and increased use of fields

This pattern of evolution can be explained discussing the evolution of computing devices. In this case, the evolution began with the mechanical computing devices called 'arithmometers'. They were followed by various electrical computing devices called 'electric calculators' in which the individual mechanical components were replaced by electrical components powered by electricity and the introduction of a field (in this case – an electrical field). The next stage of evolution was marked by the introduction of digital/electronic computing devices called 'computers'. In computers, the mechanical or electrical components are substituted by tubes and later by transistors. The size of such components is significantly smaller than the size of the mechanical or electrical computing devices. Also, the computing operations are moved from the macro-level of arithmometers to the micro-level of the individual transistors. In addition, an electromagnetic field is added following the pattern of the increased use of fields. This manifold evolution is represented in Figure 2.2, where the use of micro-level is shown as the diminishment of objects, and the increased use of fields is represented by the changing colour.

2.1.3 Increased complexity then simplification (reduction)

When a given engineering system is developed and first built, it usually contains only the most necessary components, which are required to provide the most desired function. With time, however, more additional functions are added to increase the attractiveness of a given product. That process results in the increased complexity of the system. At one point of the evolution, a complexity threshold is reached and a qualitative change occurs. The system

Intelligent agents   9

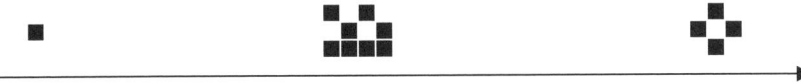

Figure 2.3 Increased complexity then simplification.

is significantly simplified and many secondary functions and related components are simply eliminated. This process is illustrated in Figure 2.3.

The evolution of the Palm personal data management device is a good example here. The first several models offered more and more functions followed by the introduction of a line of basic and simplified models with only very few major functions. A similar evolution of IAs can be observed. The first IAs were relatively simple, but they gradually grew in complexity. Finally, the concept of a swarm emerged, in which instead of a single complex agent a large number of simple agents is used. Considering the patterns of evolution, it might be expected that at least two lines of evolution would evolve. The first one will contain ever more complex and highly specialised agents working independently while the second one will deal with many simple agents intended only to provide a single function each. Their configurations, however, will be created to meet demands of their environment and tasks.

2.1.4 Evolution towards increased dynamism and controllability

Increased dynamism of a system refers to its greater flexibility that may allow it to operate in a variety of environments or to provide functions even outside the initially intended functional envelope. This increased dynamism should be associated with the increased controllability, that is, the ability of the user to control its behaviour in a more precise way. The increased dynamism may be described as an evolution from a static (time independent) system to a system with changeable components and finally to a system with variable components. A good example is the evolution of structural systems. First, they were built as unique systems with the individual components designed and fabricated for specific locations. The next stage was the introduction of mass produced structural shapes, which can be used in a given structure to redesign it and to exchange various original members to meet additional functional requirements. Finally, in the case of intelligent structures, the individual members can change their behaviour and adapt it to the changing environment, for example, during an earthquake or a hurricane. At the beginning of the described evolution, the user had no means to control behaviour of a structural system. In the second stage of this evolution, he/she had some indirect control while in the last stage the structural system is fully controllable and the user can change its behaviour at any time (Figure 2.4).

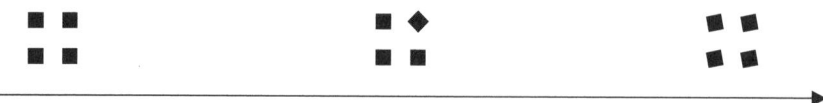

Figure 2.4 Evolution towards increased dynamism and controllability.

In the case of IAs, the described evolution towards the increased dynamism and controllability is already occurring. The first IAs were built for specific applications and all their components were custom-designed and implemented. Next, IAs with universal components emerged, and now the concept of adaptive IAs is the subject of research. The issue of the degree of possible adaptation of agents is still a research issue. Such adaptive agents will respond to changing environments through the transformation of their architecture and through the modifications of their individual components. Such agents can be described as self-developing systems and they will offer great promise in engineering applications, if ever built. Their design can be compared to integrated design in which both the design concept (agent's architecture) and its detailed design (design of the individual components) are considered and optimised for a specific class of applications.

2.2 Agent-based systems

After we have seen a general picture of evolution of IAs, in this section we step back and look more thoroughly at what agents are and where they are used. This forms a basis for a more formal analysis of IAs in Section 2.3.

2.2.1 Agents

The term *agent* was first proposed at MIT by John McCarthy and Oliver G. Selfridge in the 1950s (Bradshaw, 1997). Agent is usually understood as a system or device acting on behalf of another one. For example, a software agent is a computer system to which a user can delegate tasks. Agents are expected to operate continuously over relatively long time periods and to behave at least in a semi-autonomous way. Agents do not wait for instructions; they initiate actions and guide the interaction. An agent is seen more as an organism or an autonomous entity with which some interaction is possible, whereas a program is more simply a tool. Actually, most of an agent's features can be implied from this description. Being a self-contained unit, an agent must not be fully dependent on the external control, but rather take initiative on its own. Agents perform different actions like querying, asking, sensing to find the information they need and undertaking actions to alter the environment or communicate to other agents.

Whereas an ordinary program would probably have a sequence of instructions, an agent would rather have a constant loop of possibly 'intelligent' interactions with the environment. Not surprisingly, many programs that are not called agents by their authors, like some UNIX daemons or computer viruses, might be counted into this category (Franklin and Graesser, 1996). It is also possible to put a non-agent program inside an agent 'wrapper', which would control the program. Such procedure is sometimes called *agentification* (Jennings and Wooldridge, 1998b).

As we mentioned in (Skolicki and Arciszewski, 2003a), how an agent differs from a regular program can be seen not only at the technical, implementation level, but also at the social level. This idea was earlier noted by Bradshaw in the book on Software Agents (1997), where he differentiates between two definitions of agent – by ascription (agents are the programs which behave like agents) and by description (agents have particular attributes). Agents are programs that are perceived to work independently. The perception of agents may depend on the environment in which they operate and the functions they perform. For example, even a standard document editor could be called an agent. Once we turn it on it works in a loop interchanging actions with the user. When the user hits a key the editor decides which font to use, where to put the letter, etc. and displays the letter. This encourages the user to type consecutive keys. Of course, such an agent is not particularly intelligent. A related example of a spell checker was given by Franklin and Graesser (1996). We will address the issue of an agent's intelligence soon.

Several researchers (Bradshaw, 1997) claim that a system should be called an agent, if there is no simpler explanation for the behaviour of the system. Theoretically, we could see every system to be an agent. However, many of them, like simple electric circuits or straightforward batch programs, can have simpler, low-level explanations. On the other hand, there are some other complex systems such as people, robots or animals, which are naturally described as agents, because analysing low-level mechanisms may be too difficult to predict their behaviour.

One important characteristic of an agent system is that its behaviour is usually difficult to predict without actually running the system, due to the changing interaction with the environment and with the other agents. In this context, an agent system can be seen as a complex system – a system, the behaviour of which cannot be predicted by a simpler model. However, this partial unpredictability of agent systems makes the usage of them questionable in safety-critical applications. Therefore, agent systems are often implemented as systems suggesting the solutions, but not actually automatically performing the actions.

Agent systems have an important property of modularity. Wherever it is difficult to maintain the whole system as one unit and it is more straightforward to design it as a group of co-operating subcomponents, the agent approach seems to fit the place. In some way the agent approach provides the next level of abstraction, in the same way as procedures and objects provide it for other systems (Meyer, 1998).

Also, for systems, which cannot be totally integrated, agent approach can be an effective solution. The activeness of agents and their ability to communicate among one another may serve as a basis for exchanging data and knowledge between different systems. The agent platform serves to distribute data, control, expertise and resources.

Designing agent systems requires proper division of roles among agents. These can be a difficult task in itself, but becomes even more complicated when we have to take into consideration existing programs as building blocks. Also, more granular division is more difficult, because agents become simpler and the interaction between them becomes the main source of system's intelligence. Such interaction is difficult to design and again, in many cases only the behaviour of a given particular system can be observed. Therefore, as mentioned in Section 2.3 of this chapter, evolutionary computation (EC) plays a major role in designing finely grained agent systems.

2.2.2 A review of agent-based systems

Agents have been applied in a variety of domains, resulting in a wide spectrum of systems. Wooldridge divides applications of agents into two broad categories. The first ones are *distributed systems*, the others being *personal software assistants*. We are providing a short description of agents in selected areas, following Jennings and Wooldridge (1998b) and Wooldridge (2002).

Applications in workflow and business process management make sure that the organisation is working smoothly, in terms of data processing and managing the flow of documents. An example here is the ADEPT system, developed for British Telecom (BT) (Jennings *et al.*, 1998). Agents in the system used legacy components already existing in the company. The system supported negotiations between departments as well as among individuals.

Agents can be used for distributed sensing. The initial work was done by Lesser and Corkill (1983). The idea is to gather data from distributed sensors. There can be different types of sensors (acoustic, radar, visual, etc.). If the sensors co-operate, they may respond more accurately, more reliably and faster to the environment. An interesting possible application of a distributed system is in security systems where the sensors/agents embedded into a structure (e.g. a tall building, a bridge or a dam) observe the object and inform about any misbehaviour.

Another application for agents is for retrieving information. With the present day size of databases and internet resources in the first place, we are overwhelmed with the amount of data. Agents can serve several roles. First of all, they can make the process of retrieving information more approachable for the user, since they can communicate in a more natural way. An agent given a task of finding information on a certain topic would start searching different knowledge bases and resources and, at the end, would present a consistent summary to the user. Second, agents can be used for a personalised search. They can be told to customise the data to the needs of the user. Mail filtering agents would fall into this category, as well. Furthermore, agents can be autonomous and intelligent enough to understand what the user really wants (as opposed to a mechanistic search based on the word the user is entering), maybe by using the context of what the user was doing before. Such understanding of the meaning of the query would produce specialised, possibly much more relevant results. Finally, many agents could co-operate with each other, exchanging information, because no existing agent can possibly be an expert and have access to resources from every domain. Sun (2004, Chapter 9 in this book) discusses the use of agents for information search and retrieval.

Electronic commerce agents specialise in supporting customers in online trading. Again, instead of leaving the user alone with understanding the content of some shop or auction pages, they try to 'understand' the intentions of user and guide him along the process of buying a product. The two main areas in Electronic Commerce are agents specialising in shopping and agents helping with auctions. The first ones may contact many places and negotiate to get the best deal (Doorenbos *et al.*, 1997). The latter ones would ask the user for information about the minimum price, maximum bid, etc. and bid in the name of him or her (Chavez and Maes, 1996).

Agents used in human–computer interaction are designed to make the task more natural, easier and possibly high-level for the user. Instead of finding out the details of achieving some simple but time-expensive task, the user delegates the task to the agent, which takes over the responsibility for carrying it out, as described by Negroponte in his book (1995). Also such agents should rather behave proactively and co-operatively (but not bother you too often!), than wait for all instructions coming from the user. To do so, they may need to model the behaviour of the user, to understand and predict his or her needs.

Virtual environment is a natural place to use agents. Regardless of whether it is a virtual reality system for designing or a 3D game, we would expect to meet agents that behave in a way similar to humans. This task is not trivial, since it may require simulating emotions, which for obvious reasons is difficult to achieve. Such agents could also be used for an interactive cinema. Finally, dynamic and adaptable virtual world can be created using agents as the basis for all elements of the world (Smith *et al.*, 2003).

Modelling the behaviour of societies and observing complex interactions within such systems is another area of application for agents (Moss and Davidsson, 2001). Depending on how sophisticated is the model of every agent, we can model different aspects of societies, possibly also including beliefs and aspirations of single individuals. Such social models can be useful to study the expected outcome of new policies and laws as well as behaviour of people in critical situations like escaping from a building during an emergency situation. Other applications of agents include such various domains as process control and manufacturing, air traffic control or patient monitoring and health care.

Finally, Ugwu *et al.* (2003) report ongoing research investigating agent application for on-site logistics management. The project is developing multi-agent systems (MAS) framework for the on-site logistics material management in demonstrating construction supply chain. Their paper espoused the contextual issues related to agent-based collaborative working, highlighting the need to solve the problems engendered by the complex interactions that characterise distributed collaborative working in the construction sector. Thanks to the usage of agents, it is possible to see construction organisation as an open system. The research investigating agent frameworks for an on-site logistics also builds on the previous Agent-based Collaborative Design of Light Industrial Buildings (ADLIB) project (Anumba *et al.*, 2003, see Chapter 5 for detailed discussion of the ADLIB research project).

2.3 Intelligent agents

Taking decisions and being autonomous requires from agents to somehow 'understand' the situation. Therefore, every agent, except maybe for very rudimentary ones, possesses some amount of knowledge, some ability to learn or adapt, or change the behaviour. Agents that are capable of behaving in an 'intelligent' way are called IAs. Of course, no one has one final definition of what intelligence is, in general. No wonder then, that the definition of an IA is a bit blurry.

A standard test for intelligence of programs was given by Alan Turing (1950). The simplified version of the test is done as follows: Imagine that you can communicate with a real person through a terminal. Similarly, you communicate with the program being tested through the second terminal. If you cannot distinguish between the person and the program, then the program is thought to be intelligent. To date, no computer program (nor agent) passed this test.

It is currently believed that to achieve more intelligent behaviour from agents, they have to have knowledge about various common-sense objects, to be able to interact easier with users and to understand more unstructured or informal commands. There have been many attempts to build a system of concepts, called an *Ontology*, for agents. One of the most well-known examples is the CYC project, started over 20 years ago by Lenat (1995). The project aims at entering the entire everyday knowledge into computers. However, only recently the Cycorp Co., which works on developing the (k-b), has started releasing the product to the public.

An IA, called *Disciple*, has been developed at George Mason University (Tecuci, 1998). The agent builds ontologies for the domains about which it is supposed to reason. The reasoning is based on a set of rules, which can be adjusted, mostly by generalisation and specialisation operations.

2.3.1 Evolution of definitions

As we have seen, the word *agent* has been used by many researchers in different contexts and various levels of programs' complexity are required to call them agents. However, no matter how diverse agent systems are, they share some common characteristics. A trial to give a consistent definition and categorisation of agents was done by Franklin and Graesser (1996).

We can identify different categories of agent complexity in the history of computing. Some programs can barely be called agents, since their behaviour is very limited and the tasks which the user can delegate to them is strictly restricted. They are driven by the user and the only decisions they take are basically about the technical details of the task which we pose them with. Nevertheless, they serve the goal, and thus we can count them into the agent family. A good example here would be *gopher* (Jennings and Wooldridge, 1998b). If it happens that a program must have a stronger autonomy and must decide about more conceptual aspects of a given decision situation, then we have the next category of agents. They still serve as a tool for a specific goal. When we think about a classic example of air ticket booking, we see that choosing a route may require more complex algorithms and nested loops of queries. The third category of agents would consist of agents that are able to set the goals on their own and which would truly be able to take initiative when interacting with a user. Very often such agents would have to learn and/or adapt in order to suit the task.

Whereas many autonomous and reactive systems may be considered to be agents, only the most complex ones (like the third category above) would be candidates for the subclass called *IAs*. As mentioned in the previous section, such agents not only have to interact, but the interaction must be done in a way that looks intelligent. Therefore, the reasoning of IAs is not based on a simple look-up table, but rather involves complex mechanisms and usually requires the agent to build some internal model to understand the stimuli and infer proper reaction. For example, personal assistants have to 'look over shoulder' of the user (Casasola and Gauch, 1997), to generate a proper hint for the next actions and learning/adapting seems very important for them. In the case of 'software agents', the word *intelligent* is often omitted, since intelligence can be seen as a necessary feature of every software agent (Morales-Morell, 1997).

It is worth noting that as Computer Science and IT progresses, tasks which were previously considered to require human intelligence, are now considered to be easy for computers. Examples here are voice and handwriting recognition, chess playing, shopping, advising, etc. The 'horizon' of truly intelligent systems has moved further and it seems that it can be a Holy Grail of Artificial Intelligence, which we search but will never find.

Our understanding of IA is presented in Skolicki and Arciszewski (2003a) and stresses the need of an agent's ability to learn. Intelligence, adaptation and learning are strongly related and it seems to us that they can hardly exist without each other. This is to some extent confirmed in the paper, where we showed that most of IAs we analysed had a learning ability. Our definition of IA is as follows:

> An intelligent agent is an autonomous system situated within an environment. It senses its environment, maintains some knowledge and learns upon obtaining new data and, finally, it acts in pursuit of its own agenda to achieve its goals, possibly influencing the environment.
>
> (Skolicki and Arciszewski, 2003a)

2.3.2 Attributes describing intelligent agents

Depending on the application, number of agents, environment, function and other factors, agents of very different complexity exist. Some of them are simple and possibly work as components of a large system; some of them are complex, trying to come up with an intelligent behaviour using advanced methods of reasoning. Many of the complex agents are able to learn and adapt. They have some kind of knowledge repository, which helps them understand the situation and take actions. The agents which are able to have a mental state or emotions are called *strong* agents. Others are called *weak*. Although there were some attempts to build agents which would have emotions, on the basis of so called belief, desire, intention (BDI) agents, there was not much success in actually imitating emotional behaviour (Padgham and Taylor, 1997; Velásquez, 1997).

As we mentioned (Skolicki and Arciszewski, 2003a), agents can be characterised by the following features:

- *Autonomy, decision-making* – an agent should be at least partially independent, have some goals (internal state) and be able to make decisions (Jennings and Wooldridge, 1998a; Leitao and Restivo, 2000).
- *Proactivity* – means that agent should make suggestions as of the next actions (Jennings and Wooldridge, 1998a).
- *Social ability, communication and co-operation* – an agent is able to communicate with other agents or humans and interact with them to achieve a goal, possibly in a mixed-initiative way (Jennings and Wooldridge, 1998a; Leitao and Restivo, 2000).

- *Monitoring, perceiving* – an agent should sense and observe the environment within which it is located (Russel and Norvig, 1995; Leitao and Restivo, 2000).
- *Acting, operational control* – an agent should perform chosen actions which may change the environment or agent's state (Russel and Norvig, 1995; Leitao and Restivo, 2000).
- *Reactivity, responsiveness* – an agent should respond to the changes in the environment (Jennings and Wooldridge, 1998a). This may encapsulate the two previous features.
- *Knowledge, ontology* – an agent needs to have an understanding of the environment to perform the task (Nwana and Ndumu, 1999). It can use some kind of ontology (or a concept map) to store and maintain its knowledge.
- *Learning, adaptability* – an agent should enhance its behaviour during its life (Tecuci, 1998; Nwana and Ndumu, 1999).
- *Continuity* – some researchers require an agent to be able to run a process continuously (Franklin and Graesser, 1996).
- *Mobility* – others require an agent to be able to change its location (Jennings and Wooldridge, 1998c).

Of course, we do not require a system to have all these features to call it an agent. However, most of the agent systems will possess a substantial subset of these characteristics.

At this point of understanding of IAs, we have developed a summary of features which agents could possess and have proposed a set of attributes for agent description. The set consists of 27 binary attributes. We aimed at having a large number of these attributes to have the maximum discernability power. The attributes may sometimes overlap and they possibly do not cover all possible dimensions of agent systems. However, we have tried to create our set in a fairly extensive way. The attributes were based on characterisation of agents given in the literature, again in a more extensive research done by Jennings and Wooldridge (1998a–c). We have arbitrarily grouped the attributes into several classes, depending on the types of attributes. A list of attributes is provided here together with the explanations, as it appears in Skolicki and Arciszewski (2003a).

Interaction range (sensing and acting)

1. *Information* – this attribute specifies whether agents store information locally or have access to some common, shared memory.
2. *Perspective* – this attribute specifies whether agents can only observe a part of the environment (usually close to them) and/or communicate only with some other agents or can probe the whole environment and all other agents.

3 *Resources* – this attribute denotes whether every agent has its own resources and is independent of the others or the agents share some resources (like space, for example).
4 *Goal* – agents can have individual goals or they can have one, common goal.
5 *Autonomy* – this attribute specifies whether a given agent acts on its own, without external stimuli or it interacts with humans.
6 *Competitiveness* – this attribute specifies whether agents co-operate or compete with other agents to reach the goal.
7 *Mobility* – in some environments, the agents can move or change their perspective through communication with other agents. This attribute takes a 'mobile' value in this case. Otherwise, it is 'steady'.

Interaction depth (reasoning)

8 *Roles* – this attribute denotes whether the agents have some specific tasks assigned a priori or assigned while the system is running.
9 *Reaction* – this attribute describes whether the agents react immediately and in a simple way (they are reactive) or they perform more sophisticated actions (deliberative).
10 *Speed* – if the agent must respond in a specified time to the external real-world input, it must operate in real time. Otherwise we assign simulated time value.
11 *Empathy* – this attribute denotes whether the agents are complex enough to try to model other agents (and predict their behaviour).
12 *Transparency* – this attribute determines whether the internal state of the agent, possibly with its line of reasoning is visible outside or whether the agent communicates only predetermined states.
13 *Commitment* – some agents postpone obtaining a detailed solution or keep several solutions from which to choose later. In such cases the attribute takes the least commitment value. Otherwise, most commitment is assigned.

Learning and knowledge

14 *Language* – this attribute specifies whether the agent learns or modifies the language while running.
15 *Trust level* – this attribute specifies whether contradictions are possible in the system and whether the system has some way to deal with it.
16 *Ontology usage* – this attribute denotes whether the agents use some kind of ontology to represent concepts from the domain. If the agents use some complex database but it does not resemble an ontology, the 'simple' value is assigned.
17 *Adaptability* – this attribute describes whether the agents or the system adapt or learn something.

18 *Initial familiarity* (with other agents) – sometimes the relations and communication paths between agents are set before running and are stable. In this case we assume agents 'know' each other initially. Alternatively, the agents can dynamically discover and interact with other agents.

Structure

19 *Stability* – the attribute specifies whether the agents have a stable architecture or whether they can reconfigure at runtime.
20 *Layers* – sometimes agents are hierarchically organised and thus constitute a higher order system, often with top agents delegating the tasks to the underlying ones. Also, in some cases the agents can communicate through higher level agents serving as managers. Such cases determine the hierarchical structure. In other cases, the agents are all equal, negotiating with each other on a peer level, making a one-layer system.
21 *Diversity* – if all agents are the same (at least initially), they are homogeneous, otherwise they are heterogeneous.
22 *Reusability* – if parts of the agents are easily reusable for other purposes (being portable modules), or even whole agents can easily serve different tasks, we assign reusable value to the attribute. Otherwise, we have a one-purpose system.
23 *System flexibility* – if it is easy to add agents or components to the system, we call such a system 'open'. Otherwise, it is 'closed'.
24 *Agent flexibility* – if the agent architecture is easily expandable (i.e. components can be easily added or dropped), the agent is 'open'. Otherwise, it is 'closed'.

Quantity

25 *Amount* – this attribute specifies whether there is one or more agents in the system.
26 *Swarm* – this attribute determines whether there is a large number of agents that they may be considered as a swarm or whether there is only a limited number of relatively easily distinguishable agents.
27 *Emergence* – in some cases a large number of relatively simple operations sewn together may result in a higher level, macroscopic behaviour. The Emergence attribute attempts to capture this feature.

2.3.3 Classification of known intelligent agents

Being primarily interested in IAs for design, we have recently analysed all the agents that were described in the proceedings of the First International Workshop on Agents in Design. The workshop was a natural consequence of recent interest in the possible usage of agents in engineering and it hosted papers by researchers from various universities from all over the world (Gero and Brazier, 2002). Because it was the first conference on agents in design and there was no any strong bias towards any particular subarea, we saw a wide range of topics. This diversity seemed perfectly suited to the aim of analysing global trends in agents for design. We tried to see the line of evolution of agent systems and to predict their future development. We cannot prove formally any statistical properties of the sample and thus the results cannot be taken to be 100% correct. However, we have a feeling that our results are a good indication of the current direction of evolution of agents.

The total number of agent systems analysed was 17. Each case was studied thoroughly and we arbitrarily assigned binary values to all the attributes presented earlier. The rule of thumb was that a simpler behaviour was represented by '0', whereas a more complex one was described by '1'. This rule was however inverted for 'speed', 'amount', 'swarm' and 'emergence' attributes. The goal was to determine if we can observe a clustering of simple and multiple agents as one group (swarm agents) vs. big, slow, complex and singular agents as the second group. We encountered several such cases when there was not enough information or the information was too vague to be able to decide on the correct value. In these situations we assigned '0.5' as the value.

We performed several steps of analysis first, for every agent we calculated the average attribute value (across all the attributes). We wanted to see whether we could discover a distinguishable difference between swarm and non-swarm agents. However, it was not the case and in fact the distribution of these averages was close to a Gaussian one (Figure 2.5). The R^2 value of 0.97 in the Q–Q plot (Figure 2.6) confirmed that statement.

In the second step, for each attribute we calculated an average value for all the agents, which is shown in Figure 2.7. This time we wanted to see if any attribute fell statistically far from 0.5. We identified such cases, which

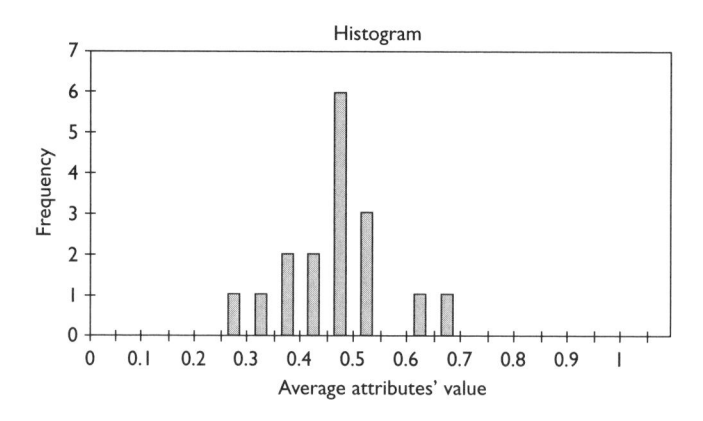

Figure 2.5 The average value of all attributes measured for every agent.

Intelligent agents 21

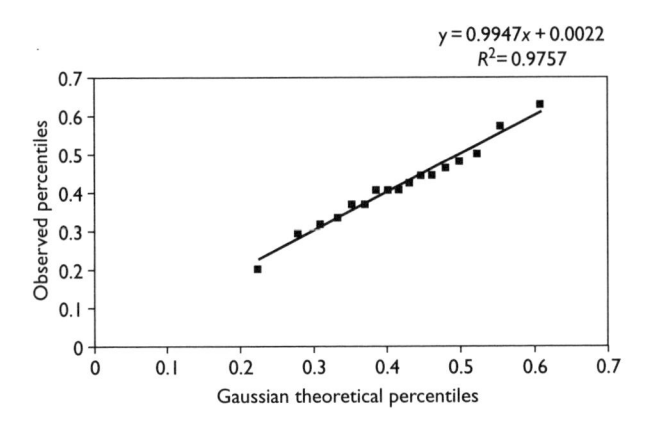

Figure 2.6 A Q–Q plot against a normal distribution.

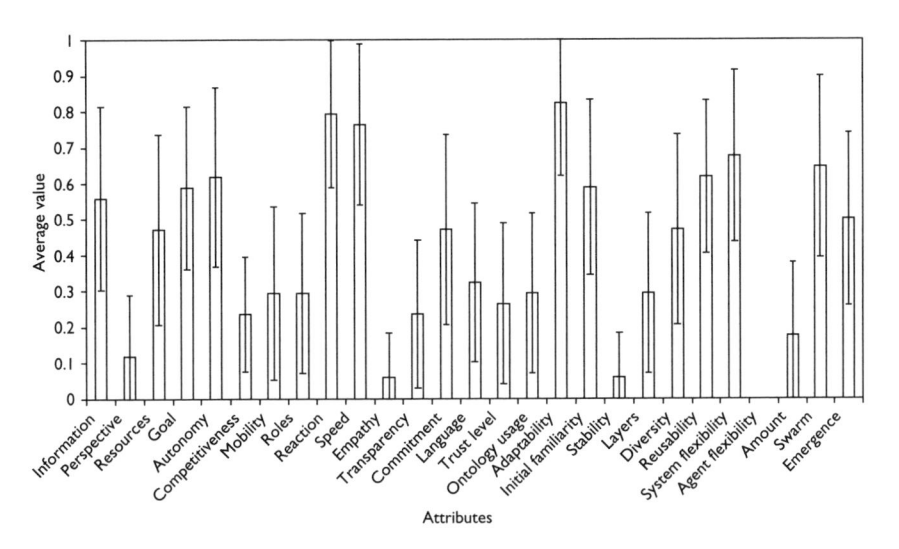

Figure 2.7 Average attributes' values.

enabled us to make the following statements about the current stage of agents in design (Skolicki and Arciszewski, 2003a):

a   A low average value of *Perspective* attribute suggests that in most cases the agents are built to act locally with limited ability to probe the environment and with high level of encapsulation. This is not surprising because that is consistent with the nature of agent-based systems.
b   A low value for the *Competitiveness* attribute reveals that the agents mostly co-operate rather than compete. New research in competitive co-evolutionary algorithms could raise the value.

c   A high level for the *Reaction* attribute shows that the agents are modelled to have a relatively high level of sophistication. Research in swarm environments could significantly change the situation.

d   Probably because of the reasons mentioned in the previous statement, the agents usually work in a 'simulated time' (*Speed* attribute), meaning that they do not have to immediately react to the input from the real world. Transferring the systems into robotics would probably change this situation.

e   A very low value of *Empathy* suggests that although the agents use elaborate techniques, they do not put much effort into modelling other agents, probably because the benefits are smaller than the costs. With growing complexity of systems, however, this feature could be required later.

f   Similarly, it seems that the agents do not show their internal state outside (*Transparency* attribute). This may be the result of encapsulation and the nature of agent systems.

g   Most agents assume the information as true (*Trust level* attribute). This is not the case, however, in the real world and also in robotics. Hence, research in this direction should be useful.

h   A high level for the *Adaptability* attribute shows that most agents or systems acquire knowledge while running. This is what we claim is necessary for an IA, and therefore this value is not surprising for us (note that this attribute has the highest value) and supports our previous statements.

i   A very low value for *Stability* and *Agent flexibility* shows an interesting direction for research. One can easily imagine agents evolving in runtime as they adapt to the environment (the first attribute) or adding and dropping components as needed – either before the system starts or also during runtime.

j   In majority of agent systems considered more than one agent was used (*Amount* attribute). We believe that it is the right direction and consistent with the philosophy behind agent systems evolution in which the interaction plays a major role. It is worth noting that even in the case of a single agent, the user usually plays the role of a second agent and thus the interaction is also important.

Finally, we managed to discover some differences between swarm and non-swarm agents, by separating all the agent systems into two groups according to the *Swarm* attribute and calculating the averages of particular attributes. This comparison is shown in Figure 2.8. Although not all the observations could be confirmed statistically, the general pattern of differences was visible. Swarm agents seem to be characterised by the following points:

- Swarm agents always act on a more local perspective
- They tend to share resources to a greater extent

Intelligent agents 23

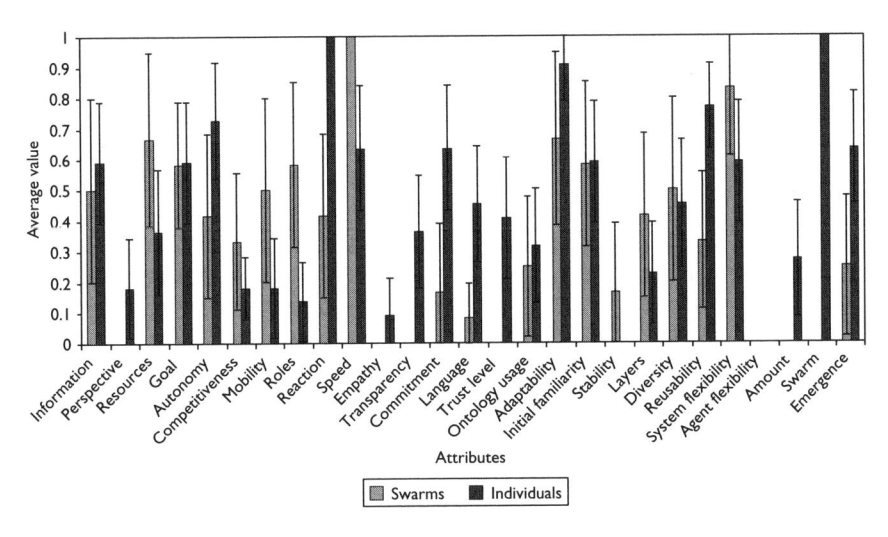

Figure 2.8 Swarm agents vs. other agents.

- They have less autonomy
- They compete more
- They are more mobile
- They tend to discover roles dynamically in runtime
- They are simpler (more reactive)
- Surprisingly they tend to act in a 'simulated time'
- They are less transparent
- They have a fixed language
- They assume the information to be true
- They seem to be less reusable.

2.3.4 Sapient agents: seven definitions

As research on agents becomes more mature, one can observe that even IAs show various levels of complexity; some have deeper mechanisms of reasoning or they 'understand' the environment better by using a knowledge-intensive approach. Their complex and difficult to predict behaviour can be attributed to their 'wisdom', which is different from knowledge in the form of a simple collection of decision rules, driving other, less sophisticated agents.

During a recent conference KIMAS'03 a special session was devoted to the so-called 'Sapient Agents' (SAs), with the goal of defining them more precisely. In this session, the authors presented a paper in which seven ways to distinguish between SAs and IAs were proposed (Skolicki and Arciszewski, 2003b).

Table 2.1 IAs vs. SAs

No.	Context	Intelligent agents	Sapient agents
1	Knowledge representation	Only decision rules	A knowledge system
2	Emergence	No recognition of emerging patterns	Recognition of emerging patterns
3	Exploitation/Exploration	Conducts only exploitation	Capable of conducting exploration
4	Evolutionary computation	Classic Evolutionary Algorithm (tactical decisions driven by current population)	Algorithm maintaining diversity, long-term benefits in complex and dynamical environments (strategic decisions driven by global understanding)
5	Time	Capable of making only short-term decisions	Capable of making long-term decisions
6	Domain dependence	Limited adaptive behaviour	Capable of abstracting knowledge and of adaptive behaviour
7	Chaos	Unaware of attractors	Capable of avoiding or finding attractors

In Table 2.1 a summary of our observations is provided. We claimed that because of a wide range of existing agents, it would be useful to identify a subclass of them and call them SAs. Such agents would be the most complex ones, with deep understanding of their domain and capable of taking safe decisions which would benefit in longer term, but which could possibly act counter-intuitive and even harmful in a short term. SAs would be required to see the 'big picture' of the changes and not restrict themselves to analyse local details. We also stated that for an agent to be sapient, it is crucial that it possesses some meta-knowledge (which is knowledge about knowledge) and that its representation of the environment and reasoning process was multi-layered or hierarchical as opposed to possibly big, but flat set of simple concepts.

Several researchers at the conference session took a different approach. They claimed that there are some agents possible, which do not fall into the category of IAs at all. Such agents would possess a true 'intelligence', as opposed to IAs, which in fact only perform smart actions. SAs would therefore be a next generation of agents, qualitatively different from the ones existing nowadays. It is interesting to note that possibly such agents would require completely new computing paradigms. For example, Cottam *et al.* (2003) mention quantum effects as a possible missing link to build SAs.

2.4 Intelligent agents and evolutionary computation

One of the most difficult tasks in creating agents is the design and implementation of the behaviours that control agents. The difficulty arises because it is difficult to anticipate a priori all of the situations that an agent might encounter. In general, agent environments are under-specified for a variety of reasons including uncertainty and time-varying properties. Under such circumstances, notions of 'optimal' behaviour are replaced by notions of 'robust' behaviour. The key difference is the ability to cope with unanticipated events in a plausible way rather than 'crash' because of a locally optimal but brittle control program.

One approach to this problem that has proven to be successful is to use EC techniques to evolve via simulation robust control programs and then insert these evolved programs into actual problem-solving agents. We describe this approach in more detail in this section.

2.4.1 Principles of evolutionary computation

The field of EC is a relatively new field of computational research and applications that traces its roots to the 1960s when modern digital computers became readily accessible to scientists and engineers. The focus of this field is on the design and implementation of Evolutionary Algorithms (EA) for the purpose of solving difficult computational problems.

An EA is an algorithm based loosely on the Darwinian notion of a population of individuals evolving over time by means of reproductive variation and selection (survival of the fittest). An example of a simple EA is:

Randomly generate an initial population of individuals and evaluate their fitness.

Do until some stopping criterion is met:
 Select some individuals to be parents.
 Produce some offspring and evaluate their fitness.
 Select some individuals to die.

Return the most fit individual produced as 'the answer'.

Such algorithms only make computational sense if the 'individuals' represent candidate solutions to the problem under study, and the fitness of an individual represents the quality of a candidate solution. For example, suppose we are presented with a difficult function optimisation problem for which there is no closed form expression of the function. For these 'black box' optimisation problems, the only available solution strategy is to query the black box for various combinations of the N argument values and use this sample information to draw inferences as to where the functions optima lie. If we let individuals represent N-dimensional vectors of possible argument values and define their fitness to be the value of the function returned by the black box, then the EA performs a parallel, adaptive search of the specified parameter space and generally returns a near-optimal solution in a relatively short period of time.

There are, of course, many other details that need to be specified before such an EA approach can be used, including the size of the population, how solutions are represented internally and how offsprings are produced. The size of the population reflects the degree of parallelism of the search process and can vary from very small (less than 10) to quite large (thousands) depending on the size and complexity of the problem being solved. The simplest solution spaces to represent in an EA are those specified as a fixed set of N parameters. In this case individuals are quite naturally represented as N-dimensional vectors that specify a particular combination of values for the N arguments.

In this case there are correspondingly simple ways that offsprings are produced. Asexual (single parent) reproduction consists of cloning a parent vector and then mutating (perturbing) one or more of the vector values. In this way offsprings are similar (but not identical) to their parents. Sexual (two parent) reproduction generates offspring that inherit parameter values from both parents, resulting in offspring that exhibit new combinations of existing parameter values.

This reproductive variation is the means by which an EA explores new areas of the solution space. The counter-balancing force is some form of fitness-biased selection that is used to select parents and determine survival. From a Computer Science point of view, the stronger the selection pressure (e.g. choose only the top 10%), the 'greedier' the EA search process is in the sense that it will more rapidly converge but likely to a suboptimal solution. The art of EA design is to find an appropriate balance between reproductive exploration and selective exploitation.

The simple EAs have proved to be highly effective for many difficult optimisation problems, particularly when the fitness of a set of parameters can only be determined by running a complex simulation. An excellent example of this is the use of EAs to explore a parameterised conceptual design space (Arciszewski and De Jong, 2001). In this case, the fitness of a conceptual design is determined by transforming the conceptual design into a detailed design and submitting it to a commercial design evaluation package.

2.4.2 Representing agent behaviours

If we want to use an EA to evolve agent behaviours, then as discussed in the previous section, we need to represent the space of behaviours to be explored and define a notion of behavioural fitness. The simplest approach is to represent behaviour space as an N-dimensional parameter space and use the approach discussed earlier where an agent's behavioural fitness is determined by its performance in a simulated environment. While this seems a bit simplistic, it is the approach taken by many machine learning algorithms (including the learning of neural network weights).

A more fundamental, but more complex approach is to represent agent behaviours as programs to be executed and evolve them over time. This implies the use of a programming language that is 'evolution friendly' in the sense that it is amenable to the creation of new and interesting programs via reproductive variation. This requirement generally rules out traditional programming languages like C and Java, but still leaves some realistic possibilities such as finite state machines, cellular automata, rule-based languages and functional languages such as LISP. The EC literature documents successful applications in each of these areas.

2.4.3 Evolving single-agent behaviours

The simplest approach to evolving single-agent behaviours is to have each individual in the population represent an entire program whose behavioural fitness can be evaluated by injecting that program into an agent in a simulated environment. The key defines effective reproduction operators in the sense that they produce useful and interesting offspring from useful and interesting parents. For asexual reproduction this means that mutations must result in behavioural 'perturbations'. For sexual reproduction this means that useful parental behaviours are combined in new and interesting ways in the offspring.

Although this sounds like a daunting task, there has been considerable success using a variety of behavioural representations including the ones noted earlier, namely, finite state machines, cellular automata, rule-based representations and LISP.

2.4.4 Evolving multiple-agent behaviours

A much more challenging task for both humans and computers is to design a team of agents capable of performing collaborative tasks such as surveillance or rescue operations. The simplest approach is to assume that the teams are homogeneous in the sense that they are all functionally identical and have the same behavioural program. This does not mean that each team agent behaves identically since agents react to their immediate environment, which differs from agent to agent. However, it allows one to use the single-agent approach stated earlier to evolve these collective behaviours.

A more difficult task is to evolve teams of heterogeneous agents in the sense that each team agent may have different capabilities and hence different control programs. In this case the most successful EA approach to date is to use a co-operative co-evolution approach in which different agent types are co-evolved in separate populations. In this way each population can use the single-agent approach described earlier. However, fitness evaluation now consists of selecting candidate agents from each of the populations and assessing in simulation how well they collectively solve problems.

Both of these approaches have been successfully applied to a number of difficult team-oriented problems.

2.5 Final comments

This chapter is intended to provide a general overview of IAs in the context of their potential applications in Civil Engineering. IAs are becoming an important emerging technology whose evolution can be, at least partially, predicted and must be carefully monitored to maximise its Civil Engineering benefits. When appropriately developed and adapted for the Civil Engineering applications, IAs could have a significant impact on the increased design and construction productivity and could ultimately lead to qualitative changes in the ways Civil Engineering is practised today.

As we know, the progress in Civil Engineering is mostly dependent on our ability to produce novel designs and to develop new construction methods and tools. In this context, the real challenge for IAs researchers is to develop a conceptual and computational foundation, which will ultimately allow IAs to produce novel solutions in design or in other Civil Engineering areas. Unfortunately, at this time we still have insufficient understanding of engineering creativity and cognition in the computational context and a lot of research is required to address the challenge. Specifically, two areas should be investigated further. First, the area of knowledge representation should be studied and both that is, static and dynamic forms of knowledge representation, which would allow the exploitation as well as exploration of the design space with strong potential for finding novel solutions. Second, various solution-generation mechanisms should be investigated. In this chapter we briefly discussed a particular approach to the use of EC as a mechanism for generating IAs. This leads to a number of interesting extensions including the possibility of evolving cellular automata that represent some interesting and novel solutions.

The technology of IAs still does not exist in a mature form and our understanding of its limitations and potential is constantly evolving. Unfortunately, still much more research is necessary, both in the areas of Computer Science and Civil Engineering. However, an international network of researchers is developing, significant progress is being made and the first practical applications of this approach are soon expected.

References

Anumba, C. J., Ren, Z., Thorpe, A., Ugwu, O. O. and Newnham, L. (2003) 'Negotiation within a multi-agent system for the collaborative design of light industrial buildings', *Advances in Engineering Software*, 34(7): 389–401.

Arciszewski, T. and De Jong, K. A. (2001) 'Evolutionary computation in civil engineering: research frontiers', in B. H. V. Topping (ed.) *Proceedings of the*

Eighth International Conference on Civil and Structural Engineering Computing, Vienna.

Bradshaw, J. (1997) *Software Agents,* AAAI Press, Menlo Park, USA.

Casasola, E. and Gauch, S. (1997) 'Intelligent information agents for the World Wide Web', Technical Report ITTC-FY97-11100-1, Information and Telecommunication Technology Center, The University of Kansas, KS.

Chavez, A. and Maes, P. (1996) 'Kashbah: an agent marketplace for buying and selling goods', in *Proceedings of the First International Conference on the Practical Application of Intelligent Agents and Multi-Agent Technology (PAAM-96),* London, UK, pp. 75–90.

Clarke, D. W. (2000) 'Strategically evolving the future: directed evolution and technological systems development', in T. Arciszewski (ed.) *Innovation: the Key to Progress in Technological and Society (Special Issue on International Journal Technological Forecasting and Social Change),* 64(2&3): 33–153.

Cottam, R., Ranson, W. and Vounckx, R. (2003) 'Sapient structures for sapient control', in *Proceedings on the International Conference – Integration of Knowledge Intensive Multi-agent Systems: KIMAS'03,* Boston, MA.

Doorenbos, R., Etzioni, O. and Weld, D. (1997) 'A scaleable comparison-shopping agent for the world wide web', in *Proceedings of the First International Conference on Autonomous Agents (Agents 97),* Marina del Rey, CA, pp. 39–48.

Franklin, S. and Graesser, A. (1996) 'Is it an agent, or just a program? A taxonomy for autonomous agents', in *Proceedings of the Third International Workshop on Agent Theories, Architectures and Languages,* Springer-Verlag, Berlin.

Gero, J. S. and Brazier, F. M. T. (eds) (2002) Agents in Design, Preprints of the First International Workshop on Agents in Design, University of Sydney, Sydney.

Jennings, N. R., Norman, T. J. and Faratin, P. (1998) 'ADEPT: an agent-based approach to business process management', *ACM SIGMOD Record,* 27(4): 32–9.

Jennings, N. R. and Wooldridge, M. J. (eds) (1998a) *Agent technology: Foundations, Applications and Markets,* Springer-Verlag, Berlin.

Jennings, N. R. and Wooldridge, M. J. (1998b) 'Applications of Intelligent Agents', in N. R. Jennings and M. J. Wooldridge (eds) *Agent Technology: Foundations, Applications and Markets,* Springer-Verlag, Berlin.

Jennings, N. R. and Wooldridge, M. J. (1998c) 'Applying agent technology', in N. R. Jennings and M. J. Wooldridge (eds) *Agent Technology: Foundations, Applications and Markets,* Springer-Verlag, Berlin.

Leitao, P. and Restivo, F. (2000) 'A framework for distributed manufacturing applications', in *Proceedings, Advanced Summer Institute International Conference, ASI' 2000,* Bordeaux, France, pp. 75–80.

Lenat, D. B. (1995) 'Cyc: a large-scale investment in knowledge infrastructure', *Communications of the ACM,* 38(11): 33–48.

Lesser, V. R. and Corkill, D. D. (1983) 'The distributed vehicle monitoring testbed: a tool for investigating distributed problem solving networks', *The AI Magazine,* 4(3): 15–33.

Meyer, J. J. (1998) 'Agent languages and their relationship to other programming paradigms', in J. P. Muller, M. P. Singh and A. S. Rao (eds) *Intelligent Agents,* Springer-Verlag, Berlin, pp. 309–16.

Morales-Morell, A. (1997) 'Intelligent Agents', Technical Paper, HCI Course, from http://citeseer.nj.nec.com/morales-morell97intelligent.html [Accessed: 6 October 2002].

Moss, S. and Davidsson, P. (eds) (2001) 'Multi-agent-based Simulation', Lecture Notes in Computer Science, Volume 1979, Springer-Verlag, Berlin.

Negroponte, N. (1995) *Being Digital*, Hodder and Stoughton, London.

Nwana, H. S. and Ndumu, D. T. (1999) 'A perspective on software agents research', *The Knowledge Engineering Review*, 14(2): 1–18.

Padgham, L. and Taylor, G. (1997) 'A system for modelling agents having emotion and personality', in W. Wobcke, L. Cavedon and A. Rao (eds) *Intelligent agent systems: theoretical and practical issues*, Lecture Notes in AI 1209, Springer-Verlag, Berlin, pp. 59–71.

Russell, S. J. and Norvig, P. (1995) *Artificial Intelligence: A Modern Approach*, Prentice-Hall, New Jersey.

Skolicki, Z. and Arciszewski, T. (2003a) 'Intelligent agents in design', in *Proceedings of ASME 2003 Design Engineering Technical Conferences and Computers and Information in Engineering Conference*, Chicago.

Skolicki, Z. and Arciszewski, T. (2003b) 'Sapient agents: seven approaches', in *Proceedings, International Conference – Integration of Knowledge Intensive Multi-agent Systems: KIMAS'03*, Boston.

Smith, G. J., Maher, M. L. and Gero, J. S. (2003) 'Designing 3D virtual worlds as a society of agents', in M.-L. Chiu, J.-Y. Tsou, T. Kvan, M. Morozumi and T.-S. Jeng (eds) *Proceedings of the Tenth International Conference on Computer Aided Architectural Design Futures*, 'Digital Design: research and practice', Kluwer, Tainan, Taiwan, pp. 105–14.

Tecuci, G. (1998) *Building Intelligent Agents: An Apprenticeship Multistrategy Learning Theory, Methodology, Tool and Case Studies*, Academic Press, San Diego, CA.

Turing, A. M. (1950) 'Computing machinery and intelligence', *Mind*, 59(236): 433–60.

Ugwu, O. O., Kumaraswamy, M. M., Ng, T. S. and Lee, P. K. K. (2003) 'Agent-based collaborative working in construction: understanding and modelling design knowledge, construction management practice and activities for process automation', *The Hong Kong Institution of Engineers (HKIE) Transactions*, [*Emerging Technology in the 21st Century*, Tenth Anniversary Issue] 10(4): 81–7.

Velásquez, J. (1997) 'Modeling emotions and other motivations in synthetic agents', in *Proceedings of Fourteenth National Conference on Artificial Intelligence (AAAI-97)*, MIT/AAAI Press, Providence, RI, pp. 10–15.

Wooldridge, M. J. (2002) *An Introduction to Multi-Agent Systems*, John Wiley & Sons, Ltd, Chichester, UK.

Chapter 3

Multi-agent systems in construction

An overview

Z. Ren, C. J. Anumba and O. O. Ugwu

3.1 Introduction

The concept of distributed artificial intelligence (DAI) originates from the real world where many cases are inherently distributed in space, function, knowledge, expertise or information (Durfee *et al.*, 1989). The notion of DAI provides a natural metaphor to match such distribution. It represents a new way of analysing, designing and implementing complex software systems. The key advantage of DAI is its responsibility for enacting various components of the business process, which is delegated to a number of autonomous agents. These agents act collectively as a society and collaborate to achieve their own individual goals as well as the common goal of the society to which they belong (Ugwu *et al.*, 1999a). Agents are inherently modular and can be constructed locally for each resource, provided they satisfy some high-level protocol of interaction.

The agent-based view in DAI offers a powerful repertoire of tools, techniques and metaphors that have the potential to considerably improve the way in which people conceptualise and implement their systems (Jennings *et al.*, 1998). DAI applications such as: information access, information filtering, electronic commerce, workflow management, intelligent management and various negotiations are becoming ever more prevalent. The common point in these different types of system is the key abstraction used – (distributed) agents. The significance of agents is that they provide a natural means for performing the above tasks over a distributed environment.

Multi-agent systems (MAS) emanate from the traditional field of DAI. In MAS, there is no central controlling party. Agents often co-operate to achieve their own individual goals, rather than to solve a common problem. MAS are suitable for domains that involve interactions between different people or organisations with different (possibly conflicting) goals and proprietary information. With the lack of centralised control, agents in MAS have to solve the problem of the relationship between each agent's behaviour and goals, and those of the global system or MAS community. Negotiation, thus, often plays a central role in agent co-operation.

Moreover, agents have reasoning abilities to infer the other agents' key features and changes in the environment.

This chapter explores the common threads that together make up the agent and MAS. The purpose is to provide an in-depth analysis of MAS, and indicate the key issues in the field. Particular focus is on the nature of autonomous agents and MAS, negotiations and learning issues in MAS.

3.2 Intelligent agents

Intelligent agents (IAs) are the basic cells of MAS. To understand MAS, the starting point is to define and understand the IA.

3.2.1 Definition

There is little agreement on the definition of the terms 'agent' and 'IA'. Clearly, they should be more than just a program, but where the boundaries lie is not clear at all. This is a manifestation of the general problem in AI of defining 'intelligence' that has led to much discussion. The result is that there are as many agent definitions as there are researchers, leading to the term being substantially overused. Some examples are:

- Brustoloni (1991) 'Autonomous agents are systems capable of autonomous, purposeful action in the real world'.
- Smith *et al.* (1994) 'Let us define an agent as a persistent software entity dedicated to a specific purpose. "Persistent" distinguishes agents from subroutines; agents have their own ideas about how to accomplish tasks, their own agendas. "Special purpose" distinguishes them from entire multifunction applications; agents are typically much smaller'.
- Maes (1995) 'Autonomous agents are computational systems that inhabit some complex dynamic environment, sense and act autonomously in this environment, and by doing so realise a set of goals or tasks for which they are designed'.
- Coen (1995) 'Software agents are programs that engage in dialogs and negotiation and coordinate transfer of information'.
- Hayes-Roth (1995) 'Intelligent agents continuously perform three functions: perception of dynamic conditions in the environment; action to affect conditions in the environment; and reasoning to interpret perceptions; solve problems, draw inferences, and determine actions'.
- IBM white paper (1994) 'Intelligent agents are software entities that carry out some set of operations on behalf of a user or another program with some degree of independence or autonomy, and in so doing, employ some knowledge or representation of the user's goals or desires'.

Although different researchers emphasise different aspects of agency, their definitions of agents suffer from one or more of these three problems (Ugwu *et al.*, 1999a):

- too broad a definition that will include things such as Unix daemons;
- too narrow a definition that prescribes a particular AI technique; or
- forming a definition in terms of equally vague terms.

Despite the different definitions, there are several broad qualities that have some measure of general agreement. Wooldridge and Jennings (1995) define an agent as 'a computer system, situated in some environment that is capable of flexible autonomous action in order to meet its design objective'. There are thus four key concepts in the definition:

- *Autonomy* – agents should operate without the direct intervention of humans or others, and have some kind of control over their actions and internal state.
- *Social ability* – agents need to be able to interact with other agents (and possibly humans) via some kind of agent-communication language.
- *Reactivity* – agents should be able to perceive their environment and respond in a timely fashion to changes that occur in it. This environment may be the physical world, a user via a graphical user interface (GUI), a collection of other agents, the Internet or perhaps all of these combined.
- *Proactiveness* – agents should not simply act in response to their environment, they should be able to exhibit goal-directed behaviour by taking the initiative.

From this and other definitions, the key feature would appear to be 'autonomy'. However, this is usually loosely defined; containing words like 'control over their own actions' or 'formulate their own goals'. An agent with freewill is a high aspiration indeed. For learning agents that deal with different situations in different ways as they learn, something that appeared like autonomous behaviour would be possible. However, few definitions actually include learning. Non-learning agents ultimately follow only the same set of instructions and/or rules at all times during their existence. Hence, for these agents, autonomy must take on a somewhat weaker meaning. A more meaningful definition may be to say an agent is autonomous if it operates without the need for the direct intervention of humans.

Nwana (1996) takes Wooldridge and Jenning's definition and reduces it to three behavioural attributes, any two of which must be possessed by a software agent. These are:

- *Autonomy* – this refers to the principle that agents can operate on their own without the need for human guidance. Agents have individual internal states and goals, and act in such a manner as to meet their goals. A key element of their autonomy is their proactiveness, that is, the ability to 'take the initiative' rather than acting simply in response to their environment.

- *Co-operation* – co-operation with other agents is paramount, and is the reason for having multiple agents in the first place. In order to co-operate, agents need to possess a social ability, that is, the ability to interact with other agents and possibly humans via some communication language.
- *Learning* – for agent systems to be truly 'smart', they would have to learn as they react and/or interact with their external environment. A key attribute of any intelligent being is its ability to learn. The learning may also take the form of increased performance over time.

Nwana's requirements for agenthood may be neatly shown as a Venn diagram in Figure 3.1. The inclusion of learning is at least an aspiration, it recognises a core quality of the intelligent behaviour. The diagram neatly provides a framework into which software agents can be currently defined. A number of other attributes for agenthood have been cited (see also Arciszewski *et al.* in Chapter 2). Some of these need to be considered for any classification of agents; others are better under the category of generally desirable qualities. In the next section Nwana's definition will be extended to include some of these attributes. Figure 3.2 shows an example of an agent architecture (Sen, 1997).

3.2.2 Agent development

Agents can be understood as an incremental extension of previous software technologies (Table 3.1). In the beginning was the *program*, a monolithic deck of machine instructions and data tied together with tangled 'goto'

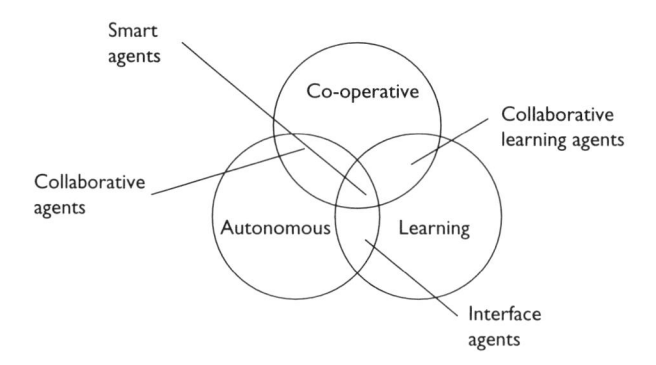

Figure 3.1 Nwana's requirements for agenthood.

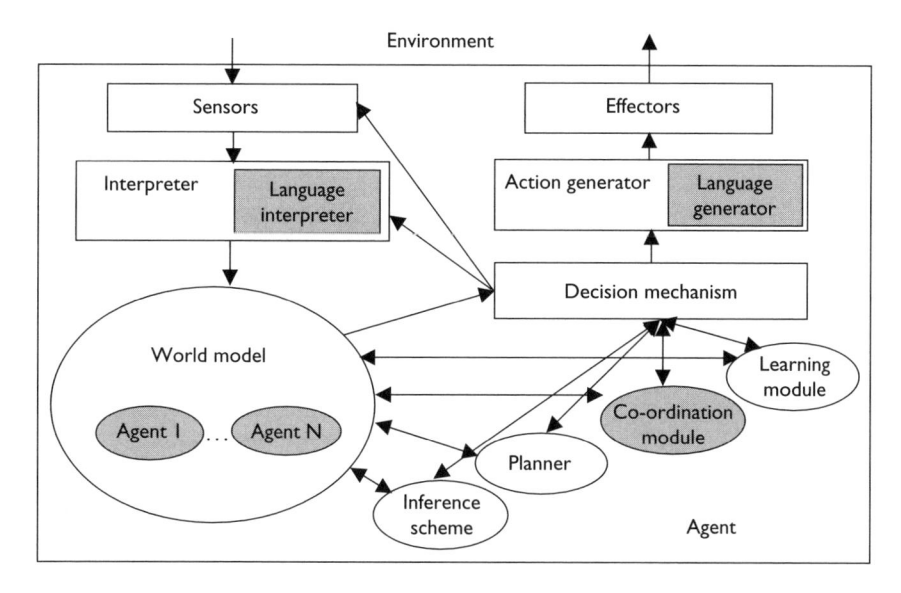

Figure 3.2 An agent architecture (shaded modules represent components particular to agents in a MAS).

Table 3.1 Agents in historical perspective

	Monolithic program	*Structured programming*	*Object-oriented (OO) programming*	*Agent-oriented programming*
How does a unit behave?	External	Local	Local	Local
What does a unit do when it runs?	External	External	Local	Local
When does a unit run?	External	External (called)	External (message)	Local (thread; goals)

Source: Parunak et al., 1997.

statements that took over the complete resources of the computer when the user fed it into the card reader. The *structured programming* movement modularised program code through constructs such as subroutines and structured loops. These constructs localised the definition of how the program would function, but relied on an external innovation and external definition of the data on which the code would operate and innovate. The next major development, *objects*, gave software modules local state as well as local code, but innovation was still determined externally, by sending a message. *Agents* add two things to (passive) objects: a local thread of control and local initiative (usually expressed as local goals). Together, these enable the agents to monitor and respond to their environment autonomously (Parunak *et al.*, 1997).

Two requirements make agents an attractive technology for a modern design environment:

- First, agents are intrinsically distributed. While their local threads can be supported on a single processor, it is also natural to distribute them across a network, supporting the distribution requirements.
- Second, the modularity of agents makes it natural to encapsulate humans as peer agents to computer processes using common language and protocols to integrate people and machines. In nature, this integration requires people to reduce the bandwidth of their communication to a level that computerised agents can handle.

Many researchers appear to confuse the following: agents and objects; distributed computing and agent-based computing; object-oriented (OO) systems, expert systems and agent-based systems. The OO computing and the distributed computing do not in themselves offer solutions to agent-based problems because distributed computing modules are usually passive and dumb. Also, their communications are usually low level while agent-based systems require high-level messages (Newell, 1982). However, OO techniques and distributed computing constitute some of the main underpinning for agent technology – modelling and implementation.

The OO techniques are good in general, but are rather low level for intelligent applications. They can be used, for instance, to implement knowledge representations, but they do not themselves provide a knowledge representation. The OO development methodologies can, however, be seen as a low-level underpinning for a multi-agent methodology. The same might be said of distributed computing methodologies and indeed, many agent-based systems are built on top of distributed platforms. However, it can be argued that if OO approaches are still relatively new, agent-based systems are even newer and less generally accepted. Again, the knowledge level is wrong. For instance, communications protocols do not operate at the high level of speech acts as one might wish for an agent-based system. More importantly, agent applications require a co-operative knowledge level (Jennings and Campos, 1997), while expert systems typically operate at the symbolic and knowledge level (Newell, 1982).

The benefit of an agent-based system would be a reduction of the semantic gap between analysis on the one hand, and design and implementation on the other, leading to a reduction in the time to design and implement, with the usual trade-off between better expandability and losses in execution efficiency and design specificity. Current methodologies emphasise top-down design, but agent-based systems adopt a different approach: top-down within the agent and bottom-up in the agent community (Wooldridge, 1997). In summary, agent-based systems research can be regarded as developing a way of looking at problems rather than a technology. Hence, agent-based systems can, and do, use OO programs, expert systems, and distributed computing technologies to implement applications and tool kits that embody this approach.

3.2.3 Agent taxonomy

The use of a taxonomy is an important approach in understanding and designing agents. There have been many attempts to produce taxonomies or classifications of agents. None of these seem to be complete and most of them become dated quickly. For example, Brustoloni's (1991) taxonomy of software agents begins with a three-way classification into regulation agents planning agents or adaptive agents. A regulation agent reacts to each sensory input as it comes in, and always knows what to do. It neither plans nor learns and so on. This yields a two-layer taxonomy.

There are many other possible classification schemes. Agents might be classified according to the tasks they perform such as: information gathering agents or email filtering agents. Agents may also be classified by the range and effectiveness of their actions or by the degree of sophistication of their internal state such as: goal driven, non-temporal agents to those with a full Beliefs, Desires and Intentions (BDI) reasoning capability. Another possible taxonomy might involve the environment in which the agent finds itself, for example software agents as opposed to artificial life agents.

Most of these taxonomies seem cumbersome and unsuited to classifying many agent types. Any taxonomy has to divide agent-space from the top and there seems to be no natural top-level divisions. Second, agents are often an *ad hoc* collection of techniques and frequently fall into two categories of a taxonomy. Given the difficulties of a formal taxonomy, a more pragmatic approach may be to list the qualities that designers may wish their agents to have, each as a separate dimension. An agent will then be represented by a point in agent-space. Table 3.2 extends Nwana's model (1996) with other desirable or useful agent qualities. Nwana's qualities are represented as core qualities. There are four agent quality types:

- *Core* – at least two need to be present for agenthood.
- *Motivational* – deals with the basis on which an agent interacts with other agents, the environment, etc. whether it is acting for itself, a group or a wider community. This is really a continuum and a point must be chosen along this line.
- *Activity* – defines the level of passivity of an agent. Does it actively seek to realise its own goals or does it wait for the environment to change around it?
- *Others* – are a collection of qualities sometimes seen as necessary for particular agent applications.

Table 3.2 Agent qualities

Quality type	Property	Meaning
Core	Autonomous	Exercises control over its own actions.
	Co-operative	Communicates with other agents, perhaps including people.
	Learning	Changes its behaviour-based on its previous experience.
Activity	Goal-orientated	It is the classical symbolic AI approach and is based on the Physical Symbol System Hypothesis, where the external world can be modelled symbolically and when processes are run on this representation the agent has its own purpose, and does not simply act in response to the environment (Newell and Simon, 1976).
	Reactive	Responds in a timely fashion to changes in the environment. In extreme cases, it represents a rejection of all symbolic world models because one method, that is, formal logic, has been found wanting. Such simple systems usually involve learning to respond to certain inputs in certain ways.
	Hybrid	Allows agents to be able to respond rapidly in a reactive and possibly learned way, while maintaining a small and consequently manageable symbolic representation of the world.
Motivation	Altruistic	Does not have private goals and acts in the interest of fellow agents or a group. These agents are really only possible in closed systems.
	Self-interested	Acts in own self-interest. Many of them are concerned with task allocation and resource sharing.
	Veracious	Will not knowingly communicate false information.
Other	Mobile	Able to transport itself from one machine to another.
	Personalisable	Easily retrained to perform a task the way a particular user wants it done (Foner, 1993).
	Graceful degradation	If failure occurs most of the task can still be accomplished, instead of failing to accomplish any of the task (Foner, 1993).

Source: Adapted from Ugwu *et al.*, 1999a.

Classes of agents can then be defined by some combination of the core qualities, motivation, activity and the extra properties. No agent will have all these qualities. For example, a poker-playing agent would not want to be veracious. Equally, a proactive, autonomous agent within an open system (e.g. the Internet) would not necessarily want to be altruistic. Some of these qualities are discussed in detail later.

3.2.3.1 Activity

Muller (1998) classifies agents according to the influential threads of agent research, that is, reactive agents, deliberative agents and interacting agents. Each of these threads focuses on one important property of an agent. Reactive agents focus on reactive and real-time behaviour. Deliberative agents focus on the ability to act in a goal-directed manner. Interacting agents focus on the ability of co-operative social behaviour.

REACTIVE AGENTS

Reactive agents are built according to the behaviour-based paradigm, have no (or at most a very simple) internal representation of the world, and provide a tight coupling of perception and action. Such simple systems usually involve learning to respond to certain inputs in certain ways. Such agents can learn by observation of the user, but cannot formulate plans to achieve distant goals. There are two important points that the reactive approach has introduced to AI generally and agent research in particular (Wooldridge and Jennings, 1995).

a   Intelligence is not exclusively to be found in large, centralised singular systems, but can be an emergent property of the interaction of many unintelligent units.
b   Most everyday activity is really routine, once learned it can be accomplished in a routine way with little variation.

Valuable though these insights are, the requirements placed on many agents means there has to be some symbolic representation. This is because they usually have to interact with humans who will ask explicit questions such as 'why?' and 'how?' They are also often expected to interact with a great variety of other agents with vastly different requirements.

DELIBERATIVE AGENTS

Deliberative agents are agents in the symbolic artificial intelligence tradition that have a symbolic representation of the world in terms of categories such as beliefs, goals or intentions and that possess logical inference mechanisms to make decisions based on their world model. In essence this states that the external world can be modelled symbolically and when processes are run on this representation (essentially theorem proving) it is capable of general intelligent action. Some key problems of which are:

a *Speed of execution* – a logic-based system uses formal rules of inference to deduce if some action or event is a logical consequence of its current knowledge/beliefs, etc. The theorem proving search process is an NP-complete problem (non-polynomial – its time complexity is exponential). The effect of this, even in moderately dynamic and unpredictable environments, is that the world changes faster than a plan-based agent can reason about actions to take to achieve its goals and execute those actions.

b *Common-sense reasoning* – this is perhaps the second most important practical shortcoming of plan-based agents. In most applications, common-sense reasoning involving notions such as time, space and causality is handled in an *ad hoc*, application-specific manner. Indeed, the issue of common-sense reasoning has defied formal treatment in AI, and presents perhaps the major stumbling block to the development of reasonably smart agent systems.

c *Reasoning about other agents' desires and intentions* – a reasoning agent in a co-operative setting, in addition to the communications overheads to ensure co-operation, agents have to reason about the goals, plans and beliefs of other agents, in order, for example, to minimise duplication of effort. This is itself an NP-complete problem and, for practical purposes, it dramatically reduces the size of the manageable knowledge-base (k-b) of each agent.

INTERACTING AGENTS

Until recently, interacting agents were not classified separately. Interacting agents are able to co-ordinate their activities with those of other agents through communication and, in particular, negotiation. Interacting agents have been mainly investigated in distributed AI; they may have explicit representations of other agents and may be able to reason about them. So far, the focus has been on the co-ordination process itself and on mechanisms for co-operation among autonomous agents rather than on the structure of these agents. A detailed discussion about interacting agents is contained in Section 3.4.

Besides the reactive agents and deliberative agents, Ferguson (1992) classifies hybrid agents. Hybrid systems allow the agent to be able to respond rapidly in a reactive and possibly learned way, while maintaining a small and consequently manageable symbolic representation of the world. Its reactive quality means the nuts and bolts of how it interacts with the world do not need to be represented symbolically. In this way the symbolic system can concentrate on planning, and dealing with unexpected situations. Thus, it is still capable of functioning reactively while waiting for symbolic decisions to be made. The Touring Machines hybrid (Ferguson, 1992) is a good example of this approach.

3.2.3.2 Motivation

Agents can be classified as self-interested and altruistic according to their motivation for action.

ALTRUISTIC AGENTS

Some domains are inherently suited to the use of altruistic agents. For example a single factory scheduling problem where each work-cell is represented by an agent. If the cells do not have private goals the agents will need to act in the interest of the company, that is, altruistically. There are many multi-agent collaboration projects that use altruistic agents within the literature, some illustrative examples follow. Negotiation in the examples is generally within a small group of 'friend' agents as stated in the following sections.

Pleiades The complete system is designed for a limited environment (called the InfoSphere) and is a set of Internet-based heterogeneous information resources by Sycara and Zeng (1994). *Pleiades* has been used to create the Visitor Hosting System where the agents co-operate in order to manage a visitor's schedule. Agents arrange appointments and meetings with other agents and accordingly formulate plans for the visitor. Agent skills include the following: knowledge of how to gather information, knowledge of other agents it must co-ordinate with and strategies for conflict resolution. The architecture has no central planner and hence agents must all engage in co-ordination by communicating to others their constraints, expectations and other relevant information.

ADEPT This project uses collaborating agents to help in business decision-making where the information needed is often spread through several companies in different databases (Jennings *et al.*, 1996). This system attempts to provide access to business processes and information on request. Due to the distributed nature of business information collaborative agent technology provides a good basis upon which to build a system of this kind. The requirements in providing quotations cover such things as obtaining credit references, designing the system and costing it.

SELF-INTERESTED AGENTS

There are many examples of self-interested agents. Many of them concern task allocation and resource sharing where the agents are self-interested, some examples follow:

a *Electronic marketplaces* – agents representing different enterprises, buying and selling (Rosenschein and Zlotkin, 1994).
b *Information retrieval* – information servers can form coalitions for answering queries (Fischer *et al.*, 1996; Tsvetovatyy and Gini, 1996).
c *Air traffic control* – aeroplanes belonging to different airlines need to share the limited resources at airport, the control mechanism needs to give priority to planes with less fuel on board (Sandholm and Lesser, 1997a).
d *Distribution problems* – package delivery companies may co-operate to reduce expenses (Sen and Durfee, 1997).

In these examples, the agents are self-interested and try to maximise their own benefits. They typically involve less than a dozen agents. There has been less work done with self-interested agents in open systems. One of the main reasons is that it is much harder to negotiate with other agents who might not be telling the truth. Trust during negotiation is one of the key issues. The majority of self-interested agents are based on Game Theory or Economic Theory. A further discussion of Game Theory and Economic Theory is discussed in Chapter 4.

3.2.4 Agent co-ordination

Although there are single-agent systems like information retrieval agent, most agent-based systems contain more than one agent. Therefore, co-ordination is central to agent-based systems to ensure a community of individual agents acting in a coherent, harmonious and expected way. The approach to agent co-ordination reflects how people view the real-world problems and model them in agent-based systems. Agent co-ordination has been studied by researchers in diverse disciplines such as: organisation theory economic theory, social psychology, anthropology and sociology. Several typical ways of structuring and co-ordinating agents have been developed. Some examples are listed in the following sections.

ORGANISATIONAL STRUCTURING

This is one of the most common and also the simplest interaction mechanisms. It usually exploits a hierarchical structure, implemented as client/server or master/slave, where the master could gather information from the agents of the group, create plans and assign tasks to individual agents in order to ensure global coherence. A typical use might be for resource allocation such as Werkman (1990). A more practical application may be an assembly line. Here all agents are set up to interact only with several other agents, all sharing the same explicit end goal (i.e. the manufacture of a product). However, these types of systems allow few of the benefits of DAI, as it assumes that there is one agent with a global view of the full task. In real and complex situations agents are likely to possess knowledge that the central agent does not know of, and feeding this back to the central agent is difficult. In addition, it is very difficult for peer agents with different goals to resolve their difficulties as all must go before a central arbiter with imperfect knowledge.

CONTRACTING

The Contract Net Protocol (CNP) (Smith, 1980) is one of the most commonly used protocols. It is different from a master/slave system in that a manager agent will break a problem into the component problems and then announce each task. Contractors then table bids to the manager. The manager then reviews the bids and awards the contract. Contract nets are best used when the problem can be broken down via a well-defined hierarchical nature into a set of tasks. This means that the main planning has to be done centrally.

MULTI-AGENT PLANNING

This usually involves a central arbiter who will review all potential plans of individual agents. This agent then checks for conflict and rejects or revises as appropriate. An example of this is multi-agent planning for air traffic control (Cammarata *et al.*, 1983). It is appropriate to have a central agent with ultimate responsibility in such a safety-critical area. The system works because within the problem space there are many possible solutions that at the most cause only minor inconvenience to other aircraft, hence conflict is less common. In other domains where there is more chance of a conflict of interests between agents, direct negotiation is more efficient.

PEER-TO-PEER NEGOTIATION

All the preceding organisational models can and do involve negotiation but it generally plays a minor part, as it will be negotiation between agents of differing ranks. Peer-to-peer negotiation effectively means that there is no structure and the agents must communicate directly with other agents to achieve its (or the group's) goals. Thus, it is most commonly adopted in MAS.

A detailed analysis of these forms of organisation is discussed in the following sections.

3.3 Multi-agent systems

DAI is designed and implemented as several interacting agents. The goal of DAI is clear enough and has been proven in many prototypes: creating a system that interconnects separately developed agents, thus enabling the ensemble to function beyond the capabilities of any singular agent in the set-up (Nwana and Ndumu, 1999). DAI systems have received considerable attention for two main reasons (Russell and Norvig, 1995).

- First, they have useful properties such as parallelism, robustness and scalability. Therefore, they are applicable in many domains which cannot be handled by centralised AI systems. In particular, they are well suited for domains which require resolution of interest and goal conflicts, integration of multiple knowledge sources and resources, time-bounded processing of very large data sets, or on-line interpretation of data arising from different geographical locations.
- Second, they are in accordance with the insight gained in disciplines such as AI, psychology and sociology that intelligence is tightly and inevitably coupled with interaction.

In other words, DAI is ideally suited to representing problems that have multiple problem-solving methods, multiple perspectives and/or multiple problem-solving entities. Such systems have the traditional advantages of distributed and concurrent problem solving, and also have the additional advantage of sophisticated patterns of interactions. Examples of common types of interactions include: co-operation (working together towards a common aim); co-ordination (organising problem-solving activity so that harmful interactions are avoided or beneficial interactions are exploited); and consensus (coming to an agreement which is acceptable to all the parties involved) (Jennings *et al.*, 1998). It is the flexibility and high-level nature of the interactions which distinguishes DAI from other forms of software and which provides the underlying power of the paradigm.

Traditionally, research into systems composed of multiple agents was carried out under the banner of DAI, and has historically been divided into two main camps: Distributed Problem Solving (DPS) and MAS. More recently, the term MAS has come to have a more general meaning, and is now used to refer to all types of systems composed of multiple autonomous components.

3.3.1 Distributed problem solving

DPS considers how the task of solving a particular problem can be divided among a number of modules (or nodes) that co-operate in dividing and sharing knowledge about the problem and about its evolving solutions. In a pure DPS system, all interaction strategies are incorporated as an integral part of the system. Agents' interactions are guided by co-operation strategies meant to improve their collective performance. Conflict among the agents in these environments may arise while each tries to achieve its own sub-task, but their overall task is the same. For example,

AIR TRAFFIC CONTROL

Cammarata *et al.* (1983) develop co-operation strategies for resolving conflicts between the plans of a group of agents. They apply these strategies to an air traffic control domain, in which the aim is to enable each agent to construct a flight plan that will maintain a safe distance with each aircraft in its vicinity and satisfy additional constraints. Agents involved in a potentially conflicting situation choose one of the agents involved in the conflict to resolve it. The chosen agent acts as a centralised planner to develop a multi-agent plan that specifies the conflict-free flight paths that the agents will follow. The decision of which agent will do the planning is based on different criteria, for example, 'most-informed' agent or 'most-constrained' agents.

THE DISTRIBUTED VEHICLE MONITORING TASK

In this domain, a set of agents is distributed geographically, with each being capable of sensing some portion of an overall area to be monitored. As vehicles move through its sensed area, each agent detects characterised sounds from those vehicles at discrete time intervals. By analysing the combination of sounds heard from a particular location at a specific time, an agent can develop interpretations of what vehicle might have created these sounds. By analysing temperate sequences of vehicle interpretations, and using knowledge about mobility constraints of different vehicles, the agent can generate tentative maps of vehicle movements in its area. By communicating tentative maps to one another, agents can obtain increased reliability and avoid redundant tracking in overlapping regions (Durfee, 1988).

INDUSTRIAL PROCESS CONTROL

ARCHON, a software platform for building DAI, and an associated methodology for building applications, has been developed to facilitate industrial process control. The system addresses two major problems. One is concerned with providing the necessary control and level of integration to help the subcomponents of an industrial process to work together. Another is concerned with decomposing the overall application goal(s) and with distributing the constituent tasks throughout the community. The system has been applied in several process control applications such as: electricity transportation management and particle accelerator control. It is one of the world's earliest field-tested DAI systems (Cockburn and Jennings, 1996). Further examples of DPS can be found in Georgeff (1983), Huhns *et al.* (1987), Durfee (1988), Levesque *et al.* (1990), Lesser (1991), Jennings (1995), Grosz and Kraus (1996), Parunak (1996) and Tambe (1997).

3.3.2 Multi-agent systems

Research in MAS is concerned with the behaviour of a collection of autonomous agents aiming to solve a given problem. There is no global control, no globally consistent knowledge, and no globally shared goals or global success criteria among these agents (Hewitt, 1985). MAS offer a way to relax the constraints of centralised, planned, sequential control, although not every MAS takes full advantage of this potential. They offer production systems that are decentralised rather than centralised, emergent rather than planned and concurrent rather than sequential (Parunak, 1996).

Jennings *et al.* (1998) summarise the characteristics of MAS as:

- each agent is individually motivated and attempts to maximise its own utility;
- each agent has incomplete information or capabilities for solving the problem, thus each agent has a limited viewpoint;
- there is no global system control;
- data is decentralised; and
- computation is asynchronous.

There is an increasing interest in MAS research because of its ability to provide robustness and efficiency; the ability to allow inter-operation of existing legacy systems; and the ability to solve problems in which data, expertise or control is distributed. This could be achieved through ontology-driven enterprise integration at data and process levels (Ugwu *et al.*, 2003). MAS have been applied in many cases where the need for interaction and negotiation exists, such as WWW searches, e-commerce, supply chain management (SCM), design/project management, knowledge management, project co-ordination, computer networks, operating systems, multi-enterprise manufacturing or multi-robot systems. Some examples are listed in the following sections.

ADJUSTING AGENT AUTONOMY IN SUPPLY CHAIN MANAGEMENT

A supply chain is a network of suppliers, factories, warehouses, distribution centres and retailers. SCM manages the co-operation of these system components, which correlate with each other through chain activities to implement system functionality. Software agents (Lin et al., 1998; Chen et al., 1999) are introduced as an entity with goals, action and domain knowledge situated in the environment. Each functionality of the chain is implemented by one kind of autonomous, intelligent, proactive and adaptive agent. These agents co-operate in a dynamically changing and open environment. A mixed negotiation process is developed for the specific problem where SCM's functional agents contain self-adaptive rules for their reasoning procedure, meanwhile human involvement is also considered. Obonyo et al. (2004), Schnellenbach-Held et al. (2004) and Udeaja and Tah (2004) discuss specific applications of agents and MAS in the construction supply chain in Chapters 7, 8 and 11 respectively.

AGENT-BASED INTERNATIONAL CRISIS NEGOTIATION

In this case, a strategic negotiation model of alternative offers is developed to facilitate the negotiations among the three parties of a hostage crisis. In this model, agents are self-motivated, rational and autonomous, each with its own utility function. Both parties can opt out, and while one loses over time, the other gains (up to a point). Other specific issues are the conflicting objectives and the utility functions of parties, and the impact of each item on bargaining behaviour in a crisis. The study provides strategies for a wide range of situations, which satisfy the criteria: symmetrical distribution, simplicity, instantaneous, efficiency and stability (Kraus and Wilkenfeld, 1993).

AGENT-BASED PROJECT MANAGEMENT

Integrated project management (IPM) means that design and project planning are interleaved with plan execution, allowing both the design and plan to be changed as necessary. This requires the right change to be propagated through plan and design. This study develops a *Redux* model to facilitate the IPM, where agents communicate their goals and decisions to the *Redux server*, uses a set of rules to maintain the consistency of goals and decisions, and propagate the effects of design changes in a collaborative design environment. The main problems addressed in the *Redux server* include conflict resolution when design constraints are violated, and communicating any design changes to appropriate team members whose decisions will be affected by such changes. Such timely communication will reduce the impact of making design decisions on the basis of obsolete data/information, and this level of message propagation involves some reasoning by the agent that generates such changes (Petrie et al., 1998). Further examples of MAS can be found in Rosenschein and Zlotkin (1994), Sandholm and Lesser (1995), Decker et al. (1997), Kraus et al. (1995), Kraus (1996, 1997), Jennings et al. (1998), Coen (1995), Nwana and Ndumu (1999) and Shen et al. (2001).

With the wide range of applications, some key aspects of MAS have gained considerable attention from researchers such as: interactions between self-interested agents, negotiation protocols and learning approaches of agents. For example, self-interested agents interact in a shared environment, which represents many real-world problems. Such domains include both systems where agents are adversarial to each other (e.g. bargaining parties) as well as domains where agents are indifferent to each other. In the former case, research concentrates on issues like modelling the knowledge and behavioural strategies of opponents, learning to exploit an opponent's weakness and developing interaction rules by which agents can arrive at equilibrium configurations (Sen, 1997). In the latter case, the key research problems are designing social laws, conventions and protocols by which each agent can achieve its own goal without significantly affecting the chances of others achieving their goals, then the whole society exhibits desirable behaviour – in other words, locally good behaviour implies globally good behaviour (Jennings et al., 1998).

The goal is to design a mechanism for self-interested agents such that if agents follow this, the overall system behaviour will be acceptable. For example, economic-based approaches, and market mechanisms in particular, are becoming increasingly attractive to MAS researchers both because of the ready availability of underlying models, and their potential applicability in Internet-based commerce. In such approaches, agents are often assumed to be self-interested utility maximisers. The areas where economic-based approaches have been applied to MAS research to date are resource allocation; task allocation and negotiation. The next section will focus on the general negotiation mechanisms in MAS.

3.4 Negotiation in multi-agent systems

In systems composed of multiple autonomous agents, negotiation is a key form of interaction that enables groups of agents to arrive at a mutual agreement regarding belief, goal or plan. Particularly, because the agents are autonomous and, in many cases, are self-motivated, they must influence others to convince them to act in certain ways (Beer et al., 1999). Negotiation is used more specifically for conflict resolution and avoidance (Cammarata et al., 1983; Sycara, 1988; Adler et al., 1989; Klein, 1991), task allocation (Davis and Smith, 1983; Durfee and Montgomery, 1990; Zlotkin and Rosenschein, 1994) and resource allocation (Adler et al., 1989; Sathi and Fox, 1989; Conry et al., 1992) and hence for the coherence of the agent society. The potential benefits of agent negotiation include saving time and money, efficiency for computationally intense negotiations searching for optimal results, and the ability to incorporate multiple negotiation strategies for changing environments.

3.4.1 Overview

Although negotiation is highly important for the modelling of MAS, there is no clear and common definition of what negotiation is in the agent world, and no formal theory of agent negotiations (Muller, 1996). Jennings *et al.* (1998) describe negotiations between agents as:

> The main characteristics of negotiation in terms of agent applications can be stated as: the presence of some form of conflicts that need to be resolved in a decentralised manner, by (self-interested) agents, under conditions of bounded rationality, and incomplete information. Furthermore, the agents communicate and actively exchange proposals and counter-proposals. Conflicts between agents may be about the limited available resources, conflict beliefs between agents, or disagreements on issues, such as price. In the first case, the negotiation becomes an optimisation problem; in the second case, at least one of the agents has to change its beliefs; and in the third case, the negotiation is a bargaining situation. It is often difficult to see what exactly one wants to achieve when resolving the situation.

However, many other researchers like Smith (1980) and Shen *et al.* (2001) consider negotiation as a more general approach to agent co-ordination. It is not necessary that there is a conflict between agents as a condition of negotiation. Negotiation could be simply for the purpose of a better solution to a problem. Perhaps, a basic definition is that of Bussmann and Muller (1992): '...negotiation is the communication process of a group of agents in order to reach a mutually accepted agreement on some matter'. This section will discuss agent negotiation in a broad sense (i.e. consider negotiation mechanisms for both DPS and MAS).

Despite the ambiguous term and the wide range of negotiations – from situations involving task and resource allocation to situations involving agent to agent bargaining, the negotiations are intended to improve the global state of affairs or to achieve individual agents' objectives such as: minimising the time to find a solution; minimising the total resource usage in doing so; maximising the quantity of the result; trying to achieve Pareto optimality (i.e. the outcome maximises the product of the agents' utilities) or trying to reach a Nash equilibrium (Shen *et al.*, 2001).

To negotiate with one another, each individual agent should be able to make proposals counter-proposals, accept or reject proposals and generate arguments in support of their adopted stance on the negotiating subject. Preist (1999) extends this with some points relating specifically to competitive negotiation:

- First, all parties should be clear about the belief sets of each of the agents involved in the negotiation process. At the very least a shared

ontology of the domain being negotiated over, or a way of establishing one needs to exist.
- Then they must be able to make contact with potential negotiators and recognise the type of 'game' in progress. Questions, such as the number of negotiators and the bargaining power of individuals, need to be considered and an appropriate strategy based on this information selected.
- During the negotiation, it requires both a means to evaluate the relative value to the agent of different offers made during negotiation and a means to estimate the relative value to other agents of offers the agent may potentially make.
- Finally, in some situations, the ability to exchange constraints on acceptable offers and to reason with these constraints may make negotiation more efficient.

The important issues in agent negotiation include the application domain, system architectures and infrastructures, interaction protocols, and the architecture and strategies adopted by individual agents. Muller (1996) summarises both the agent and group negotiation issues into three major categories (Table 3.3):

- *Negotiation language category* – research is concerned with the communication primitives for negotiation, their semantics and their usage in terms of a negotiation protocol. This category also comprises the investigation of the structure of the negotiation topics.
- *Negotiation decision category* – algorithms to compare the negotiation topics and correlation functions for them are discussed. The definition of utility functions and the representation and structure of the agents' preferences are fixed and negotiation strategies also fall into this category.
- *Negotiation process category* – general models of the negotiation process are investigated and the global behaviour of the negotiation participants is analysed.

Table 3.3 Categories for agent negotiation

Negotiation language category	Negotiation decision category	Negotiation process category
Protocols	Utility	Procedure
Primitives	Matching	Behaviour
Semantics	Preferences	—
Object structure	Strategies	—

Source: Muller, 1996.

Many other researchers (Jennings *et al.*, 1998; Nwana and Ndumu, 1999; Shen *et al.*, 2001) have different prospects about the major features of agent negotiation. For example, Shen *et al.* (2001) outlines the three main features of agents negotiation as: language used by participating agents, protocol followed by agents as they negotiate, and the decision process each agent uses to determine its position, concessions and criteria for agreement. Nwana and Ndumu (1999) emphasise the importance of a common set of speech acts, a common service ontology and a common set of prescriptive conversation policies (i.e. protocols for negotiating agents).

3.4.2 Agent negotiation mechanism

Many factors need to be considered prior to designing an agent negotiation mechanism (Ren *et al.*, 2002):

- First, autonomous agents, such as a computer system, work differently from human negotiators. Rule-based (mechanical) negotiation theories are essential for the agent negotiation system, whilst some behavioural theoretical models, like learning approaches, are also applicable to the system.
- MAS negotiation mechanism should consider some highly desirable properties such as the ability to guarantee convergence, Pareto-optimality or equilibrium. The specific application scenario and objective should be reached. Meanwhile, factors like efficiency and symmetry should also be considered. For example, the designed negotiation mechanism should prevent agents from spending too much time on negotiation, and therefore not keeping to their timetables for satisfying their goals. This could be achieved by assigning appropriate time-based penalties into the negotiation process.
- Finally, many other specific factors will also influence the design and implementation of a negotiation mechanism such as:
  a information source (complete vs. incomplete);
  b influence processes to change position (rationality in term of value, risk or utility);
  c level of shared goals among agents (all sharing the same goals vs. self-interested agents);
  d number of agents (a few vs. very large number);
  e type of agents (automated agents vs. system composed of people and automated agent; reactive vs. learning);
  f communication and computation costs (the availability and costs of communication vs. the computation capability and costs); and
  g agent organisations (peer-to-peer vs. superintendent to subordinate).

Different approaches have been adopted to develop agent negotiation systems and techniques, which can be categorised as either environment-centred or agent-centred. Environment-centred designers ask the question: How can the rules of the environment be designed so that the agents in it, regardless of their origin, capabilities or intentions, will interact productively and fairly? Developers of agent-centred negotiation mechanisms focus on the question: What is the best strategy for an agent to follow, given an environment in which the agent must operate?

Given the ubiquity and importance in many different contexts, negotiation mechanism covers a broad range of phenomena, and encompasses multifarious approaches. Despite this variety, agents' negotiations are generally composed of two phases: a communication phase where information relevant to the negotiation is communicated to participating agents, and a bargaining phase where 'deals' are made between individuals through relaxation of initial goals, mutual concessions, lies or threats (Adler *et al.*, 1989). Agent negotiation mechanisms mainly address these two aspects which cover three main broad areas: negotiation objects, negotiation protocol and negotiation strategy.

NEGOTIATION OBJECTS

Negotiation objects are the range of issues over which agreement must be reached. Particularly, to what kinds of agreements can the agents come? The object may contain single issue, such as price or multiple issues relating to price, timing, quality, etc. Also relevant are the allowable operations on these objects. In the simplest case, the structure and contents of the agreement are fixed, and negotiation amounts to accepting or rejecting the offer. The next level, however, offers flexibility to change the values of the issues in the negotiation object through counter-proposal, changing the structure of the negotiation object and so on. Finally, participant might be allowed to dynamically extend the structure of the negotiation objects (Jennings, 2000).

NEGOTIATION PROTOCOLS

Given a set of possible deals, negotiation protocols set the rules by which the agents will come to a consensus. In a broad and general sense, negotiation protocols cover the permissible types of participants (e.g. the negotiators and their relevant parties), the negotiation states (e.g. accepting bids, negotiation closed), the events that cause state transitions (e.g. no more bidders, bid accepted), and the valid actions of the participants in a particular state (e.g. which can be sent by whom, to whom and when) (Muller, 1996).

In a more specific sense, negotiation protocols set the stage for the negotiation process. They contain the basic rules for a negotiation process and communication since negotiation involves exchange of messages, protocol structures called conversations and definite classes of dialogue. The simplest dialogues are found in contract-net approaches where they are limited to exchange involving offers, bids and grants of contract. More complex dialogues are found in human types of negotiation, when trying to change other agents' beliefs. Reeds (1998) identifies five types of dialogue: persuasion, negotiation, inquiry, deliberation and information seeking. For example, persuasion dialogue covers the case of conflicting beliefs. Negotiation dialogue differs procedurally from persuasion in an important respect, in that coherence between beliefs is not demanded, and the relevant beliefs of the participants may well remain at odds after negotiation.

There are two approaches to dialogue. One is an unmediated approach which involves bilateral or multilateral, communication between all of the agents. That mechanism may not scale well. In the mediated approach, agents submit messages to some institution implementing the mechanism. The process may be iterative, with the institution providing some feedback based on previous messages received. The process terminates under conditions prescribed by the mechanism rules (Wellman and Wurman, 1998).

NEGOTIATION STRATEGIES

Given a set of possible deals and a negotiation protocol, negotiation strategies provide the decision-making apparatus by which participants attempt to achieve their objectives. It determines from which possible alternative actions the agent will choose at each step. Each agent's strategy will strongly depend on the type of application it is involved in. For example, an agent may update its local states and perform its function based on certain predefined rules if the agent works in a collaborative way. Or it may select to use a rational strategy to maximise its utility if the agent works in a competitive way.

The relative importance of the negotiation protocols and strategies varies according to the negotiation and environmental context. In some circumstances, the negotiation protocol is the domain concern. For example, the system designer may determine that the negotiation is best organised using a particular form of action. This mechanism, design choice, constrains the type of operations that can be performed on the negotiation object and prescribes the behaviour of the agent's decision-making models. In other cases, however, the agent's strategy is the domain concern. The protocol does not prescribe an agent's behaviour, and there is scope for strategic reasoning to determine the best course of action. In such a case, the relative success of two agents is determined by the effectiveness of their individual strategies.

3.4.3 Negotiation models

To make agent systems work in the complex environment context, different negotiation models have been developed from various aspects, either for DPS or for MAS problems. For example, negotiation has been examined under a game-theoretic approach in which every agent knows all relevant information about other agents (Genesereth *et al.*, 1986; Zlotkin and Rosenschein, 1989; Kraus and Wilkenfeld, 1991), and under conditions where agents are hostile and completely unwilling to share private information (Sycara, 1988). Various points along the co-operation/hostility continuum are examined in (Zlotkin and Rosenschein, 1989, 1990). Negotiation can occur among peers (Cammarata *et al.*, 1983; Rosenschein and Zlotkin, 1994), through a mediator or arbitrator (Sycara, 1988), or hierarchically through an organisation (Davis and Smith, 1983; Durfee and Montgomery, 1990).

By focusing on the design of a negotiation mechanism, the following section examines a few important negotiation protocols and related strategies in both DPS and MAS domains. Jennings *et al.* (1998) classify the current agent negotiation models as AI, game theory and psychology-based negotiations. Shen *et al.* (2001) make a more thorough classification, which includes contract, plan, market, game theory and AI-based negotiations. With the quick development of agent negotiation systems, many models are specific to their application environments, and thus not easily classified.

3.4.3.1 Contract-based negotiation

Smith (1980) introduces a simple negotiation mechanism among co-operative agents in DPS environment, called CNP, in which an agent having some work to subcontract broadcasts an offer and waits for other agents to send bids. After some delay, the best offers are retained and contracts are allocated to one or more contractors who process their subtasks. The CNP provides for co-ordination in task allocation, with dynamic allocation and natural load balancing. This approach is quite simple and can be efficient.

However, the CNP fails to capture many intuitive and important aspects of the negotiation process. For example, bidders cannot counter-propose better options, they cannot modify any of the service agreement parameters, and the emphasis in devising a complete specification is placed solely with the task manager. Also, when the number of nodes is large, the number of messages on the network increases, which can lead to a situation where agents spend more time processing messages than doing the actual work or worse, the system stops through being overloaded with messages.

Various improvements and extensions to the basic CNP have been proposed. In the most basic approach, the choice of a contractor is done by comparing bids corresponding to a particular offer, using whatever mechanisms are relevant to the problem. For example,

- Malone et al. (1988) developed a Distributed Scheduling Protocol (DSP) by overlaying CNP with an economic model. They introduce a motivation framework in the terms of economic theory, and provide a more theoretical term in which to discuss the task-sharing algorithm, while the autonomy agents willingly bid for tasks without explicit motivation in Smith's (1980) original work. DSP includes two primary dimensions: (a) contractors select manager's tasks in the order of the tasks' numerical priorities and (b) managers select contractors that satisfy the minimum requirements to perform the job.
- Sandholm (1993) presents a modified version of the CNP for competitive agents in the transportation domain. It provides a formalisation of the bidding and the decision awarding processes, based on marginal cost calculations that are based on local agent criteria. More importantly, an agent will submit a bid for a set of delivery tasks only if the maximum price mentioned in the tasks' announcement is greater than what the deliveries will cost that agent.
- A more general approach proposed by Sandholm and Lesser (1997b) is to develop protocols with continuous levels of commitment based on a monetary penalty method, where commitment varies from the original bounded to breakable as a continuum by assigning a commitment breaking cost to each commitment separately. This cost can increase with time, decrease as a function of acceptance time of the offer, or be conditions on events in other negotiations or the environment.

3.4.3.2 Plan-based negotiation

Plan-based negotiation is based on co-operation protocols and strategies for resolving conflicts among the plans of a group of agents. As an important negotiation technique, it suffers from limitations inherent in centralised or distributed multi-agent planning. Durfee and Lesser (1991) developed a Partial Global Planning (PGP) approach, which requires agents to exchange descriptions of intermediate situations and results, enabling them to check for potential task overlaps and decide which agent should do what work. PGP is a flexible and dynamic approach to co-ordination that does not assume any particular distribution of sub-problems, expertise or other resources. Agent interactions take the form of communicating plans and goals at an appropriate level of abstraction. These communications enable a receiving agent to form an expectation about the future behaviour of a sending agent, adjusting its own local planning appropriately, thus improving agent predictability and network coherence.

Conry et al. (1988) proposed a negotiation protocol called multi-stage negotiation for co-operatively resolving resource allocation conflicts. They were specifically concerned with negotiation strategies for distributed constraint satisfaction problems, where a group of agents have a goal, but each agent has only limited resources. The local constraints give rise to a complex set of global and inter-dependent constraints. This investigation is done in the context of the monitoring and control of a complex communication system. Their implementation involves developing algorithms for multi-agent planning, taking the inevitable conflicts into consideration.

Kreifelt and Von Martial (1991) developed a negotiation approach, where negotiation consists of two stages: first, agents plan their activities separate, and then second, co-ordinate their plans. A separate co-ordination agent carries out the co-ordination of all the agents' plans. The negotiation protocol is described in terms of agent states, message types and conversation rules. The problem of this approach is that it does not actually present a negotiation model but just prescribes one, and it is really left to the agents to achieve consensus.

3.4.3.3 Market-based negotiation

The goal of market-based negotiation is to resolve a distributed resource allocation problem. Agents are classified as producers and consumers of goods and services. Equilibrium is reached when the price of goods is such that all resources are being used up. In this situation, a particular agent may want to acquire goods, but it could be limited by budgetary constraints. Thus, it will make offers based on the current price of goods and its own preferences. It has an internal utility function, and its goal is to increase utility, which corresponds to the hypothesis of rational behaviour. Producers have a specific production technology, and seek to maximise their profits. Given a set of prices, the trading process involves a sequence of offers in which each consumer states how much of each resource it wants to purchase. If the demand differs from the supply, then prices will have to be adjusted by the producers (Shen *et al.*, 2001).

Mullen and Wellman (1996) adopt a market-based negotiation for a digital library service. In this approach, an alternative information service is treated as competing with economic activities. Given a measure of priorities over the end-user services provided, the various agents effectively compete to provide the highest level of service using the minimal computational resources. One central capability of the agent is thus to be able to reach agreements on suitable compensation. The goal is to achieve an efficient overall allocation of resources towards the optimal provision of services to users.

One of the drawbacks of the market-based approach using prices as a primary controlling mechanism is that the convergence process may be slow, involving a large number of offers (and computations). A new approach focusing on resources rather than prices was introduced by Ygge and Akkermans (1998). It appears to be complex but efficient. One appealing feature of the new approach is that, in each iteration, the computed scheme is feasible although not optimal. In project *KASBAH* (Chavez *et al.*, 1997), negotiation is done between one buyer and one seller, there are no globally maintained prices and the information that each agent can have is limited, which is more in line with actual web applications. Lee (1998) proposes a strategy of risk re-distribution, trying to minimise the amount of computation in each agent. Other approaches based on probability theory have been proposed by Ekenberg *et al.* (1995).

Considering the natural link between the market-based approach and the CNP, the latter has been incorporated with various market-based strategies depending on the type of contract an agent can consider such as the de-commitment of contract discussed by Sandholm and Lesser (1995). Various types of strategies exist; some of them have been studied by Matos *et al.* (1998): time dependent, resource dependent or behaviour development.

3.4.3.4 Game theory-based negotiation

Although various negotiation mechanisms have been developed based on the above approaches, these negotiation models are mainly used to resolve conflicts in the DPS domain. They are far from satisfying the requirements of real-world problems. Hence, many negotiation mechanisms have been developed for MAS. Most of them (Roth, 1979; Zlotkin and Rosenschein, 1991, 1994; Kraus and Wilkenfeld, 1993; Kraus *et al.*, 1995; Pena-Mora and Wang, 1998; Ugwu *et al.*, 1999a,b) are based on game theory.[1]

There is a particular match between game theory and agent-based systems. One of the major assumptions of game theory is that the players are rational. When game theory is applied in analysing human negotiation, the problem regularly faced is that human beings do not always act rationally, and frequently do not have consistent preferences over alternatives. On the other hand, agents, being *pre-programmed* in their behaviour, make concrete the notion of 'strategy' which plays a central role in game theory – the idea that a player adopts rules of behaviour before starting to play a given game, and that these rules entirely control its responses during the game (Rosenschein and Zlotkin, 1994). Moreover, the game solutions, like equilibrium points, stable strategies and Pareto-optimal solutions are also the basis of the MAS negotiation solutions.

Game theory has some other essential assumptions, such as utility maximisation, complete knowledge, isolated negotiation, inter-agent comparison of utility, symmetric ability, binding commitments and no explicit utility transfer. These assumptions are also the basis for game theory-based negotiations. Zlotkin and Rosenschein (1991, 1994) explain that their work on MAS followed the general direction that treats negotiation in the spirit of game theory, while altering game theory assumptions which are irrelevant to MAS. The main point is that by appropriately adjusting the rules of the game by which the programs must interact, the private strategies of agents can be influenced. Certain strategies simply become the best for an agent to adopt given the rules of the game (Rosenschein and Zlotkin, 1994). A number of examples are presented in the ensuing paragraphs.

Zlotkin and Rosenschein (1991) study the problem of incomplete information in MAS negotiation where agents need to reach an agreement on task allocation. The incomplete information is either about the opponent's goals or about the value of its goals. By adopting certain game-theoretic techniques to model communication and promises, they introduce a mechanism that they called 'one negotiation phase' in which agents simultaneously declare private information before beginning the negotiation. They also identify situations and protocols where agents have incentives to tell the truth in 'one negotiation phase' and cases where it is beneficial for agents to lie. In their study, the process of negotiation was severally restricted and it assumed that each agent knew the complete payoff matrix associated with the interaction.

Zlotkin and Rosenschein (1994) extend the previous work. They explore various situations in different negotiation domains and the possible negotiation protocols and strategies where they divide negotiation domains into three: task orientated domains (TODs), state orientated domains (SODs) and worth orientated domains (WODs), with each domain being a generalisation of the previous one.

- TODs are the simplest case where an agent's activity is defined in terms of the set of tasks it has to achieve. It is assumed that all resources are available to the agent, the benefit of negotiation being the redistribution of tasks amongst a group of agents. There is no possibility of deadlock as all agents can proceed with their original task list and be no worse off. This is clearly limited.
- SODs deal with problems where agents wish to change their environment from an initial state to some goal state. The classic AI Blocks World problem is a good example. There is the possibility of conflict and deadlock, as agents may have different goals, and to satisfy all may be impossible or require more effort from each agent than if they were alone in the world. In this situation agents must be able to make concessions.
- WODs are domains where agents attach a worth to each potential state. This allows much more flexible goals to be set and allows concessions to be made on these goals. An example would be agents in an electronic marketplace where the goal for a seller may be to obtain the highest price for x within time y. There is again the possibility of conflict and deadlock, but now within a more complicated bargaining environment.

Within these domains, Zlotkin and Rosenschein (1994) explore three different types of situation:

- co-operative, in which the cost to both agents of a joint plan that achieves all goals is less than or equal to any individual plan that would achieve a single agent's goal;
- compromise, in which at least one agent will have to pay more to achieve its goal jointly than it would to achieve them individually but, given the inevitable presence of another agent, a deal can be made that achieve the goals; and
- conflict, in which at least one agent has to pay an unacceptably high cost for any joint plan and, therefore, no deal can be made.

Kraus (1996) studies the multi-interactions among self-motivated agents where agents do not have complete information. By adopting different game theory techniques, she develops a strategic model that not only takes into consideration the passage of time during the negotiation (i.e. time constraints), but also includes belief systems. Using a distributed mechanism, agents negotiate and can reach efficient agreements without delays. She also assumes that the set of possible agreements is limited and that there is full information, but she makes fewer restrictions on the negotiation procedure. The study provides strategies for a wide range of situations, which satisfies the criteria: symmetrical distribution, simplicity, instantaneous, efficiency and stability.

Although game theory provides a theoretical basis for MAS negotiation mechanism, it is not a perfect solution. Problems like complete information assumption, inflexibility and inadequacy or inappropriateness for real-world situations are all barriers to the application of game theory in MAS. As a result, many researchers have attempted to incorporate other negotiation mechanisms like the bargaining theoretic approach, with game theory.

3.4.3.5 AI-based negotiation

Several common AI approaches have been adopted in developing agent negotiation mechanisms such as: case-based reasoning (Sycara, 1987), negotiation search (Lander and Lesser, 1991, 1993) and (k-b) approach (Werkman, 1990).

Sycara (1987) presents a model of negotiation, called *Persuader*, that operates in the domain of labour negotiation. It involves three agents (i.e. a union, a company and a mediator), and is inspired by human negotiation. It models the iterative exchange of proposals and counter-proposals in order for the parties to reach an agreement. Agents can modify other agent's beliefs, behaviours and intentions via persuasion. Each agent's multi-dimensional utility model is private knowledge. Belief revision is used to change the agent's utilities so that an agreement can be reached via persuasive argumentation. A case-based reasoning mediator is incorporated into the model with the idea that human negotiators make decisions with reference to past negotiation experiences.

Lander and Lesser (1991, 1993) develop a negotiation system, called TEAM, that explores negotiation search for conflict resolution among heterogeneous and reusable agents in the domain of DPS. In TEAM, an extended search is carried out by an agent recognising a conflict in order to find another solution in its local search space that avoids the conflict, while relaxation is used to expand the local solution space. In the negotiated search, loosely coupled agents interleave the tasks of: local search for a solution to some sub-problem; integration of local sub-problem solutions into a shared solution; information exchange to define and refine the shared search space of the agents, and assessment and reassessment of emerging solutions. A selected solution is acceptable if it satisfies the requirements imposed by each agent.

Werkman (1990) proposes a (k-b) model of incremental negotiation in DPS domain. This scheme uses a shared knowledge representation, which allows agents to negotiate in a similar manner to co-operating experts with a common background of domain knowledge. Essentially, it explores a blackboard having partitions for requested proposals, rejected proposals, accepted proposals, communications and shared knowledge. The negotiation in this model follows a procedure of proposal, evaluation and counter-proposal. An arbitrator agent is introduced to help agents to resolve possible deadlocks, by reviewing their negotiation dialogue and using their mutual information network to generate alternative proposals. This is done using issue relaxation techniques or some intelligent proposal generator approach. This approach may fail to achieve resolution in which case the arbitrator may set time limits or use other techniques.

3.4.3.6 Psychology-based negotiation

Behaviour negotiation theory is an important approach in analysing negotiation because it reflects the human negotiators' psychological responses during negotiation. Socio-psychology aspect is always a central point in analysing human negotiations. Bussmann and Muller (1992) present a cyclic negotiation model in the DPS domain based on Gulliver's (1979) eight phases of negotiation process. This model addresses the limitation of other negotiation proposals and models such as market, game theory and AI approaches. The cyclic nature of the model also addresses the thorny issue of conflict resolution. The general strategy is that the negotiation begins with one, some or every agent making a proposal. Next, the agents evaluate and check the proposals against their preferences, and criticise them by listing any of their preferences violated by the proposal. The agents then update their knowledge about the other agent's preferences and the negotiation cycle resumes with a new proposal or proposals in the light of this newly learned information. Conflicts between the agents are handled in a concurrent conflict resolution cycle.

3.4.3.7 Other approaches

Besides these negotiation approaches, many other negotiation mechanisms have been developed to address different aspects of negotiation in different environments. These include the heuristics approach (Kraus and Lehmann, 1995), the flowchart approach (Polat *et al.*, 1993) and the argument-based approach (Sierra *et al.*, 1998).

One of the more important ideas is the argument-based negotiation. Sierra *et al.* (1998) proposed an argument-based negotiation in the domain of business process management. The emphasis is on how the agents can justify their negotiation stance, and how they persuade one another to change their decisions. Negotiators provide arguments to support their stance. Thus, in addition to generating proposals, counter-proposals and critiques, the negotiator is seeking to make the proposal more acceptable by providing additional meta-level information in the form of arguments for its position. The nature and types of the arguments can vary enormously. However, common categories include threats, rewards and appeals. Whatever its precise form, the role of the supporting argument is either to modify the recipient's region of acceptability or its rating function over this region. In so doing, arguments have the potential to increase the likelihood and/or the speed of agreements being reached. In the former case, this is by persuading agents to accept deals that they may previously have rejected. In the latter case, this is achieved by convincing agents to accept their opponent's position on a given issue (and to cease negotiating over it).

3.4.4 Negotiations in distributed artificial intelligence and multi-agent systems

Although the earlier discussions have, explicitly or implicitly, addressed the working domain of each negotiation mechanism, it is necessary to clarify the differences between the negotiation mechanism in these two domains.

The negotiations in DPS domain are often termed as 'co-operative negotiation'. They are often used in situations where agents have 'a global goal/single task envisioned for the system' (Smith and Davis, 1983). System designers impose an interaction protocol and a strategy for each agent. The main question is what social outcomes follow given the protocols and assuming that agents use the imposed strategies. Systems like Contract Net, Distributed Scheduling and some market-based approaches are typical examples of negotiation in DPS.

The negotiations in the MAS domain are often regarded as 'competitive negotiation'. They are often used in situations where 'agents of disparate interests attempt to make a group choice over well-defined alternatives' (Rosenschein and Zlotkin, 1994). The agents are provided with an interaction protocol, but each agent will choose its own strategy to maximise its own good without concern for the global good. The protocols need to be designed using a non-co-operative strategic perspective. The main question is what social outcomes follow given a protocol that guarantees that each agent's desired local strategy is best for that agent and thus the agent will use it. This approach is required in designing robust non-manipulation MAS where agents are constructed by separate designers and represent different real-world parties. Compared with negotiation in DPS, such self-interest naturally prevails in negotiations among independent businesses or individuals. In building computer support for negotiation in such settings, the issue of self-interest has to be dealt with.

To improve the quality of the MAS negotiation mechanism, the adoption of proper negotiation theoretic models is crucial. Such models should match the nature of negotiation objects, suit the application environment and reflect the characteristics of MAS. The negotiation cannot be tackled by technological or economic methods alone. Instead, the successful solutions are likely to emerge from a deep understanding and careful integration of both.

3.5 Learning in multi-agent systems

A major problem in the development of MAS is the difficulty for a developer to foresee all potential situations an agent could encounter and specify agent behaviour optimally in advance. Therefore, it is widely recognised in the agent community that one of the more important features of high-level agents is their capability to adapt, to learn and to modify their behaviours. Weiss (1993) concludes that the two important reasons for learning in MAS are: to be able to endow artificial MAS with the ability to automatically improve their behaviour; and to get a better understanding of the learning processes in natural MAS.

3.5.1 Overview

As a group, agents work in an open, complex and dynamic environment, which results from a number of factors such as (extended from Prasad and Lesser, 1999):

- *environment uncertainty* – it is impossible to define all conditions before the systems start to work;
- *dynamic environment* – the system exists in an environment whose conditions vary over time;

- *communication constraints* – every communication link has limited parameters such as range and bandwidth and a certain noise level;
- *degree of clustering* – in the case of a larger number of agents, it is advantageous to divide them into groups according to their functions; however, this functional grouping is limited;
- *time stress* – the time for decision-making is not infinite, especially in real-time systems the question of quick response plays a vital role;
- *option multiplicity* – it represents the number of planning options available to each agent;
- *density of the solution space* – it represents the ratio of acceptable, conflict-free plans to the number of potential plans;
- *complexity of interactions between agents* – an agent's activities might lead to a change in the other agents' decisions; and
- *varying goals, abilities, preferences, skills and levels of knowledge of individual agents.*

Agents in such a system face uncertainties due to their partial views of the other agents and the environment. Incomplete information about the progress, characteristics, expectations or preferences of the other agents, and generated partial results may lead to global incoherence and degradation in system performance (Weiss, 1996). To effectively utilise opportunities, agents need to learn about other agents and adapt their local behaviour based on group composition and dynamics.

As a result, there has been considerable research on MAS learning for particular applications such as: the prisoner's dilemma (Sandholm and Crites, 1995), predator/prey (Nagayuki et al., 2000), agent co-ordination (Prasad and Lesser, 1999), engineering design (Grecu and Brown, 1998a,b) and negotiation (Zeng and Sycara, 1998), with varying degree of success. The following sections explore the major characteristics of agents, and approaches to agent learning, with a focus on the learning during negotiations.

3.5.2 Issues of learning

Learning in MAS can be defined operationally to mean the ability to perform new tasks that could not be performed before or to perform old tasks better as a result of changes produced by the learning process. In a stronger and more specific meaning, 'multi-agent learning' refers only to situations in which several agents collectively pursue a common learning goal. In a weaker and less specific meaning, 'multi-agent learning' additionally refers to situations in which an agent pursues its own learning goals, but is affected in its learning by other agents, their knowledge, beliefs and intentions (Weiss, 1996). Agents learn in a communal way; their learning is influenced by individual goals and preferences, exchanged information, shared assumptions, commonly developed viewpoints, environment, social and cultural convention and norms which regulate and constrain their behaviours and interaction. To conduct effective learning in MAS, several important issues need to be addressed before and during the learning process such as: the objectives and focuses of learning, the key components of learning and the approach to learning.

3.5.2.1 Objective of learning

The learning objective is highly dependent on the application domains and the goals of each individual agent. An agent in DPS may learn from the others and the environment, and adapt its behaviour for better co-operation or solution of a specific problem; whilst an agent in MAS often learns for its own benefit. Therefore, the learning objectives and learning approaches in different systems are often different.

The items which an agent expects to learn from others could be the preferences, utility functions, risk attitudes, tasks, strategies, sequences of actions or plans, specific domain knowledge, prediction of decisions, types of conflicts and so on. For example, in a MAS negotiation domain, an agent may expect to achieve the objectives through learning as described in following sections.

CHANGING OWN BELIEFS AND LEARNING ABOUT OTHERS' BELIEFS

Agents can hold different beliefs about the same fact. As a result of the knowledge and information exchanged during the negotiation, an agent can change its beliefs, if the new belief is supported by more powerful evidence. The change of beliefs through learning can imply significant changes in the proposal generated by an agent. As beliefs are seen as directly related to preferences, learning in this direction can influence the preferred order of an agent's decisions in a multiple solution situation. Assuming that an agent has acquired knowledge from the external environment, it can influence other agents to revise their beliefs as a result of the negotiation and thus propagate the external influence (Grecu and Brown, 1994).

Meanwhile, an agent can make inferences about the other agents' beliefs, and analyse their intentions and further negotiation strategies, so that it can determine its negotiation strategies accordingly. Such learning plays an extremely important role for many agents. Generally, the key negotiation features, like agents' utility functions, risk attitudes and reservation values represent agents' beliefs.

LEARNING NEGOTIATION STRATEGIES

Depending on different negotiating situations, the negotiating agents not only need to understand the opponents' beliefs and intentions, but also are urged to know the opponents' negotiation strategies. Or in some other cases, when both parties' objectives are relatively apparent, negotiation results depend on the negotiation strategies taken by the negotiating agents. However, to understand the negotiation strategies is one of the most complex problems even if the agent has some knowledge about the opponent's beliefs. In multi-step negotiations, agents could change their strategy dynamically, and it is very hard to tell whether and when a change of strategy occurs (Cross, 1977). By analysing the difference between the predicted response and the actual response, an agent can adjust its perceptions about the opponents' strategies.

Grecu and Brown (1994) emphasise the importance of learning negotiation strategies for the learning agent itself. The results of negotiation could be used to evaluate the quality of specific negotiation steps. This could reinforce the agent's drive to use the same sequence of actions in a similar situation or, alternatively, to weaken that drive. More generally, a negotiation history can be used to extract and to compile particularly useful sequences of negotiation actions. Agents might recognise the applicability of a strategy at moments where the existing commitments prevent the execution of the initial steps of that strategy. Otherwise, the strategy might prove to be inefficient if used continuously during negotiation. In addition, an agent's observation of the other agents' actions, guided by its strategies, may eventually lead to the synthesis of new strategies.

LEARNING CONFLICT PATTERNS

Based on participation in different conflict resolution processes, an agent can learn to recognise and classify different types of conflict. This implies learning about the context leading to conflicts and learning the characteristics of conflict. The first factor is important for avoiding conflicts, the second one for taking negotiation decisions. The negotiation model proposed by Klein (1991) uses a hierarchy of possible types of conflicts between agents. An inductive learning approach could be used to automate the construction of such a classification scheme and to facilitate adding new conflict types.

Although these learning cases, especially the first two, represent some of the essential learning objectives in MAS negotiation systems, the learning objectives could be various depending on different application domains such as: negotiation issues, domain knowledge or environment issues.

3.5.2.2 Key elements in learning

Learning in a multi-agent environment can be a very complex process, considering that agents learn mutually, and the environment continuously changes. Moreover, agents' actions and strategies are often not directly observable, the action taken by the learning agent can strongly bias the range of behaviours that are encountered (Grecu and Brown, 1998a). However, no matter how complex the learning process is, there are always three essential components in the process, on which the learning agent bases its learning. These are agents' expectations, feedback information, and credit assignment to evaluate the feedback information and decision-making.

EXPECTATIONS

Expectations are the bases for agent learning. In a general MAS domain, expectations represent an agent's beliefs that events will occur in a pre-defined way. Expectations encode the agent's current knowledge of an event and the global environment in which it operates, and represent a basis for action in a partially observable and partially computable world (Grecu and Brown, 1998b). Expectations also reflect the criteria through which the agent relates to the environment and the expectations of an agent guide its decision-making during the operation process. In a negotiation domain, an agent's expectations determine what and how much the agent expects to get from the others or what the others will do.

On the other hand, an agent's expectations are limited in its anticipatory power due to the constraints imposed on perceiving the other agents and the environment. During the system operation process, an agent plans to perform a task according to its expectations, but may fail to do that because an event does not occur or because other agents respond in an unexpected manner. It means the agent's expectations are violated, either in a good or a bad sense. The violation of the expectation indicates that the agent's knowledge about other agents, events or environment has limited validity. If it is not noticed or taken into consideration, the agent will repeatedly fail in outlining and implementing its actions. A continuous learning process makes it possible for an agent to modify its expectations to be more realistic during the negotiation process.

Hu and Wellman (1996) characterise an agent's belief process in terms of conjectures about the effect of their actions. A conjectural equilibrium is then defined, in which all agents' expectations are realised, and each agent responds optimally to its expectations. They present a MAS where an agent builds a model of the response of others. Their experimental results show that depending on the starting point, the agent may be better or worse than had it not attempted to learn about/from the other agents.

FEEDBACK

The availability of information is a primary requirement for the learning process. Sound and unbiased feedback information provides a learning agent with resources about the other agents' perceptions, the properties of events and the system's working environment. A feedback can originate from direct communication with other agents or indirectly, mediated by intermediary agents or without communication, directly through the learning agent's observations of the effects of its decisions and other agents' actions. A feedback can be biased by the path to the receiver and may also contain different information or be from several sources therefore its effect depends on how these sources are used: filtered, independently or in a combined manner. This feedback can also be affected and reduced by processes such as conflicts and backtracking, or hidden through social decision-making schemes.

EVALUATION CRITERIA

Evaluation criteria are closely related to the expectations of the learning agent. A basic problem that any learning system is confronted with is how the agent evaluates the feedback from the others as a response to the agent's last decisions or actions. In a more general sense, it is a problem of properly assigning credit for overall performance changes to each of the system activities that contributed to those changes (Weiss, 1996). The selection of performance criteria shapes the direction in which the system will evolve. The evaluation problem can be usefully decomposed into two sub-problems: the assignment of credit for an overall performance change to external actions, and the assignment of credit for an action to the corresponding internal decisions.

3.5.3 Methods of learning

The approach to learning is the most essential problem for any MAS learning problem. Most of the current approaches are extended from machine learning methods. A few researchers have made detailed analyses:

- Winston (1997) has drawn a big picture of the existing learning methods in MAS (Figure 3.3). Most of the efforts in AI focus on the first two classes of methods, with recent emphasis on the second class.
- Carbonell (1989) makes a summary of the major machine learning paradigms, which have a high potential of being applied in MAS. They are:
    i *inductive learning* – acquiring concepts from sets of positive and negative examples;
    ii *analytic learning* – explanation-based learning and certain forms of analogical and case-based learning methods, deductive methods and analytic methods;
    iii *genetic algorithms* – classifier systems; and
    iv *connectionist learning methods* – non-recurrent 'backdrop' hidden layer neural networks.

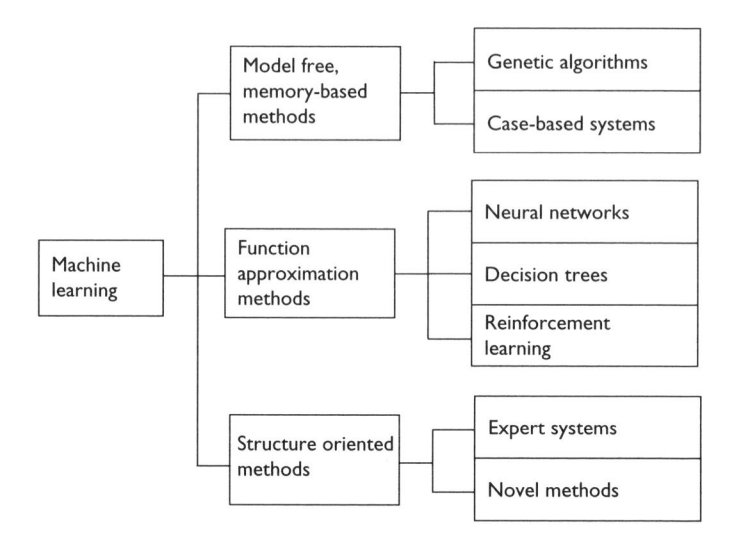

```
Machine learning
├── Model free, memory-based methods
│   ├── Genetic algorithms
│   └── Case-based systems
├── Function approximation methods
│   ├── Neural networks
│   ├── Decision trees
│   └── Reinforcement learning
└── Structure oriented methods
    ├── Expert systems
    └── Novel methods
```

Figure 3.3 The classification of learning methods for MAS.
Source: Modified from Winston, 1997.

- Jennings et al. (1998) conclude that the previous work related to learning in MAS is limited and much of this work relied on techniques derived from reinforcement learning, genetic algorithms and classifier systems.
- Grecu and Brown (1998a) list the learning approaches in the MAS engineering design domain as: explanation-based learning, concept/induction learning, knowledge compilation, multi-strategy learning, case-based learning, reinforcement learning, generic algorithm learning and neural networks. Some of these are briefly discussed here.

  i Reinforcement learning is based on the idea that the tendency to produce an action should be reinforced if it produces favourable results, and weakened if it produces unfavourable results. It uses few computation resources per example but requires a large number of examples.

  ii The concept learning model developed by Shaw (1996) closely follows the paradigm of collective group induction. Agents develop different hypotheses of concept description by seeing a set of training instances. Agents differ in the rules they use to generate their hypotheses. Hypotheses are periodically integrated into group hypotheses that achieve a better accuracy than the individual agent hypotheses.

  iii Many multi-strategy learning approaches use an initial learning phase to achieve acceptable levels of performance in agents. The agents' knowledge is then used to seed and evolve an agent population with refined performance. Gordon and Subramanian (1994) describe an agent learning approach in an embedded adversarial multi-agent setting. The first learning phase uses compilation to operationalise the agent's task knowledge. The second phase refines the rule sets representing the agents' task strategies through a genetic algorithm approach.

iv The genetic algorithm approach has led to a significant amount of experimental and theoretical results. Grefenstette and Daley (1996) describe experiments with co-evolutionary approaches that are similar to an ecological environment where species evolve during the interactions with each other. Their goal is to design behaviour strategies for intelligent robots in MAS environments. Haynes and Sen (1997) analyse crossover operators and fitness functions that allow rapid evolution of agents with good task performance.

- Weiss (1996) has undertaken a comprehensive study about the possible learning approaches in MAS according to the learning method and learning feedback. According to the learning method or strategy by a learning entity, the following methods are usually distinguished. A major difference between these methods lies in the amount of learning effort required by them, increasing from top to bottom:

  i *rote learning* – direct implantation of knowledge and skills without requiring further inference or transformation from the learner;
  ii *learning from instruction and by advise taking* – operationalisation, transformation into an internal representation and integration with prior knowledge and skills – of new information like an instruction or an advice that is not directly executable by the learner;
  iii *learning from examples and by practice* – extraction and refinement of knowledge and skills like a general concept or a standardised pattern of motion from positive and negative examples or from practical experience;
  iv *learning by analogy* – solution-preserving transformation of knowledge and skills from a solved to similar but unsolved problem; and
  v *learning by discovery* – gathering new knowledge and skills by making observations, conducting experiments and generating and testing hypotheses or theories on the basis of the observational and experimental results.

According to the learning feedback that is available to a learning entity, learning can be distinguished as:

a *supervised learning* – the feedback specifies the desired activity of the learner and the objective of learning is to match this desired action as closely as possible;

b *reinforcement learning* – the feedback only specifies the utility of the actual activity of the learner and the objective is to maximise this utility; and
c *unsupervised learning* – no explicit feedback is provided and the objective is to find out useful and desired activities on the basis of trial and error and self-organised processes.

In all three cases, the learning feedback is assumed to be provided by the system environment or the agents themselves. This means that the environment or an agent providing feedback acts as a teacher in the case of supervised learning and as a 'critic' in the case of reinforcement learning; in the unsupervised learning, the environment and the agents just act as passive 'observers'.

Also, learning in MAS can also be analysed according to other criteria such as:

- the purpose and goal of learning (i.e. learning to improve a single agent's skills and abilities or to improve the agent system's coherence and co-ordination as a whole);
- the categories of learning (i.e. only one of the agents gets involved in the learning process or all available agents are involved); and
- an agent's involvement in a learning process (the involvement of an agent is essential or not for achieving the pursued learning goal).

These learning paradigms emerged from quite different scientific roots, employ different computational methods, and often rely on different ways of evaluating success. Some of the learned elements have generality, which makes them applicable to any type of negotiation partner, some will apply only to agents of a particular type, while other learned elements represent information useful only when they negotiate with specific agents. Furthermore, different agents do not necessarily adopt the same learning method or the same type of learning feedback. In the course of learning, an agent may employ different learning methods and types of learning feedback. Finally, learning can occur not only during the negotiation process, but also afterwards. The negotiation history together with some evaluation techniques for the agent's past actions can be used to classify agents as more or less successful.

3.5.4 Challenging research issues

Although agent learning has gained a high interest in the MAS research community, the current learning approaches are very limited. For example, most current approaches to learning in MAS have been using purely reactive architectures where agents base their actions just on the current situation and not on the previous history, and there is no notion of deliberate planning towards an explicit goal (Kaelbling *et al.*, 1996). Such approaches are less than optimal in complex domains where (k-b) planning and complex co-ordination is necessary. In other words, although these learning techniques can be successful in restricted domains, they strip agents of the ability to adapt in a domain-dependent fashion, based on background knowledge of the respective situation. This ability is crucial in complex domains where background knowledge has a large impact on the quality of the agents' decision making.

Besides the learning techniques, Weiss (1996) points out some other aspects which challenge agent learning in MAS:

- requirements for learning in MAS;
- principles and concepts of learning in MAS;
- models and architectures of MAS capable of learning;
- extension and transformation of single-agent learning approaches to MAS learning approaches;
- parallel and distributed inductive learning in MAS;
- multi-strategy and multi-perspective learning in MAS;
- learning in MAS as organisational self-design; and
- theoretical analysis of learning in MAS.

3.6 Multi-agent system in the construction industry

This section addresses two major aspects of the application of MAS in the construction industry. One is the necessity, possibility and advantages of MAS in solving construction problems, another is how to apply MAS in the industry. A few other applications are also described.

3.6.1 *Why multi-agent systems?*

One of the most important driving forces behind MAS research and development is the technology push of a growing standardised communication infrastructure – Internet, WWW, Knowledge Query and Manipulation Language (KQML) and eXtensible Markup Language (XML). The Internet is particularly important as it enables separately designed agents, often belonging to different organisations, to interact in an open environment in real time and carry out transactions safely. Another reason is the strong application pull for computer support for negotiation at the operative decision-making level.

MAS are not required merely to produce modularity (though they reduce complexity), extra speed (though this may be an effect of their inherent parallelism), reliability (though they provide redundancy), flexibility or re-usability (Newell, 1982). In the same way, they are not required simply because a problem is too large for a centralised single agent due to resource limitation, neither because of the sheer risk of a centralised system, nor merely for reasons of efficiency, heterogeneous reasoning, etc.

Problems requiring multi-agent solutions include the following (Sen, 1997; Nwana and Ndumu, 1999):

- Problems requiring the interconnecting and inter-operation of multiple, autonomous, 'self-interested' existing legacy systems, for example, expert systems or decision-support systems.
- Similarly, problems whose solutions draw from distributed autonomous experts. Agents in a MAS are often designed based on functionality: such systems are modular in design and hence can be easier to develop and maintain.
- Problems that are inherently distributed in nature (e.g. distributed sensor interpretation, co-ordination of self-interested agents, etc.).
- Problems whose solutions require the collation and fusion of information, knowledge or data from distributed, autonomous and selfish information sources.

A key benefit for the application of MAS in the construction industry is that MAS provides a decentralised approach to modelling the fragmented construction engineering and management problems. Such fragmented problems, though widely recognised, are difficult to be solved using other approaches. The idea of incorporating MAS into the construction industry provides a novel approach to tackling distributed problems. Some other key benefits could be: effective decomposition of large-scale problems; improved collaborative and concurrent working; and easier and cheaper access to specialist information. The use of agent-based systems is expected to result in increased competitiveness of the construction industry as the decentralisation of complex, large-scale problems and the collaborative input to their resolution, will lead to better quality, more economic, safer and more optimal solutions (Ugwu *et al.*, 1999b; Anumba *et al.*, 2001).

3.6.2 Some applications

Given the potential benefits that MAS could bring to the industry, several research projects have been conducted, attempting to resolve the conflicts between project participants, to facilitate decision-support at various stages of a construction project or to reach a better solution for problems in construction engineering. The two main areas being addressed here are engineering design and negotiations.

3.6.2.1 Agent-based design

Much of the work in the area of automated design has been done on building design support. Focus is on co-ordinating the activities and information flow between designers who are often geographically distributed, and on modelling agents to facilitate workflow between architectures and engineering in a collaborative design environment. Some of the developed systems are described as follows:

- Fenves *et al.* (1994) developed an Integrated Building Design Environment (IBDE) project. Agents in IBDE are classified into two groups: generators and critics. The generators typically are (k-b) systems that contribute to the development of the emerging design such as the '*ARCHPLAN*' (developing the building design concept), the '*CORE*' (generating layouts of the service core) and the '*STRPES*' (configuring the structure system). The critics do not contribute directly to the design descriptions, but evaluate the current description and make recommendations for redesign such as constructibility assessment by the '*CONSTRUCTION CRITIC*' and structure evaluation by the '*STRUCTURAL CRITIC*'. Each generator and critic agent is described according to the role it plays in the overall project. As an information-processing unit, each agent is first described in terms of its principal inputs and outputs, then is further described by the problem-solving paradigm it adopts, and how it transforms inputs into solutions.
- Chiou and Logcher (1996) have implemented an agent-based system for design. Agents have areas of specialist knowledge and perform design and checking tasks. They interact directly with a user who is responsible for design changes. However, there is no provision for direct negotiation between agents during the design process, so the number of designs evaluated remains small. Hence, convergence to a near optimal design depends on the user and does not make full use of the available computational power which can allow for the evaluation of many slightly differing designs automatically.
- Heckel *et al.* (1996) developed an Agent Collaboration Environment (ACE). The ACE supports collaboration amongst members of the design team by providing the infrastructure for a community of co-operative design agents that assist the users. In the framework, agents are organised into business processes to reflect the various functional tasks and workflow between different design teams in an organisational setting. Agents communicate with each other using libraries of design objects such as beams, columns and footings. Agents are reactive, but have the capability to run in the background and advise users on design issues such as code violations. Heckel *et al.* (1996) metaphorically described the ACE as 'a gathering place of several "experts" represented as agents. Each agent has particular expertise in a relatively narrow domain, but the groups of agents are all working on a common problem' with each agent embodying a particular expertise that

must be acquired from source. The primary role of an ACE agent is as 'design assistants that use heuristic rules and a powerful checklist facility to automate routine design tasks, enhancing productivity and design quality'.

- Radeke (1997) describes the Global Engineering Network Intelligent Access Libraries (GENIAL) project. The objective of the project is to facilitate large-scale collaborative engineering by establishing a common semantic infrastructure that will enable heterogeneous software systems to communicate and exchange information. An example of such system–system communication is automated database transactions at enterprise-level. This will enable enterprises from different engineering sectors to combine internal knowledge with engineering knowledge accessed on-line and world-wide via the global engineering network. It is expected that the project will lead to domain-specific tool kits for acquisition, retrieval and presentation of internal and external engineering knowledge.
- The DESSYS (2000) project is part of a wider research program, Virtual Reality Design Information System. DESSYS investigates the deployment of multiple software agents to improve collaborative decision-making in multi-disciplinary architectural design environment. This research covers knowledge modelling for a decision-support system in geotechnical design. It also investigates some of the important issues for decision-support in construction. The first issue is on different techniques of transferring knowledge and how the knowledge is used in the process of making decisions in structural-design. The second issue deals with formalising expert-knowledge in such a way that it can be used as decision-knowledge in the early phases of a multi-disciplinary building-process. The final phases deal with developing a (k-b) agent, and validating the prototype.

3.6.2.2 Negotiation

Negotiation is the core of many MAS in construction because it is often unavoidable between different project participants with their particular tasks and domain knowledge whilst they interact to achieve their individual objective as well as their group goals. Furthermore, the importance of negotiation in MAS is likely to increase. One reason is the growth of fast and inexpensive standardised communication infrastructures, over which separately designed agents belonging to different organisations, can interact in an open environment in real time, and safely carry out transactions. Second, there is an industrial trend to be able to respond to larger and more diverse orders. Such ventures can realise the resource allocation efficiently and are easier to adapt to a dynamically changing economic environment. The following are negotiation algorithms developed in construction.

- Ugwu et al. (1999b) developed an ADLIB model in which agents are used to model construction design project, discussed in detail in Chapter 5. Each agent needs to satisfy its minimum specification requirement, while it may give up something for the other participants to reach their minimum requirements. The relationship between the agents lies between fully co-operative and fully hostile. Different design teams co-ordinate their requirements through negotiations. The Monotonic Concession Protocol (MCP) is adopted in their negotiations, whilst each agent adopts a gradient concession strategy given the MCP protocol.
- Pena-Mora and Wang (1998) develop a CONVINCER model to facilitate the negotiation of conflict resolution in large-scale civil engineering projects. Essentially, the model is based on game theory, where a player's actions are based on the premise that the decision of any player can affect the payoff of all players. Also, the model analyses various conflict situations and settlement solutions. The negotiation model contains two steps. First, each negotiator expresses his interests in the conflict. Once the interests have been expressed correctly, the influence of positions of conflicting interests on the overall negotiation outcome is evaluated using game theory. By following this collaborative negotiation approach, they build an agent, which facilitates or mediates the negotiation of conflicts in projects.
- Kim et al. (2000) develop a project schedule co-ordination model through compensatory negotiation – a framework wherein a project can be rescheduled dynamically by all of the concerned project participants based on their resource profiles. It is intended to lead to individually rational and globally optimal solutions in agent-based project schedule co-ordination domains. The compensatory negotiation approach allows an agent to transfer utility to other agents for compensation of the disadvantageous agreements through a multi-linked negotiation process. In their study, utility is defined as the difference between benefit (i.e. profit gain for the task from a possible option) and cost (i.e. cost incurred for succeeding task due to the possible option).
- Oliveira et al. (1997) develop a MACIV model. The project aims to design and implement a MAS, enabling decentralised management of the different resources in a construction company. In their system, negotiation follows a six-step process, that is, announcing, task evaluation, selection phase, market manipulation and price adjustment. The last two steps are repeated until all the agents except one get out of the process, or some timeout arrives. Each agent makes decisions according to its estimate of the cost for the execution of each task in terms of the travel cost, depreciation cost, operation cost, operator cost and profits. A co-ordinator agent is introduced in the system.

- Ren *et al.* (2002) develop a MASCOT model which facilitates construction claim negotiation with a MAS. Unlike other MAS in construction, this model particularly focuses on the development of MAS negotiation mechanism as claim negotiation involves much more complex human and technical issues than other negotiations. A sophisticated negotiation mechanism has been developed based on Zeuthen's bargaining model which specifically addresses the participants' risk perception in conflict avoidance. This matches the obliged self-interested nature of construction claims negotiation. Agents in this model can also learn their opponent's negotiation strategies during negotiation process. Based on its previous knowledge about the opponent and the opponent's new offers during negotiation, an agent can improve its estimate about the opponent's reservation value and adjust its negotiation strategy. Furthermore, the MASCOT negotiation protocol also considers some important practical issues of claims negotiation such as the involvement of the client agent and the expanded solution searching. A detailed discussion can be found in Chapter 6.

A common drawback in these negotiation algorithms is that they often fail to address the complex issues of negotiations in construction. For example, although the CONVINCER model focuses on the negotiation resolution of construction conflicts, each party's decision-making is finally determined by a simple game theory rule, which can hardly represent the real negotiation situations. A more sophisticated negotiation mechanism needs to be developed for construction claims negotiation – the focus of Chapter 6.

3.6.3 Problems and challenges

Although MAS provide many potential advantages, they also face many difficult challenges such as: the information discovery problem, the communication problem, the ontology problem, the legacy software integration problem, the reasoning and co-ordination problem, and monitoring problem (Nwana and Ndumu, 1999). To design and implement MAS for industrial problems, developers need to address a number of questions (Gasser and Huhns, 1989; Jennings *et al.*, 1998):

- how to formulate, describe, decompose, and allocate problems and synthesise results among a group of IAs;
- how to enable agents to communicate and interact. What communication languages and protocols to use. What and when to communicate;
- how to ensure that agents act coherently in making decisions or taking action, accommodating the non-local effects of local decisions and avoiding harmful interactions;

- how to enable individual agents to represent and reason about the actions, plans and knowledge of other agents in order to co-ordinate with them. How to reason about the state of their co-ordinated process (e.g. initiation and completion);
- how to recognise and reconcile disparate viewpoints and conflicting intentions among a collection of agents trying to co-ordinate their actions;
- how to effectively balance local computation and communication. More generally, how to manage allocation of limited resources;
- how to avoid or mitigate harmful overall system behaviour, such as chaotic or oscillatory behaviour; and
- how to engineer and constrain practical MAS. How to design technology platforms and development methodologies for MAS.

All these questions are concerned with the key points of the development of the MASCOT model, and will be addressed (more or less) in this project. For example, the first four questions address the critical issues such as how to model the industrial problems in a MAS environment, and how to build the agent co-ordination mechanism based on the identified industrial problem. These questions need to be answered during the conceptual model development process. The answers of the later questions are crucial for the quality of the developed system; and ensure that the system will work properly.

Jennings *et al.* (1998) puts particular emphasis on two major technical impediments to the widespread adoption of agent technology:

- the lack of a systematic methodology enabling designers to clearly specify and structure their applications as MAS and
- the lack of widely available industrial-strength tool kits for building MAS.

The former means that most existing applications have been designed in a fairly *ad hoc* manner – either by borrowing a methodology and trying to shoehorn it to the multi-agent context or by working without a methodology and designing the system based on intuition and past experience. This is the biggest and hardest problem for the widespread application of MAS. What is required is a means of analysing the problem of working out how it can be best structured as a MAS, and then determining how the individual agent can be structured. A clear, non-*ad hoc* methodology is vital for further development of MAS.

The latter impediment means that most MAS projects expend significant development effort building up basic infrastructure before the main thrust of agent and inter-agent development can commence. Again, this is an unsustainable position. The position can be alleviated to a certain extent by exploiting existing technologies as and where appropriate – rather than re-inventing the wheel as often happens at the moment. However, greater support is still needed for the process of building agent-level features. Thus, a tool kit is required, providing facilities for: specifying an agent's problem-solving behaviour, specifying how and when agents should interact, and visualising and debugging the problem-solving behaviour of the agents and of the entire system.

Solutions to the above problems are intertwined. For example, a different modelling scheme for an individual agent may constrain the range of effective co-ordination regimes; different procedures of communication and interaction have implications for behavioural coherence; different problem and task decompositions may yield different interactions. The application of MAS in the construction industry face all these problems, and have the additional difficulties arising form the flexible and sophisticated interactions between autonomous problem-solving components required by the specific industrial problems. The solutions to these problems need to be formed in the context of solving real or quasi-real-world problems.

3.7 Summary

This chapter explored the main concepts and characteristics of IAs and MAS. It first presented the definitions, taxonomies, development process and organisations of IAs; it then discussed the major characteristics, disciplines, classifications and applications of agent-based systems. Both DPS and MAS were introduced, particularly the latter and a number of applications were examined. Section 3.4 explored the basic principles of agent-negotiation and various agent-negotiation mechanisms. Section 3.5 analysed the key components of agent learning; discussed the major agent-learning approaches; and pointed out the problems and challenges for developing learning approaches. Finally, this chapter discussed the potential benefits, application methods, implementation areas and possible problems for the application of MAS in construction.

Here is a list of some of the main points included in this chapter.

- In general, MAS represent a melting pot of ideas orienting from such areas as distributed computing, OO systems, software engineering, artificial intelligence, economics, sociology and organisational science. At its core is the concept of autonomous agents interacting with one another for their individual and/or collective good. This basic conceptual framework offers a natural and powerful means of analysing, modelling and resolving many real-world problems.
- Negotiation, the most important agent-collaboration mechanism, plays a central role in MAS. The development of various agent-negotiation mechanisms and their applications in different environments reveal the potential and possibility to facilitate construction claims negotiation.

This has been implemented and tested in the MASCOT prototype system discussed in Chapter 6. The study of negotiation mechanisms provided valuable experience for the development of MASCOT negotiation protocol and strategies, and pointed out the direction in which the MASCOT system focused.
- An agent's learning ability increases its negotiation power in an MAS. The analysis of the components and processes of agent learning suggested what and how an agent would learn from other agents or the environments. The philosophical differences between different learning methods suggested which method was more suitable for a particular system, and which had more potential to be applied in the MASCOT system.
- The study of the research projects conducted on the applications of MAS in the construction industry to facilitate engineering designs and negotiations indicated how MAS have been framed to solve different industrial problems. This is particularly important for the development of the MASCOT system in the construction industry environment (see Chapter 6).

The next chapter (Chapter 4) explores the major negotiation theories in some detail.

Note

1 Here, the game theory-based negotiation actually covers the economic theory-based negotiation. A detailed discussion is in Chapter 4.

References

Adler, M. R., Davis, A. B., Weihmayer, R. and Worrest, R. W. (1989) 'Conflict resolution strategies for nonhierachical distributed agents', in L. Gasser and M. Huhns (eds) *Distributed Artificial Intelligence*, Volume II, Pitman Publishing, London, pp. 139–61.

Anumba, C. J., Newnham, L., Ugwu, O. O. and Ren, Z. (2001) 'Intelligent agent applications in construction engineering', in Singh (ed.) *Proceedings of ISEC, 2001 Hawaii – Creative Systems in Structural and Construction Engineering*, Balkema, Rotterdam, pp. 161–6.

Beer, M., d'Inverno, M., Jennings, R. N., Luck, M., Preist, C. and Schroeder, M. (1999) 'Negotiation in multi-agent systems', *The Knowledge Engineering Review*, 14(3): 285–9.

Brustoloni, J. C. (1991) 'Autonomous agents: characterisation and requirements', *Carnegie Mellon Technical Report CMU-CS-91-204*, Carnegie Mellon University, Pittsburgh.

Bussmann, S. and Muller, H. J. (1992) 'A negotiation framework for co-operating agents', in *Proceedings of CKBS-SIG*, Dark Centre, University of Keele, UK, pp. 1–17.

Cammarata, S., McArthur, D. and Streeb, R. (1983) 'Strategies of distributed problem solving', in *Proceedings of the Eighth International Joint Conference on Artificial Intelligence*, pp. 767–70.

Carbonell, J. M. (1989) 'Introduction: paradigms for machine learning', *Artificial Intelligence*, 40(1–3): 1–9.

Chavez, A., Dreilinger, D., Guttman, R. and Maes, P. (1997) 'A real-life experiment in creating an agent marketplace', in *Proceedings of the First International Conference on the Practical Application of Intelligent Agents and Multi-agent Technology*, London, UK.

Chen, Y., Peng, Y., Labrou, Y., Cost, S., Chu, B., Yao, L., Sun, R. and Willhelm, B. (1999) 'A negotiation-based multi-agent system for supply chain management', http://citeseer.nj.nec.com/268420.html

Chiou, J. D. and Logcher, R. D. (1996) Testing a Federation Architecture in Collaborative Design Process – Final Report, Report no. R96-01, CERL.

Cockburn, D. and Jennings, N. R. (1996) 'ARCHON: a distributed artificial intelligence system for industrial applications', in G. M. P. O'Hare and N. R. Jennings (eds) *Foundations of Distributed Artificial Intelligence*, John Wiley & Sons Ltd, Chichester, UK, pp. 319–44.

Coen, M. H. (1995) http://citeseer.nj.nec.com/coen95sodabot.html

Conry, S. E., Meyer, R. A. and Lesser, V. R. (1988) 'Multistage negotiation in distributed planning', in A. H. Bond and L. Gasser (eds) *Readings in Distributed Artificial Intelligence*, Morgan Kaufumann Publishers Inc., San Mateo, Los Altos, CA, pp. 367–84.

Conry, S., Kuwabara, K., Lesser, V. and Meyer, R. (1992) 'Multistage negotiation in distributed constraint satisfaction', *IEEE Transactions on Systems, Man and Cybernetics*, 21(6): 1462–77.

Cross, J. G. (1977) 'Negotiation as a learning process', in I. W. Zartman (ed.) *The Negotiation Process: Theories and Application*, Sage Publications, London, pp. 29–54.

Davis, R. and Smith, R. G. (1983) 'Negotiation as a metaphor for distributed problem solving', *Artificial Intelligence*, 20: 63–109.

Decker, K., Sycara, K. and Williamson, M. (1997) 'Middle-agents for Internet', in *Proceedings of the Fifth International Joint Conference on Artificial Intelligence (IJCAI-97)*, Nagoya.

DESSYS (2000) http://www.ds.arch.tue.nl/Research/Agents/DessysIntro.stm

Durfee, E. H. (1988) *Coordination of Distributed Problem Solvers*, Kluwer Academic Publishers, Dodrechht (Hingham, MA).

Durfee, E. H. and Lesser, V. R. (1991) 'Partial Global Planning: a co-ordination framework for distributed hypothesis formation', *IEEE Transactions on Systems, Man and Cybernetics*, 21(5): 1167–83.

Durfee, E. H. and Montgomery, T. A. (1990) 'A hierarchical protocol for co-ordinating multi-agent behaviours', *AAAI*, Boston, MA, 86–93.

Durfee, E. H., Lesser, V. R. and Corkill, D. D. (1989) 'Trends in co-operative distributed problem solving', *IEEE Transactions on Knowledge and Data Engineering*, KDE, 1(1): 63–83.

Ekenberg, L., Boman, M. and Danielson, M. (1995) 'A tool for co-ordinating autonomous agents with conflicting goals', in *Proceedings of the First International Conference on Multi-agent Systems*, AAAI Press/The MIT Press, Cambridge, MA, pp. 89–93.

Fenves, S., Flemming, U., Hendrickson, C., Maher, M. L., Quadrel, R., Terk, M. and Woodbury, R. (1994) *Concurrent Computer-integrated Building Design*, PTR Prentice-Hall, NJ.

Ferguson, I. A. (1992) 'Touring machines: autonomous agents with attitude', *Technical Report no. 250*, Computer laboratory, University of Cambridge, UK.

Fischer, K., Muller, J. P., Heimig, I. and Scheer, D. (1996) 'Intelligent agents in virtual enterprises', in *Proceedings of the First International Conference on the Practical Application of Intelligent Agents and Multi-agent Technology (PAAM 96)*, London.

Foner, L. (1993) 'What's an agent, anyway? a sociological case study', *Agents Memo 93-01*, MIT Media Lab, Cambridge, MA.

Gasser, L. and Huhns, M. N. (eds) (1989) *Distributed Artificial Intelligence* II, Pitman, London.

Genesereth, M. R., Ginsberg, M. L. and Rosenschein, J. S. (1986) 'Co-operate without communication', in *Proceedings of AAAI-86*, Philadelphia, Pennsylvania, PA, pp. 51–7.

Georgeff, M. P. (1983) 'Communication and interaction in multi-agent planning', in *Proceedings of the Third National Conference on Artificial Intelligence (AAAI-83)*, Washington, DC, pp. 125–9.

Gordon, D. and Subramanian, D. (1994) 'A multi-strategy learning scheme for agent knowledge acquisition', *Informatic*, 17: 331–44.

Grecu, D. L. and Brown, D. C. (1994) 'Learning by design agents during negotiation,' in *Proceedings of the Third International Conference on AI in Design – Workshop on Machine Learning in Design*, Lausanne, Switzerland.

Grecu, D. L. and Brown, D. C. (1998a) 'Guiding agent learning in design', in *Proceedings of the Third IFIP Working Group 5.2 Workshop on Knowledge Intensive CAD*, Tokyo, Japan, pp. 237–50.

Grecu, D. L. and Brown, D. C. (1998b) 'Dimensions of learning in design', *Artificial Intelligence for Engineering Design, Analysis and Manufacturing*, Cambridge University Press, UK, pp. 117–22 (Special issue on Machine Learning in Design).

Grefenstette, J. and Daley, R. (1996) 'Methods for competitive and cooperative co-evolution', *AAAI-96 Spring Symposium on Adaptation, Co-evolution and Learning in Multi-agent Systems, AAAI Technical Report SS-96–01*, AAAI Press, Menlo Park, Canada.

Grosz, B. and Kraus, S. (1996) 'Collaborative plans for complex group actions', *Artificial Intelligence*, 86(2): 269–57.

Gulliver, P. H. (1979) *Disputes and Negotiation: A Cross-culture Perspective*, Academic Press Inc., San Diego, CA.

Hayes-Roth, B. (1995) 'An architecture for adaptive intelligent systems', *Artificial Intelligence* [Special Issue on Agent and Interactivity].

Haynes, T. and Sen, S. (1997) 'Crossover operators for evolving a team', in John R. Koza, Kalyanmoy Deb, Marco Dorigo, David B. Fogel, Max Garzon, Hitoshi Iba and Riek L. Riolo (eds) *Proceedings of the Second Conference on Genetic Programming*, Morgan Kaufmann, San Francisco, CA, pp. 162–7.

Heckel, J., McGraw, K. D., Morton, J. D. and Lawrence, P. (1996) 'The agent collaboration environment, an assistant for architects and engineers', in *Proceedings of Conference on Congress on Computing in Civil Engineering*, Washington, DC.

Hewitt, C. E. (1985) 'The challenge of open systems', *Byte*, 10(4): 233–42.
Hu, J. and Wellman, M. P. (1996) 'Self-fulfilling bias in multi-agent learning', in *Proceedings of the Second International Conference on Multi-agent Systems (ICMAS-96)*, Kyoto, Japan, pp. 118–25.
Huhns, M., Mukhopadhyay, U. and Stephens, L. M. (1987) 'DAI for document retrieval: the MINDS project', in Huhns, M. (ed.) *Distributed Artificial Intelligence*, Pitman Publishing, London, pp. 249–84.
IBM Intelligent Agent (1994) http://www.networking.ibm.com/wbi/wbisoft.htm
Jennings, N. (2000) 'Automated negotiation and argumentation', http://www.ecs.soton.ac.uk/~nrj/neg-arg.html
Jennings, N. R. (1995) 'Controlling cooperative problem solving in industrial multi-agent systems using joint intentions', *Artificial Intelligence*, 75(2): 195–40.
Jennings, N. R. and Campos, J. R. (1997) 'Towards a social-level characterisation of socially responsible agents', *IEEE Transactions on Software Engineering*, 144(1): 11–25.
Jennings, N. R., Faratin, P. and Johnson, M. J. (1996) 'Using intelligent agents to manage business processes', in *Proceedings of the First International Conference on the Practical Application of Intelligent Agents and Multi-agent Technology (PAAM 96)*, London.
Jennings, N. R., Sycara, K. and Wooldridge, M. (1998) 'A roadmap of agent research and development', *Autonomous Agents and Multi-agent Systems*, 1(1): 7–38.
Kaelbling, L. P., Littman, M. L. and Moore, A. W. (1996) 'Reinforcement learning: a survey', *Journal of Artificial Intelligence Research*, 4.
Kim, K., Paulson, B. C., Petrie, C. J. and Lesser, V. R. (2000) 'Compensatory negotiation for agent-based project schedule coordination', CIFE Working Paper #55, Stanford University, Stanford, CA.
Klein, M. (1991) 'Supporting conflict resolution in cooperative design systems', *IEEE Transactions on Systems, Man and Cybernetics*, 21(6): 1379–90 (Special Issue on Distributed Artificial Intelligence).
Kraus, S. (1996) 'Beliefs, time and incomplete information in multiple encounter negotiations among autonomous agents', *Annals of Mathematics and Artificial Intelligence*, 20(1–4): 111–59.
Kraus, S. (1997) 'Negotiation and cooperation in multi-agent environments', *Artificial Intelligence Journal*, 94(1–2): 79–98 (Special Issue on Economic Principles of Multi-Agent Systems).
Kraus, S. and Lehmann, D. (1995) 'Designing and building a negotiating automated agent', *Computational Intelligence*, 11(1): 132–71.
Kraus, S. and Wilkenfeld, J. (1991) 'The function of time in cooperative negotiations', *AAAI*, 1: 179–84.
Kraus, S. and Wilkenfeld, J. (1993) 'A strategic negotiation model with applications to an international crisis', *IEEE Transactions on Systems, Man, and Cybernetics*, 23(1): 313–23.
Kraus, S., Wilkenfeld, J. and Zlotkin, G. (1995) 'Multiagent negotiation under time constraints', *Artificial Intelligence*, 75(2): 297–345.
Kreifelt, T. and Von Martial, F. (1991) 'A negotiation framework for autonomous agents', in Y. Demazeau and J. P. Muller (eds) *Decentralised Artificial Intelligence II*, Elsevier Science, Amsterdam, The Netherlands.

Lander, S. E. and Lesser, V. R. (1991) 'Customising distributed search among agents with heterogeneous knowledge', in *Proceedings of the Fifth International Symposium on AI Applications in Manufacturing and Robotics*, Cancun, Mexico.

Lander, S. E. and Lesser, V. R. (1993) 'Understanding the roles of negotiation: is distributed search among heterogeneous agents', in *Proceedings of the Thirteenth International Joint Conference on Artificial Intelligence*, Chambray, France, pp. 438–44.

Lee, L. C. (1998) 'A model of progressive multi-agent negotiation', in *Proceedings of the Third International Conference on Multi-Agent Systems*, The IEEE Computer Society, Paris, France, pp. 448–9.

Lesser, V. R. (1991) 'A Retrospective view of FA/C distributed problem solving', *IEEE Transactions on Systems, Man, and Cybernetics*, 21(6): 1347–62 (Special Issues on Distributed Artificial Intelligence).

Levesque, H. J., Cohen, P. R. and Nunes, J. H. T. (1990) 'On acting together', in *Proceedings of the Eighth National Conference on Artificial Intelligence (AAAI-90)*, Boston, MA, pp. 94–9.

Lin, F., Tan, G. W. and Shaw, M. J. (1998) 'Modelling supply-chain networks by a multi-agent system', in *Proceedings of the Thirty-first Hawaii International Conference on System Sciences (HICSS'98)*, The IEEE Computer Society, Kona, Hawaii.

Maes, P. (1995) 'Artificial life meets entertainment: life like autonomous agents', *Communications of the ACM*, 38(11).

Malone, T. W., Fiken, R. E., Grant, K. R. and Howard, M. T. (1988) 'Enterprise: a marketlike task schedule for distributed computing environments', in B. A. Huberman (ed.) *The Ecology of Computation*, North Holland, New York, pp. 177–205.

Matos, N., Sierra, C. and Jennings, N. R. (1998) 'Determining successful negotiation strategies: an evolutionary approach', in *Proceedings of the Third International Conference on Multi-Agent Systems*, IEEE Computer Society, pp. 182–9.

Mullen, T. and Wellman, M. P. (1996) 'Some issues in the design of market-oriented agents', in M. Wooldridge, J. P. Muller and M. Tambe (eds) *Intelligent Agents II* (LNAI1037), Springer-Verlag: Heidelberg, Germany, pp. 283–98.

Muller, H. J. (1996) 'Negotiation principles', in G. M. P. O'Hare and N. R. Jennings (eds) *Foundations of Distributed Artificial Intelligence*, John Wiley & Sons Ltd, New York, pp. 139–64.

Muller, J. P. (1998) 'Architectures and applications of intelligent agents', *The Knowledge Engineering Review*, 13(4): 353–80.

Nagayuki, Y., Ishii, S. and Doya, K. (2000) 'Multi-agent reinforcement learning: an approach based on the other agent's internal model', in *Proceedings of the Fourth International Conference on Multi-Agent Systems (ICMAS-2000)*, Los Alamitos, pp. 215–21.

Newell, A. (1982) 'The knowledge level', *Artificial Intelligence*, 18: 87–127.

Newell, A. and Simon, H. A. (1976) 'Computer science as empirical enquiry', *Communications of the ACM*, 19: 113–26.

Nwana, H. S. (1996) 'Software agents: an overview', *The Knowledge Engineering Review*, 11(3): 205–44.

Nwana, H. and Ndumu, D. (1999) 'A perspective on software agents research', *The Knowledge Engineering Review*, 14(2): 125–42.

Oliveira, E. D., Fonseca, J. M. and Steiger-Garcao, A. (1997) 'MACIV: a DAI based resource management system', *Applied Artificial Intelligence Journal*, 11(6): 525–50.

Parunak, H. V. D. (1996) 'Application of distributed artificial intelligence in industry', in G. M. P. O'Hare and N. R. Jennings (eds) *Foundations of Distributed Artificial Intelligence*, John Wiley & Sons Ltd, New York, pp. 139–64.

Parunak, H. V. D., Baker, A. D. and Clark, S. J. (1997) 'The AARIA agent architecture: an example of requirements-driven agent-based system design', in *Proceedings of the First International Conference on Autonomous Agents*, Marina del Rey, CA, pp. 284–91.

Pena-Mora, F. and Wang, C. (1998) 'Computer-supported collaborative negotiation methodology', *Journal of Computing in Civil Engineering*, 12(2): 64–81.

Petrie, C., Goldmann, S. and Raquet, A. (1998) 'Agent-based project management', http://citeseer.nj.nec.com/petrie98agentbased.html

Polat, F., Sher, S. and Guvenir, H. A. (1993) 'A negotiation platform for cooperating multi-agent systems', *Concurrent Engineering: Research and Applications*, 1(3): 179–87.

Prasad, M. V. N. and Lesser, V. R. (1999) 'Learning situation-specific coordination in cooperative multi-agent systems', *Autonomous Agents and Multi-agent Systems*, 2: 173–207.

Preist, C. (1999) 'Economic agents for automated trading', *Software Agents for Future Communication Systems*, Springer-Verlag, Berlin.

Radeke, E. (eds) (1997) 'GENIAL-global engineering networking intelligent access libraries', *GENIAL white paper, No: G_White_paper_V1.0_970805_SNI.doc*.

Reeds, C. (1998) 'Dialogue frames in agent communication', in *Proceedings of the Third International Conference on Multi-agent Systems*, The IEEE Computer Society, pp. 246–53.

Ren, Z., Anumba, C. J. and Ugwu, O. O. (2002) 'Negotiation in a multi-agent system for construction claims negotiation', *Applied Artificial Intelligence*, 16(5): 359–94.

Rosenschein, J. S. and Zlotkin, G. (1994) *Rules of Encounter*, MIT Press, Cambridge, MA.

Roth, A. E. (1979) *Axiomatic Models of Bargaining*, Springer-Verlag, Berlin.

Russell, S. and Norvig, P. (1995) *Artificial Intelligence: A Modern Approach*, Prentice-Hall, London.

Sandholm, T. (1993) 'An implementation of the contract net protocol based on marginal cost calculations', in *Proceedings of the Eleventh National Conference on Artificial Intelligence (AAAI-93)*, Washington, DC, pp. 256–62.

Sandholm, T. and Crites, R. (1995) 'Multi-agent reinforcement learning in the iterated prisoner's dilemma', *Biosystems*, 37: 147–66 (Special Issue on the Prisoner's Dilemma).

Sandholm, T. and Lesser, V. (1995) 'Issues in automated negotiation and electronic commerce: extending the contract net protocol', in *Proceedings of the First International Multi-agent Systems (ICMAS-95)*, San Francisco, CA, pp. 328–35.

Sandholm, T. and Lesser, V. (1997a) 'Coalitions among computationally bounded agents', *Artificial Intelligence*, 94(1) (Special Issue on Economic Principles of Multiagent Systems).

Sandholm, T. and Lesser, V. (1997b) 'Issues in automated negotiation and electronic commerce: extending the contract net framework', in M. Huhns and M. Singh (eds) *Reading in Agents*, Morgan-Kaufmann Publishers, pp. 66–73, http://books. elsevier.com/uk//computerscience/uk/subindex.asp?maintarget=&isbn=1558604952 &country=United+Kingdom&srccode=&ref=&subcode=&head=&pdf= &basiccode=&txtSearch=&SearchField=&operator=&order=&community= computerscience.

Sathi, A. and Fox, M. (1989) 'Constraints-directed negotiation of resource reallocations', in N. H. Michael and L. Gasser (eds) *Distributed Artificial Intelligence*, Volume II, pp. 163–93.

Sen, S. (1997) 'Multiagent systems: milestones and new horizons', *Trends in Cognitive Science*, 1(9): 334–9.

Sen, S. and Durfee, D. H. (1997) 'A formal study of distributed meeting scheduling', *Group Decision and Negotiated Support Systems*, 7: 265–89.

Shaw, M. (1996) 'Cooperative problem-solving and learning in multi-agent information systems', *International Journal of Computational Intelligence and Organisations*, 1(1): 21–34.

Shen, W., Norrie, H. D. and Barthes, J. (2001) *Multi-agent Systems for Concurrent Intelligent Design and Manufacturing*, Taylor & Francis, London.

Sierra, C., Jennings, N. R., Noriega, P. and Parson, S. (1998) 'A framework for argument-based negotiation', in M. P. Singh, A. Rao and M. J. Wooldridge (eds) *Artificial Intelligence*, Lecture Notes in Artificial Intelligence, Volume 1365, Springer-Verlag, Berlin, pp. 177–92.

Smith, D. C., Cypher, A. and Spohrer, J. (1994) 'Kidsim: programming agents without a programming language', *Communications of the ACM*, 37(7): 54–67.

Smith, R. and Davis, R. (1983) 'Negotiation as a metaphor of distributed problem solving', *Artificial Intelligence*, 20: 63–105.

Smith, R. G. (1980) 'The contract net protocol: high-level communication and control in a distributed problem solver', *IEEE Transactions on Computing*, 29(12): 1104–13.

Sycara, K. P. (1987) *Resolving Adversarial Conflicts: An Approach to Integrating Case-based and Analytic Method*, PhD Thesis, School of Information and Computer Science, Georgia Institute of Technology.

Sycara, K. P. (1988) 'Resolving goal conflicts via negotiation', in *Proceedings of the Seventh National Conference on Artificial Intelligence (AAAI-88)*, St. Paul, MN.

Sycara, K. P. and Zeng, D. (1994) 'Visitor-hoster: towards an intelligent electronic secretary', in *Proceedings of the Third International Conference on Information and Knowledge Management (CIKM-94)*, ACM ISBN: 0–89791–674–3, Gaithersburg, MD, USA.

Tambe, M. (1997) 'Towards flexible teamwork', *Journal of Artificial Intelligence Research*, 7: 83–124.

Tsvetovatyy, M. B. and Gini, M. (1996) 'Toward a Virtual marketplace: architectures and strategies', in *Proceedings of the First International Conference on the Practical Application of Intelligent Agents and Multi-agent Technology*, London.

Ugwu, O. O., Anumba, C. J., Newnham, L. and Thorpe, A. (1999a) 'Applications of distributed artificial intelligence in the construction industry', Research Report No. ADLIB/01, Loughborough University, Loughborough, UK.

Ugwu, O. O., Anumba, C. J., Newnham, L. and Thorpe, A. (1999b) 'Agent-based decision support for collaborative design and project management', *The International*

Journal of Construction Information Technology, 7(2): 1–16 (Special Issue: Information Technology for Effective Project Management and Integration).

Ugwu, O. O., Kumaraswamy, M. M., Ng, T. S. and Lee, P. K. K. (2003) 'Agent-based collaborative working in construction: understanding and modelling design knowledge, construction management practice and activities for process automation', *HKIE Transactions Tenth Anniversary Issue, Special Issue on Emerging Technology in the 21st Century*, 10 (4): 81–7.

Weiss, G. (1993) 'Learning to coordinate actions in multi-agent systems', in *Proceedings of the International Joint Conference on Artificial Intelligence (IJCAI'93)*, San Matco, Canada.

Weiss, G. (1996) 'Adaptation and learning in multi-agent systems: some remarks and a bibliography', in G. Weiss and S. Sen (eds) *Adaption and Learning in Multi-agent Systems*, Lecture Notes in Artificial Intelligence, Volume 1042, Springer-Verlag, Berlin, pp. 1–21.

Wellman, M. P. and Wurman, P. R. (1998) 'Real time issues for internet auctions', in *Proceedings of the First IEEE Workshop on Dependable and Real-time E-Commerce Systems*.

Werkman, K. J. (1990) 'Knowledge-based model of negotiation using sharable perspectives', in *Proceedings of the Tenth International Workshop on DAI*, Texas.

WinSton, P. H. (1997) http://www.stanford.edu/~vengerov/ideas.html

Wooldridge, M. (1997) 'Agent-based software engineering', *IEE Proceedings on Software Engineering*, 144(1): 26–37.

Wooldridge, M. and Jennings, N. R. (1995) 'Intelligent agents: theory and practice', *The Knowledge Engineering Review*, 10(2).

Ygge, F. and Akkermans, H. (1998) 'On resource-oriented multi-community market computations', in *Proceedings of the Third International Conference on Multi-agent Systems*, The IEEE Computer Society, Paris, France, pp. 365–71.

Zeng, D. and Sycara, K. (1998) 'Bayesian learning in negotiation', *International Journal of Human-Computer Studies*, 48: 125–41.

Zlotkin, G. and Rosenschein, J. S. (1989) 'Negotiation and task sharing among autonomous agents in cooperative domains', in *Proceedings of the Eleventh International Joint Conference on Artificial Intelligence*, Morgan Kaufmann, Detroit, MI, pp. 912–17.

Zlotkin, G. and Rosenschein, J. S. (1991) 'Incomplete information and deception in multi-agent negotiation', in *Proceedings of the 12th International Joint Conference on Artificial Intelligence*, pp. 225–31.

Zlotkin, G. and Rosenschein, J. (1994) 'Cooperation and conflict resolution via negotiation among autonomous agents in noncooperative domains', *IEEE Transactions on Systems, Man, and Cybernetics*, 21(6): 1317–24 (Special Issues on Distributed Artificial Intelligence).

Chapter 4

Negotiation theories

Z. Ren, C. J. Anumba and O. O. Ugwu

4.1 Introduction

Negotiation, as an important human co-operation approach, has been studied and defined by many researchers in different research domains, such as: economics, society, politics and AI systems. For example,

- Collins Cobuild English Dictionary – negotiations are formal discussions between people who have different aims or intentions, especially in business or politics, during which they try to reach an agreement.
- Zartman (1977) – negotiation is a joint decision process between two or several parties or their representatives. Negotiation tends to be a matter of finding a formula encompassing the optimum combination of interests of both parties and then of working out the details that implement these principles and affect the agreement. Negotiation is a dynamic process, on-going, involving moves and countermoves.
- Hammer and Clay (1977) – negotiation is the interaction that occurs when two or more persons attempt to agree on a mutually acceptable outcome in a situation where their orders of preference for possible outcomes are negatively correlated.
- Gulliver (1979) – negotiation is one kind of problem-solving process, in which people attempt to reach a joint decision on matters of common concern in situations where they are in disagreement and conflict.
- Rosenschein and Zlotkin (1994) – negotiation is a form of decision-making process where two or more parties jointly search a space of possible solutions with the goal of reaching a consensus (deal).
- Lesser (1998) – negotiation, the process of arriving at a state that is mutually agreeable to a set of agents, is intimately related to co-ordination.

These definitions provide different views about negotiation objective, nature, scope and elements involved.

- From a social-psychological perspective, Bartos (1977) concludes that the nature of negotiation is to resolve the conflicts between competitive individualism and co-operative collectivism. Negotiation often involves dual and mostly conflicting motivations: the individual (competitive) desire to maximise one's own utility and the collective (co-operative) desire to reach a fair solution. Negotiations can proceed smoothly only when they are guided by the collective desire for fairness or when the loss of breaking negotiation is higher than that of reaching an agreement for either party even if negotiators are selfishly motivated.
- In distributed artificial intelligence (DAI), as defined by Lesser (1998), negotiation could be a co-ordination approach for participants to gain more. There is not necessarily any conflict between the participants. This is so-called co-operate negotiation.

In most cases, negotiation participants have power over each other during this process (Young, 1991). Each party tries to reach its objective by gaining or giving some of its benefits to others. There are relatively fixed parties and flexible values in negotiation.[1] Value and behaviour are modified to alter divergent positions towards a common convergence of values (Spector, 1977). Negotiation normally includes three stages:

- Negotiation starts from the point where each party tries to maximise his own pay off.
- By exchanging information, two parties explore the nature and extent of their differences and the possibilities open to them, and seek to induce or persuade each other to modify their expectations and requirements, and then search for an outcome that is at least satisfactory enough to both parties (Gulliver, 1979). Different strategies are adopted by negotiation participants, such as: concession making, contending, problem-solving, inaction, withdrawal (Pruitt and Carnevale, 1993).
- Finally, an agreement is reached, theoretically, at an equilibrium point where the opposing interests are balanced. Practically, the final results are inevitably influenced by many factors such as: the negotiators' personal capabilities, negotiation strategies, time constraints, expectations and the relationship between the parties. Therefore, the outcome of a negotiation could be the victory of one party, a simple compromise, win–win result or a conflict.

As an important social activity, negotiation involves more complex human interactions than simple technical issues. Many important theories and principles have been developed to explain various aspects of negotiation. Pruitt and Carnevale (1993) summarise three main traditions in the study of negotiation:

- The first consists of books and manuals providing advice on various aspects of practical negotiation issues (e.g. Fisher *et al.*, 1982; Cohen, 1991; Warham, 1993; Fowler, 1996).
- The second consists of mathematical models of rational behaviour developed by economists and game theorists (e.g. Luce and Raiffa, 1957; Young, 1975). These models are ordinarily limited to a relatively narrow set of tactics, such as concession making or third-party recommendations for particular agreements. They focus on the various situations, assumptions, outcomes and bargaining processes of negotiation. The two kinds of model are sometimes combined by theorists who use the tools of rational analysis to examine the wide range of tactics used by most negotiators and third parties (e.g. Gulliver, 1979; Raiffa, 1982).
- The third, behavioural, tradition seeks to develop and test predictive and practical theory about the impact of environmental conditions on negotiator (or mediator) behaviour, human behaviours during negotiation, and the impact of these conditions and behaviours on outcomes (e.g. Walton and Mckersie, 1965; Rubin and Brown, 1975; Morley and Stephenson, 1977). The theories developed based on this approach are classified as behaviour theory.

This study aims at developing a multi-agent systems (MAS) negotiation system and focuses on the second approach since it provides theoretical foundations for computer-based negotiation systems. Meanwhile the behaviour-theoretical approach is also taken into consideration, as the concepts of behaviour models are crucial for computer negotiation systems to simulate practical negotiations.

4.2 Game theory and economic theory

Prior to making a detailed discussion, it is necessary to explain the essential terms used in this chapter. Basically, there are two different ways to classify and present the game- and economic-theoretical approaches in negotiation. Rojot (1991), accepted by most of the negotiation theorists, terms the theories developed based on these two theoretical approaches as Mechanical Theory, where the distinctions between the game- and economic-theoretical models are identified with focus on their different modelling approaches to human negotiations. The former is primarily static whilst the latter considers the dynamic features of negotiation.

On the other hand, many AI researchers (Rosenschein and Zlotkin, 1994; Sandholm, 1996; Kraus, 1997) view the game- and economic-theoretical approaches as different branches of game theory. Here, the game-theoretical approach is termed as axiom bargaining approach whilst the economic-theoretical approach is termed as strategic bargaining approach. This classification reveals the inter-relationship between these theoretical approaches (i.e. game theory is patricianly developed based on economic theory; furthermore, as far as the mathematics is concerned, the two are equivalent). However, the drawback is that such classification easily brings confusion between the strategic (normal) form of game theory and the strategic bargaining approach (see Osborne and Rubinstein, 1994; Rosenschein and Zlotkin, 1994; Sandholm, 1996; Kraus, 1997).

This chapter, with focus on the application of these theoretical models in negotiation, adopts Rojot's classification. However, it does not conflict the game theory classification, rather it is a matter of different terms as far as the negotiation modelling approaches are concerned. Furthermore, to avoid confusion, the term 'classical game theoretical approach' is used to distinguish the general 'game theoretical approach'. These are discussed in the following section.

4.2.1 The classical game-theoretical approach

Game theory seeks to get at the essentials of decision-making and the associated strategies in situations two or more parties are interdependent, and therefore, the outcome of their conflict and competition must be the product of their joint requirements and the interaction of their separate choices (Bacharach and Lawler, 1981). These parties are in a situation in which there may be many possible outcomes with different values to them. Although they may have some control which will influence the outcome, they do not have complete control over others. Each party in a game faces a cross-optimisation problem.

Game theory represents the most thorough search for a determinate solution to negotiation problems (Bacharach and Lawler, 1981). All types of negotiation can be conceptualised as different kinds of games (Brams, 1990). It focuses on the logical and hypothetical conceptualisation of problems and processes wherein many of the variables of the real negotiation are ignored. For example, all game-theoretical models are based on some important simplifying assumptions which determine how the theory works and its scope. According to Nash (1950), the fundamental assumptions are:

- both the number of players and their identity are assumed to be fixed and known to everyone;
- players are rational and expect others to be rational;
- players attempt to maximise their own gain or utility;
- players have complete information on the utility of alternative settlements to themselves and their opponents. The pay-off function for each player is fixed and known at the outset. The utility function of each player is fixed and known.

Based on these assumptions, game theory places negotiation in the context of a general theory of individual choice. Given the mutual dependence relationship in decision-making, such choice becomes a strategic issue; that is, a party must assess how to maximise his own gain in the context of potential interference from others (Gulliver, 1979). The game-theoretical models are essentially static models which attempt to deduce what strategy a player should take, taking account of the fact that an outcome is the result of the interdependent strategies and choices of two or more players. Therefore, the major objective of the game theory approach is to offer a set of rules that describe how rational actors choose the best strategy most consistent with the bargainers' conflicting interests (Shubik, 1975).

The basic elements of game theory are players, pay offs and rules of action. First, there must be a well-defined set of courses of action for each of a number of players. Second, there must be well-defined preferences for each player among possible outcomes of the game or mixtures of his outcomes. Third, the relationships must exist whereby the outcome is determined by the player's courses of action (Bacharach and Lawler, 1981).

A game can be categorised as zero sum or non-zero-sum, depending on whether there is a fixed set of available pay offs. In zero-sum games, players try to garner as much of this fixed pot of rewards as possible. Every player assumes that his opponent will 'do his worst' (Von Neumann and Morgenstern, 1947). The key point in this analysis is the proposition that a rational player in a two-person, zero-sum game can predict the preferences of his opponent accurately. There is no strategic interaction[2] in his decision-making problem (Young, 1975).

In a non-zero-sum game, players look to find opportunities such that mutual gains can be made, thereby increasing the size of the pot. In this situation, there are strategic interactions between players. To eliminate strategic interaction from non-zero-sum situations, game theory is divided into a number of separate domains distinguished by the specific assumptions they employ:

- Nash (1950) has sought to predict outcomes in non-zero-sum interactions by searching for solutions that seem particularly stable-equilibrium points. Nash shows that under some rational behaviour and symmetry assumption, players would reach an agreement on a deal that will be individual rational, Pareto-optimal and will maximise the product of the players' utility.
- Nash (1953) also sought to circumvent the 'outguessing' regress by introducing fixed decision rules, whose application yielded unique prediction concerning the distribution of pay offs between the players. These decision rules were based on specific criteria concerning the worth or power of the individual players and/or on certain desirable attributes of a solution. Games of this kind are known as co-operative games.

- Harsanyi (1977) has sought to achieve determinate solutions for non-zero-sum games by introducing the notion that each player may be able to assign 'subjective probabilities' to the choices of the other players.

Unlike the competitive games, these models not only use the idea of a solution concept where the agents' strategies form some type of equilibrium. Instead, desirable properties for a solution, called axioms of the bargaining solution, are postulated, and then the solution concept that satisfies these axioms is sought (Osborne and Rubinstein, 1990, 1994).

For example, in the Nash bargaining solution (1950), it is assumed that the space of feasible utility vectors is compact and convex. When many deals are individually rational (i.e. pay more than the fallback) to both agents, there may be many Nash equilibrium solutions. An example is that the agents are bargaining over how to split a dollar, all splits that gives each agent more than zero are in equilibrium. If Agent 1's strategy is to offer P and no more, Agent 2's best response is to take the offer as opposed to the fallback which is zero. Now, Agent 1's best response to this is to offer P and no more. Thus, a Nash equilibrium exists for any P that defines a contract that is individually rational for both agents, and feasible ($0 < P < 1$). Due to the non-uniqueness of the equilibrium, a stronger (axiomatic) solution concept such as the Nash bargaining solution is needed to prescribe a unique solution.

The first axiom is based on the view that the agents' numeric utility functions really represent only ordinal preferences among outcomes – the actual cardinality of the utilities do not matter. Therefore, the utility function can be transformed, and the resulting game should be equivalent to the original game. The second axiom requires symmetry: if the agents have symmetric bargaining positions, their outcome utilities should be equal. Third, independence of irrelevance is required. The fourth axiom requires Pareto efficiency. It turns out that there is a unique solution that satisfies the four axioms. This Nash bargaining solution selects the utility pair that maximises the product of the players' gains in utility over their fallback utility. Therefore, these axioms are a powerful tool for isolating basic problems. The attempts to formulate acceptable axioms uncover many ambiguities in the substantive aspects of the social sciences.

Game theory provides a useful benchmark and a fundamentally important methodological approach to the study of situations involving potential conflict. It provides a device to isolate key factors in negotiation with facility (Shubik, 1975). For example, some of the most important games such as *the prisoner's dilemma* and *the game of chicken* provide deep insight to negotiation. Rapoport (1983) conclude the strength of classical game theory as: 'The mathematician's elucidation of problems in game theory sometimes leads not to a solution but to a clarification, namely of what it is that the problem involves, what obstacles stand in the way of solutions, what special cases of the problem can be treated by what methods'.

Although classical game theory is expected to provide a scientific framework that would allow the prediction of outcomes as well as the explanation of a negotiator's behaviour and decisions in real-life negotiation, there does not appear to be any convincing formulation that offers reliable explanation or prediction (Gulliver, 1979). The assumptions of the theory and its highly abstract mathematical solutions raise general problems in regard to their potential validity and applicability (Zartman, 1977):

- the classical game theory formula presents a static representation of what is essentially a dynamic process in which components and their interrelationships are intrinsically subject to change, thus affecting their contribution to and the nature of the outcome;
- the classical game theory formula includes components that are not measurable nor ascertainable even in broadly acceptable approximations[3]; and
- the classical game theory is unsuitable for the analysis of the negotiation process because its assumptions identify and remove all the obstacles that bargainers have to confront.

4.2.2 The economic-theoretical approach

The economic-theoretical approach, in contrast to the classical game-theoretical approach, seeks to develop dynamic models of process, involving offers and counter-offers and interdependent concession making. There is no concern for the discovery of once-for-all strategies but rather an intention to examine how the bargainers should interact in terms of their expectations of each other. According to Young (1975), this approach is principally under conditions of bilateral monopoly which seeks to explain a jointly determined outcome in terms of the rational tendencies of the parties to reach an optimal point of intersection on their lists of interchangeable preferences.

Economic-theoretical approaches are fundamentally convergence models. They analyse the processes through which the demands of the participants converge over time towards some specific point on the contract curve. Therefore, the key element of these models is the development of a specific concession mechanism that permits the positions of the parties to converge in the course of a series of offers and counter-offers (Bacharach and Lawler, 1981).

The best known economic-theoretical approach is Zeuthen's economic welfare under condition of bilateral monopoly. In this model, the two players' bargaining problem is considered as a one-player decision process under the assumption that if none of the players concedes at a particular step, they will reach a conflict.

- The individual player compares the certain value he can obtain by accepting the other side's offer. Based on this offer, his own favoured outcome and the results of conflict, he calculates the maximum probability of conflict he would be willing to accept in preference to acquiescing on the current offer of the other side.
- Concession will be made by the side willing to accept the smaller risk of conflict at any given moment in time. A player needs only to reduce his own demand to the point where he is willing to accept a greater risk of conflict than the other (Young, 1975). Accordingly, each player must continue to concede until he is willing to accept a larger risk of conflict than his opponent.

Cross's model (1975) emphasises the role of time as an important factor in bargaining, and conceptualises the process of making concession in terms of the adjustment of expectations through learning. In this model, an individual bargainer starts his calculations with his preference ordering for the outcomes in the pay-off possibility set; a schedule of costs arising from the time that elapses before a specific contract is agreed upon; and a precise estimate is made of the other side's concession rate over time. Each bargainer proceeds to calculate the optimal level for his own initial demand on the assumption that the other player will make all of the concessions by taking into account the trade-off between improvements in the final settlement associated with higher initial demands and the increased costs which higher demands produce as they extend the time required to reach a specific agreement.

Based on these economic models, some bargaining alternative offering approaches in modern game theory have been discussed (e.g. Kreps, 1990; Osborne and Rubinstein, 1990, 1994; Rasmusen, 1994). An example is that one can again think about deciding how to split a dollar. With a finite number of offers and no time discount, the last offerer will get it all, and the other agent is indifferent between accepting a zero offer or getting a zero fallback pay off. With time discount, a finite game can be solved starting from the end. The agent that is to make the first and the last offer gets a pay off that approaches $1/(1 + Q)$ as the number of negotiation rounds approaches infinity. The term Q is the discount factor. When protocol in a non-discounted setting allows an infinite number of bargaining rounds, the solution concept is powerless because any split of the dollar can be supported in sub-game perfect Nash equilibrium. On the other hand, in a discounted infinite round setting, the sub-game perfect Nash equilibrium is unique. Specifically, the first offerer gets $(1 - Q2)/(1 - Q1Q2)$, where the first offerer's discount factor is $Q1$, and the other player's $Q2$ (Rasmusen, 1994).

Another model of sequential bargaining does not use discounts, but assumes that there is a fixed bargaining cost per negotiation round. If the agents have symmetric bargaining costs, the solution concept is again powerless because any split of the dollar can be supported in sub-game perfect Nash equilibrium. On the other hand, if the first offerer's bargaining cost is smaller than the other agent's, the first agent gets the entire dollar. If the first agent's bargaining cost is greater than the second's, the first agent receives a pay off that equals the second's bargaining cost. The second agent receives one minus this and agreement is reached on the first round.

The economic-theoretical models differ from the models of the classical game theoretical approach in several ways:

- First, the economic-theoretical models deal only with certain non-zero sum or mixed-motive situations. Thus, they focus on interactions in which there is a distinct range of possible outcomes within which each of the participants would prefer to reach an agreement than to accept an outcome of non-agreement, even though they have conflicting interests concerning the precise terms of agreement. Situations involving pure conflict and pure co-operation are outside the scope of these models.
- Second, economic-theoretical models treat negotiation as a process of convergence over time involving a sequence of offers and counter-offers. Consequently, the economic-theoretical models are dynamic models which focus on the bargaining process as well as on the ultimate outcome of bargaining, whereas the game-theoretic models are predominantly static models which concentrate on the ultimate distribution of pay offs among the parties.
- Third, the economic-theoretical models tend to emphasise the formation of expectations about the behaviour of the relevant other(s) in contrast to the models of game theory which stress either conditions that allow each player to make accurate predictions concerning the behaviour of the relevant other(s) or characteristics of a permissible solution which are sufficient to yield determinate outcomes (Young, 1975).

4.3 Behaviour theory

The applications of classical game theory and economic theory are limited by their assumptions which neglect the complex human factors in negotiation. On the contrary, behaviour theory attempts to analyse the negotiation processes in which negotiators influence each other's expectations, perceptions, assessments and decisions during the search for an outcome, thereby affecting the outcome. Much attention is given to the nature of changing expectations and negotiators' tactics, and to the significance of uncertainties of information, perception and evaluation – all matters that tend to be ignored by the mechanical theory. All this involves a closer approximation to the real world (Zartman, 1977). The major behaviour models are discussed in the following section.

4.3.1 Psychological model

This approach focuses on analysing the personality or psychological responses of the decision-makers themselves rather than the negotiation process. It seeks to explain the bargaining effectiveness in terms of variables such as the behavioural characteristics of the negotiators and their perceived and actual use of interpersonal strategies. It addresses the extent to which personality, perception, expectation, persuasion and interaction of these factors within the negotiator dyads can adequately describe and explain the process and outcomes of bargaining (Spector, 1977). Four factors are analysed in this approach:

- *Negotiator personality* identifies basic predisposition towards the opponent and motives for future actions and responses. It shapes one's perspective and expectations of particular objects and goals.
- *Perceptions and expectations* of the opponent's strengths, weaknesses, intentions, commitments and goals are likely to affect negotiation responses, the tone of interpersonal communication and the learning process.
- The use of *persuasive techniques* and their success in modifying negotiator values toward initially desired end-states should help to achieve acceptable outcomes.
- The *interaction of psychological and contextual factors* address the possibility that certain personalities become instrumental in motivating the bargaining process only under particular negotiation conditions.

The advantages of the psychological approach lie in the fact that it focuses on the fixed element of the process – the parties – and their ability or propensity to modify the variable element – that is, the values at stake. It deals with realistic aspects of negotiation using concepts that are possible to operationalise. In many practical negotiations, psychological approaches play a very important role in the success of the negotiations and the drawback of this approach is the analysis of the agent rather than the process. Further it focuses on the secondary rather than the primary element of decision-making.

4.3.2 The learning model

This approach views negotiation as a learning process in which each party is largely dependent on his experience of the results of past actions by the two parties. What has occurred previously is used as a standard of assessment by which to choose what to aim for and what to do next (Gulliver, 1979). The basis of this model is that negotiators' strategies are changeable. Negotiation strategies are contingent, contain errors, expectations will change and this will lead in turn to a modification of each party's choice of strategy.

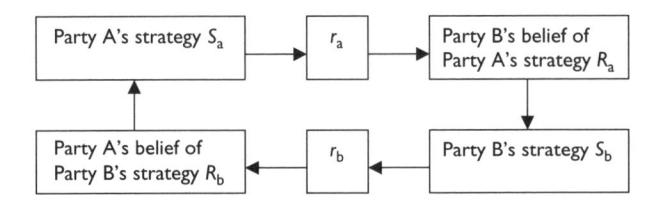

```
┌─────────────────────┐       ┌─────────────────────┐
│ Party A's strategy Sₐ│ ──rₐ→ │ Party B's belief of │
│                     │       │ Party A's strategy Rₐ│
└─────────────────────┘       └─────────────────────┘
         ↑                              │
┌─────────────────────┐       ┌─────────────────────┐
│ Party A's belief of │ ←─r_b─│ Party B's strategy S_b│
│ Party B's strategy R_b│      │                     │
└─────────────────────┘       └─────────────────────┘
```

Figure 4.1 A general learning mechanism.

Negotiators' behaviours are classified as: actual pay off demands and manipulative moves such as threats or coercive actions. The latter are not directly related to the pay off, but they do affect the overall value of the negotiation because they may influence the settlement point and costs of negotiation. The learning process involves identifying the other party's real pay off demands and manipulative moves to decide the negotiator's own strategy.

The learning mechanism establishes a dynamic interaction between the two parties' behaviours. Supposing S_a is Party A's strategy, R_a is Party B's estimate of Party A's strategy under an uncertainty $V_b \times S_a$ determines the current course of r_a, which is used by Party B in the formation of R_a. In response to R_a, Party B selects strategy S_b which determines the course of r_b, and this in turn is the basis for Party A's estimate of R_b. As a consequence of learning, S_a and R_b may be continually changing, so that what is learned in the form of R_a or R_b is a composite of a sequence of strategies rather than a single one (Cross, 1977). Such a learning process is expressed in Figure 4.1.

The learning model plays an important role in current negotiation theory since it reflects and simulates one of the most important characteristics of practical negotiations – the inference process of negotiators. For example, its importance has been recognised in the game research community as fundamental for understanding human behaviour as well as for developing new solution concepts (Cross, 1977; Jordan, 1992; Osborne and Rubinstein, 1994). Various learning mechanisms, such as: Q-learning, collective learning and Bayesian learning have been developed for intelligent agents.

4.3.3 The dual responsiveness model

Unlike the learning model, the dual responsive approach shows that negotiator's responsiveness can be based on *both his own previous concessions and the opponent's concessions*. A negotiator's response is a function of his own previous pattern of concession making as well as the opponent's concession rate (Figure 4.2). His responses are mediated by expectations, which are adjusted through the course of the conference. This approach suggests two types of monitoring the functions in negotiation: each negotiator

98 Z. Ren et al.

```
┌─────────────────────┐
│ My previous concession │
│ process             │
└─────────────────────┘
         ↕                    ┌──────────────────────────┐
                              │ My decisions of concession│
                              │ at next step             │
┌─────────────────────┐       └──────────────────────────┘
│ The opponent's previous │
│ and current         │
└─────────────────────┘
```

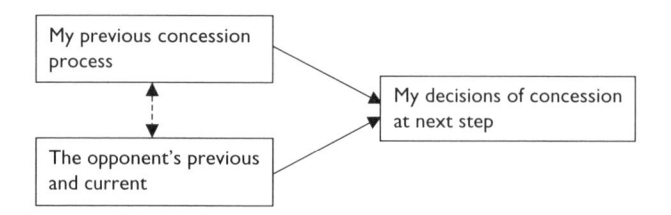

Figure 4.2 The dual responsive model.

monitors the other side for evidence of movement and monitors his own side for evidence of preferences. Unlike the learning approach, this approach calls attention to the importance of an internal dynamic in bargaining while both emphasise the importance of mutual responsiveness that is likely to occur in both directions.

4.3.4 *The joint decision-making model*

Zartman (1977) points out that the above models do not correspond to the conceptual characteristics that negotiation is a mode of the decision-making process, and do not deal with the process as it is actually practised. He concludes that negotiation tends to be a matter of finding a formula encompassing the optimum combination of interests of both parties and then of working out the details that implement these principles and affect the agreement.

Negotiators first seek a general definition of the items under discussion, conceived and grouped in such a way as to be susceptible to joint agreement under a common notion of justice. Once agreement on a formula is achieved, it is possible to turn to the specifics of items and to exchange proposals, concessions and agreements. Details are resolved most frequently in terms of the referents, which justify them and give them value rather than in their own intrinsic values. This means that convergence does not take place by inching away from fixed positions towards a middle, but rather by establishing a referent principle from which the value of the detailed item will be derived (Zartman, 1977).

The joint decision-making model differs from all the others, and views negotiation at a macro-level, providing some special advantages:

- First, it is possible to prescribe a negotiation through the achieved formula which can be made in impractical or artificial terms;
- Second, it leaves room for analysis of power as added value while others are not able to do; and
- Third, it can be used in conjunction with the other models.

4.4 Summary

This chapter, first defined negotiation and explored the nature and major elements of negotiation. Then, it discussed the two major negotiation theories: mechanical theory and behaviour theory. Section 4.2 reviewed two major theoretical approaches of mechanical theory: classical game theory and economic theory. These approaches are mainly mathematical models based on the rational behaviour assumption. The former (e.g. Nash solutions, 1950, 1953) is mainly concerned with the predictions of outcomes under certain assumptions about the players and the outcomes themselves. The latter (e.g. Zeuthen's model) focuses on analysing the negotiation process given the rational assumption.

Although these two theoretical approaches have been criticised as over simplistic relying too much on the assumptions and ignoring the social norms, relationships between negotiators, and group decision processes, these approaches uncover many ambiguous problems in negotiation. Furthermore, these approaches (especially the economic-theoretical approach) provide important theoretical models for the development of the computer negotiation systems.

Section 4.3 discussed behaviour theory and several important theoretical models including psychological model, learning model, dual-response model and joint decision model. Unlike mechanical theory, behaviour theoretical models aim to address the complex human responses in negotiation, such as how a negotiator analyses the opponent's expectations and psychological changes during negotiations, and the corresponding strategies that the negotiator will take to deal with such changes. Although behaviour models are normally difficult to simulate in computer-based negotiation systems, they represent some important concepts in practical negotiations, which developers of MAS negotiation systems need to address in their systems.

Notes

1 At this point, the focus is on the changing value of a negotiator during the negotiation. However, the participants of a negotiation could be changed (add or reduce) during the negotiation process. This is true particularly when negotiation is conducted between agents in MAS.
2 *Strategic interaction* is simply the set of behaviour patterns manifested by individuals whose choices are interdependent in negotiation (Young, 1975).
3 This is also the common limitation faced by the economic theory (i.e. the adoption of the utility functions as the criterion to evaluate a participant's pay offs and determine his negotiation strategies).

References

Bacharach, S. B. and Lawler, E. J. (1981) *Bargaining: Power, Tactics and Outcomes*, Jossey-Bass Publishers, San Franciso, CA.

Bartos, O. J. (1977) 'Simple model of negotiation: a sociological point of view', in I. W. Zartman (ed.) *The Negotiation Process: Theories and Application*, Sage Publications, London, pp. 13–28.

Brams, S. J. (1990) *Negotiation Games: Applying Game Theory to Bargaining and Arbitration*, Routledge, New York.

Cohen, R. (1991) *Negotiating across Cultures: Communication Obstacles in International Diplomacy*, Lewis, Washington, DC.

Cross, J. G. (1975) 'A theory of the bargaining process', in O. R. Young (ed.) *Bargaining: Formal Theories of Negotiation*, University of Illinois Press, Urbana, IL, pp. 191–218.

Cross, J. G. (1977) 'Negotiation as a learning process', in I. W. Zartman (ed.) *The Negotiation Process: Theories and Application*, Sage Publications, London, pp. 29–54.

Fisher, R., Ury, W. and Patton, B. (1982) *Getting to Yes: Negotiating Agreement without Giving in*, Hutchinson, London.

Fowler, A. (1996) *Negotiation Skills and Strategies, 2nd edn*, Institute of Personnel and Development, London.

Gulliver, P. H. (1979) *Disputes and Negotiation: A Cross-Culture Perspective*, Academic Press Inc, San Diego, CA.

Hammer, W. C. and Clay, G. A. (1977) 'The effectiveness of different offer strategies in bargaining', in D. Druckman (ed.) *Negotiations, Social Psychological Perspectives*, Sage Publications, London.

Harsanyi, J. C. (1977) *Rational Behaviour and Bargaining Equilibrium in Games and Social Situation*, Cambridge University Press, Cambridge, UK.

Jordan, J. S. (1992) 'The exponential covergence of bayesian learning in normal form games', *Games and Economic Behaviour*, 4: 202–17.

Kraus, S. (1997) 'Negotiation and cooperation in multi-agent environments', *Artificial Intelligence Journal*, 94(1–2): 79–98. Special Issue on Economic Principles of Multi-agent Systems.

Kreps, D. M. (1990) *A Course in Microeconomic Theory*, Princeton University Press, Princeton, NJ.

Lesser, V. (1998) 'Reflections on the nature of multi-agent coordination and its implementations for an agent architecture', *Autonomous Agents and Multi-agent Systems*, 1: 89–111.

Luce, R. D. and Raiffa, H. (1957) *Games and Decisions*, John Wiley & Sons, Ltd, New York.

Morley, I. E. and Stephenson, G. M. (1977) *The Social Psychology of Bargaining*, Allen and Unwin, London.

Nash, J. F. (1950) 'The bargaining problem', *Econometrica*, 28: 155–62.

Nash, J. F. (1953) 'Two-person cooperative games', *Econometrica*, 21: 128–40.

Osborne, M. J. and Rubinstein, A. (1990) *Bargaining and Markets*, Academic Press, New York.

Osborne, M. J. and Rubinstein, A. (1994) *A Course in Game Theory*, MIT Press, Cambridge, MA.

Pruitt, D. G. and Carnevale, P. J. (1993) *Negotiation in Social Conflict*, Open University Press, Buckingham.

Raiffa, H. (1982) *The Art and Science of Negotiation*, Belknap Press of Harvard University Press, Cambridge, MA.

Rapoport, A. (1983) *Mathematical Models in the Social and Behavioural Sciences*, John Wiley & Sons, Ltd, Chichester, UK.

Rasmusen, E. (1994) *Games & Information: An Introduction to Game Theory*, 2nd edn, Blackwell Publishers, Cambridge, UK.

Rojot, J. (1991) *Negotiation from Theory to Practice*, Macmillan, London.

Rosenschein, J. S. and Zlotkin, G. (1994) *Rules of Encounter*, MIT Press, Cambridge, MA.

Rubin, J. Z. and Brown, B. R. (1975) *The Social Psychology of Bargaining and Negotiation*, Academic Press, New York.

Sandholm, T. W. (1996) *Negotiation among Self-interested Computationally Limited Agents*, (PhD) Dissertation University of Massachusetts, Amherst, MA.

Shubik, M. (1975) *Game Theory and Related Approaches to Social Behaviour: Selections*, Robert E. Krieger Publishing Company, New York.

Spector, B. I. (1977) 'Negotiation as a psychological process', in I. W. Zartman (ed.) *The Negotiation Process: Theories and Application*, Sage Publications, London, pp. 55–66.

Von Neumann, J. and Morgenstern, O. (1947) *Theory of Games and Economic Behaviour*, Princeton University Press, Princeton, NJ.

Walton, R. E. and McKersie, R. B. (1965) *A Behavioural Theory of Labour Negotiations: An Analysis of A Social Interaction System*, McGraw-Hill, New York.

Warham, S. M. (1993) *Primary Teaching and the Negotiation of Power*, Paul Chapman, London.

Young, H. P. (1991) *Negotiation Analysis*, The University of Michigan Press, Ann Arbor, MI.

Young, O. R. (ed.) (1975) *Bargaining: Formal Theories of Negotiation*, University of Illinois Press, Urbana, IL.

Zartman, I. W. (ed.) (1977) *The Negotiation Process: Theories and Applications*, Sage Publications, London.

Chapter 5

Agent-support for collaborative design

O. O. Ugwu, C. J. Anumba and A. Thorpe

5.1 Introduction

In today's increasing competitive environment, timely delivery of data and/or information gives a construction firm some unique advantages in decision-making. The era of centralised databases has been characterised with problems of data consistency and/or quality of information. This is because different users require different dimensional views of data for decision-making in their respective functional areas. The Architecture, Engineering and Construction (AEC) sector has recognised the need for efficient management and delivery of information. In this context, information is classified as a key resource throughout the construction process. The International Standards Organisation (ISO) recognises the reliance of project teams on information, in the design, management and construction of projects. The following extracts from ISO illustrate the importance of information in construction.

> **Information** is a key resource... During the inception process, there is specific reliance on information of all kinds, the output of the process being project specific information. This is the major input to the design process, during which large amounts of reference information are also utilised. Again, the main output of the process is project specific information in the form of drawings, specifications, etc.
>
> The production process is especially dependent on information input, particularly project specific information, but also reference information relating to products, labour, regulations, etc. An important output from the production process is 'as built' information, relevant to both the use and decommissioning processes.
>
> (ISO, 1994, p. 9)

However, in practice it is a challenging task to meet the different information needs of users. The problem is exacerbated by the complex patterns of interaction and communication in a collaborative design environment, since participants can be separated in space or time. Examples of interaction include face-to-face, synchronous, synchronous distributed and asynchronous distributed. Each communication pattern poses its own specific requirements in information or data exchange. Ugwu et al. (1999a) discuss details of the communication patterns.

Collaborative design in construction is now widely seen as a prerequisite for efficient delivery of construction products. This requires that multiple distributed participants (i.e. stakeholders) work together to achieve project goals, which are often anchored in safety, quality, time, cost (with sustainability becoming an increasing requirement from project sponsors as well). The recent high profile international construction industry studies clearly allude to the expected step-gains from collaborative working in construction. Examples of such commissioned studies include the construction industry review committee in Hong Kong SAR China (CIRC, 2001), the Latham and Egan reports in the UK (Latham, 1994; Construction Task Force, 1998; M4I-URL) and the subsequent initiatives such as the Movement for Innovation (M4I) in the UK.

At the moment there exists technology to exchange information between project participants such as email systems and the web-based project portals. However, given the inherent distributed data, resources and expertise, there is need for more content support to design and construction knowledge workers. Ugwu et al. (2000c) outlined the following characteristics of distributed design environments in construction that make them suitable for the application of computing techniques such as distributed artificial intelligence (DAI) including Multi-agent systems (MAS):

- *distributed information* – design information is often distributed along functional lines and professional disciplines;
- *distributed expertise* – expertise and decision points in construction domain are distributed, and often reflect the different specialist disciplines in an organisation and/or project management structure;
- *distributed data sources* – data sources are distributed.

In the context of this chapter, agent-support for collaborative design means the use of intelligent agents (IA) that represent specialist design team members and carry out some designated tasks on behalf of their owners in a design space. Autonomy and learning are some of the core desirable qualities of an IA. An autonomous agent is one that is capable of acting (even if such action is a reaction to external stimulus within its environment) without direct intervention of humans or other agents. The agent has control over its own actions and internal state. Learning is the ability of the agents to update its knowledge base as situations in the environment changes.

Learning is desirable for developing ubiquitous agent systems. Moreover, a basic requirement for collaborative design using IA is to develop MAS models that encapsulate the organisational structures, domain product and process views, interaction protocols and social contexts of collaborative working.

Chapters 3 and 4 discussed theoretical foundations and various issues in MAS research. This chapter focuses on the use of MAS for decision-support in collaborative design in construction. It discusses development issues and application design decisions. The main goal is for readers to have a deeper understanding and appreciation of the implementation aspects of agent technology in construction, and the chapter has been structured with this as the main goal. It gives a detailed description of the Agent-based Collaborative Design of Light Industrial Buildings (ADLIB) research project that investigated issues involved in deploying agents for collaborative design in construction. Some of the issues investigated include: design process knowledge modelling and representation, developing negotiation models for peer-to-peer negotiation, interaction (messaging) protocols and co-ordination at different levels (e.g. agent–agent and agent–human levels), in a multi-agent design space. The chapter is mainly descriptive (and not prescriptive), but it is expected that readers would have a better understanding of methodological issues, knowledge modelling and representation and other technical aspects of IA and multi-agent development for decision-support in construction.

The remaining chapter is organised as follows: Section 5.2 discusses the problems and requirements for collaborative design in the context of agent-support for design processes, while Section 5.3 proposes a generic framework and reference model for agent-based collaborative design. The reference model can be adapted and extended as appropriate to solve other problems in the construction supply chain. Such extensions and applications are illustrated in Chapters 6, 8 and 11. Section 5.4 uses a case study project to place agent-support and MAS development in a practical perspective, giving a pathological analysis of the ADLIB research project that investigated MAS development and how agent paradigm can be used to streamline collaborative working process. These include description of operational context, methodological issues and modelling constructs used for business process analysis in the problem domain. Sections 5.5 and 5.6 give detailed description of how some research issues in MAS design and implementation were addressed in the domain. These include: (i) knowledge level modelling and representation of collaborative design processes; (ii) agent design and implementation issues, such as co-ordination, co-operation, interaction protocols, communication and messaging, information sharing; (iii) other aspects of agent implementation such as agent–software integration and (iv) prototype system validation with the project industrial partners. This is then followed by a summary in Section 5.6.

5.2 Agent-based collaborative design in construction – contextual issues

Collaborative working typically involves various interacting disciplines (people and processes). The success of such collaborative efforts is often predicated on good communication and co-ordination, and effective implementation of working processes and protocols. As an illustration, collaborative design of an infrastructure project such as that to provide a steel structure would involve architect, structural engineer, quantity surveyor, steel fabricator, building services, health and safety personnel. In this case, there is the need to evaluate the impact of decisions taken at a given interface and decision point on the work of other disciplines especially those downstream in the chain. As an illustration, an architect may not fully understand the impact of a given building envelope configuration on thermal load analysis and hence building services requirements. In addition, the structural engineer may not fully appreciate the impacts engendered by a given structural member configuration on constructability. However, the reality in current practice is that in most cases information is often passed too late (or not passed at all), for appropriate actions to be taken early enough. This often leads to unnecessary delay and wastage. Agent-based collaborative working makes it possible to evaluate (design) information that becomes available from different contexts and perspectives of stakeholders involved in the project. The next sub-section discusses some features of collaborative working that need to be addressed in agent research.

5.2.1 Organisational contexts and collaborative design practices

In deploying agent-based systems, a process must be performed in accordance with the organisational protocols. This means that a task should only be assigned to an appropriate team member or his representative agent. Also, different members of the project team may work at different levels of detail on specific parts of the project. As an illustration, an architect may simply specify the finishing materials, layout of structural members and dimensional requirements of a building envelope. However, the structural engineer would design the structural members based on the design loads, calculate the joint loads but not design the joint details. The steel fabricator would be responsible for designing the joints and connections and in the process, may subdivide their parts of the designing with a view to: (i) achieving design load bearing joints that achieve three-dimensional stability and (ii) designing for constructability expressed in terms of (a) manufacturability of the structural members especially for non-standard sections such as girders and (b) ease of transportation and erection of the members at the construction site which is often based on the cross-sectional details and weight of the members.

Table 5.1 Requirements of agent-support in collaborative design and project management

Collaborative working requirements	Examples of supporting agent
Information sharing and communication: • Posting (publishing design proposal) • Cross-functional communication	Utility agents (MAS platform-specific) • Yellow Pages • Broker agents
Collaborating design operations: • Co-ordination • Design • Health and safety • Costing and constructability • Building services	Task agents (configured at design time) • Interface/Architectural Agent (IAA) (publisher role) • Structural Design Agent (SDA) • Safety Agent (SA) • Costing and Constructability Agent (CCA) • Building Services Agent (BSA)
Product configuration: • Design negotiation	All task agents in a peer-to-peer negotiation, details discussed in Anumba et al., 2002, 2003

In addition, the costing specialist would then abstract the total quantities of materials and components, and use these in estimating the cost of a design proposal. Some of the computational tasks may require the use of additional resources and processes outside the responsible person or agent, resulting in further delegation of sub-tasks and problems to the appropriate specialists best able to handle it down the responsibility chain. In order to achieve this level of task and process decomposition and execution, an agent-based system must have clearly delineated roles, embody and be able to reason with knowledge describing the products, processes, organisational structuring, social contexts and sequences within which tasks are executed in the organisational model. Table 5.1 summarises the requirements for agent-based collaborative design.

5.2.2 Information sharing and management

Information sharing is a pre-requisite for collaborative working in construction and indeed any sector. Some of the specific issues involved are that: (i) collaborating entities (including humans and software) must be able to exchange information related to the problem solving, through messaging and (ii) the collaborating entities must have a common understanding of the contexts and information exchanged in the messages. The requirement is for a collaborative framework that facilitates effective co-ordination, sharing and management of information amongst the various disciplines involved in the collaborative problem-solving process.

5.2.3 Domain knowledge modelling and representation

In agent-based systems, task decomposition, role modelling and assignments capture organisational structures and social contexts of collaborative working. Thus the specification of information required by the participating agents must refer to concepts in the problem domain, so that agents share information with unambiguous understanding of the definition of the concepts associated with problem solving in the domain. This requires that the system (a society of agents) be provided with a knowledge-level representation of the problem domain that encapsulates the tasks, processes, product features, social behaviour (interactions) and other organisational constructs, as well as the rules associated with problem solving. This also means that an agent-based system must replicate the organisational patterns of interaction (e.g. during negotiation). Udeaja and Tah (2004) give an example of task decomposition for MAS support in construction materials procurement, in Chapter 11.

In order to achieve the above requirements of agent-based collaborative design, it is necessary to have robust frameworks for such agent development. The following section discusses a generic framework – agent management reference model for collaborative design.

5.3 Agent management reference model for collaborative design

This section describes the MAS management reference model at a generic level. An agent management model is essential for efficient co-ordination of the interaction and communication between dispersed agents. It then outlines some of the functional utilities within an agent development platform that facilitates the construction of MAS.

The agent management framework consists of the following components, an Agent Platform (AP), domain-specific agents and wrapper agents (Figure 5.1). The reference model can be adapted to suit various application problem domains in construction. Example applications are discussed in Chapters 6–8 and 11. The functions of these components are discussed in the ensuing sections.

The collaborative design agents are developed using an agent development platform (such as British Telecom's (BT) Zeus (BT, 2002)). This AP has some functional utilities that facilitate the design and construction of MAS. It is usually a proprietary or open-source tool that provides the following services: Agent Management System (AMS), Directory Facilitator (DF) – Yellow or White Pages, Agent Communication Channel (ACC) Router, and Internal Platform Message Transport (IPMT). Zeus, uses a TCP/IP (Point-to-Point) Messaging built on Java to achieve interoperability (BT, 2002). Other message transport Protocols include: HTTP, Knowledge Query Manipulation

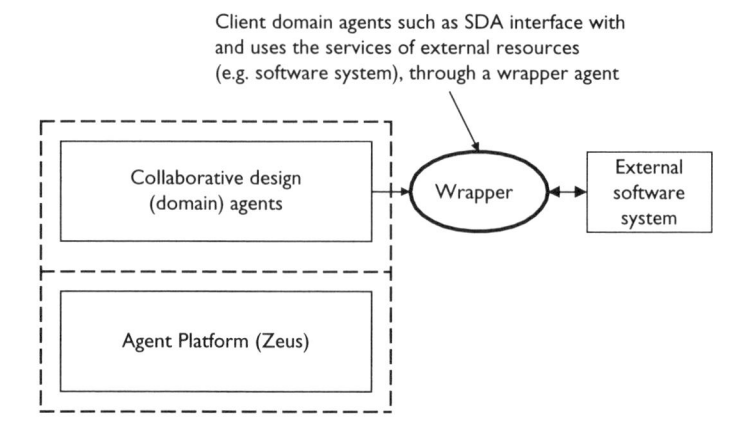

Figure 5.1 Conceptual framework showing the general agent management reference model (adapted from Ugwu et al., 1999b).

Language (KQML) and KIF. The next section gives a brief description of some of the utilities and services that are provided by the AP. These definitions are based on Foundation on Intelligent Physical Agents (FIPA) normative/ informative specifications (FIPA 1997a,b,c, 1999):

- *Agent Platform (AP)* – this is the platform on which agents reside at a given time. The AP provides a lot of utilities and services for the resident agents. These are discussed later.
- *Agent Management System (AMS)* – AMS is an agent that 'manages the creation, deletion suspension… of agents on the agent platform'. It provides a 'white pages' directory service for all agents resident in an AP. The white pages directory service in an AP is utilised by domain agents to locate their peers whenever messages are communicated.
- *Agent Communication Channel (ACC)* – ACC is an agent which uses the information provided by the AMS to route message(s) between agents within an AP and to agents resident on other platforms.
- *Directory Facilitator (DF)* – this is an agent that provides a 'Yellow Pages' directory service for the agents. DF stores descriptions of the agents and the services they offer.
- *Internal Platform Message Transport (IPMT)* – this is an abstract service provided by agent management platform to which an agent is attached at a given point in time. IPMT is platform-specific (i.e. the protocol(s) adopted is dependent on the proprietary tool). The message transport service provides for the reliable and timely delivery of messages to their destination agents. The next section discusses the case study collaborative research project.

5.4 Case study – the ADLIB project

The ADLIB project was commissioned by the Engineering and Physical Sciences Research Council (EPSRC) UK (Grant No. GR/M42169) under the Innovative Manufacturing Initiative (IMI) scheme and conducted at Department of Civil and Building Engineering, Loughborough University, UK. The research explored the issues involved in agent-based collaborative design and investigated how to automate certain design tasks using the enabling agent technology. The design domain was light industrial buildings with a focus on portal frame structures.

The research project had some unique dimensions in that it combines research on IA and MAS, with distributed collaborative working in a construction context. The fundamental objective was to look at people, process and technology issues and investigate how some of the current problems of collaborative design processes can be improved using IA and multi-agent paradigm. The ultimate objective was to develop a theoretical MAS framework/foundation and a prototype system that implements the framework. Thus the research project involved both fundamental and applied research.

The fundamental research component focused on developing *collaborative design knowledge representation* as captured in the knowledge acquisition phase of the project. This knowledge modelling resulted in the development of *ontology* for Light Industrial Buildings (LIB) domain. Another element of this research component is the development of *communication, interaction and negotiation protocols* that facilitate *automation* of certain tasks in a MAS design space. In the area of applied research, the effort focused on investigating *agent* and *MAS architectures* that are suitable for collaborative design problems in construction, and the *integration* of existing systems and other software applications. The prototype implementation integrated the MAS with FASTRAK – a commercial system for analysis, design and optimisation of portal frame structures (CSC2003-URL).

5.4.1 Research objectives

The research project had the following objectives:

- to refine the initial conceptual models and develop a functional specification for the MAS;
- to gain appropriate agent knowledge. This involved acquiring, from the project partners and other sources, appropriate specialist knowledge for the various agents and sub-agents identified in the MAS environment;
- to implement a prototype system in accordance with the functional specification, including details of the system development environment, agent communications; and
- to test the prototype system, using appropriate test cases, supplied by the industry partners.

5.4.2 Operational context of the ADLIB project – understanding business processes in collaborative design

Collaborative design in construction constitutes a class of problems in which a group of domain experts often work on different aspects of a project to find a solution. Such problems are knowledge intensive and the domain experts usually possess priceless knowledge of the problem-solving procedures that need to be captured and encapsulated within the agents in the MAS environment. This section describes an example of scenarios for the development and application of MAS in the ADLIB domain. The scenario was developed from: (i) interactions with ADLIB industrial partners during the knowledge acquisition stages of the research and (ii) a workshop organised to validate the user requirements captured.

The parties involved in a typical light industrial project such as portal frame structure include: a client, an architect, a structural engineer, an HVAC engineer, a specialist steel fabricator, a quantity surveyor and a safety advisor. A client commissions a medium to large industrial warehouse for use as a workshop, and the architect and the design team prepared an initial agreed design. The architect specifies the type of roof sheeting material. Following a change in the client's business needs, it is necessary to change the building's use to cold storage, which requires better all-round insulation. Furthermore, the local planning authority is unhappy with the roof pitch of 15° and suggests reducing it to 6° in line with similar buildings in the area. This requires a change to the pitch of the roof of the portal frame building, which in turn instigates the following:

- re-design of the structural frame with new roof pitch;
- re-assessment of heating and lightening requirements;
- re-evaluation of materials (e.g. roof light, floor slab);
- re-evaluation of the roof light design (total provision, type of material, etc.);
- re-evaluation of the constructability requirements;
- re-evaluation of the cost implications (build or whole life cost); and
- re-evaluation of the health and safety implications for building construction and maintenance.

The client drives the stated changes and there are cost implications. For example, the change in roof pitch has knock-on effects on the robustness of roof sheeting in terms of driving holes to fix sheeting (constructability related) plus the potential for leakage, type of sheeting (e.g. insulated olio type sheeting) and roof cladding.

The structural engineer uses his *existing design software* to design the structural frame. This is an important existing organisational business practice, which the potential MAS users do not necessarily have to be compelled to change. The following design parameters are generated from the design software and passed to other members of the team: member sizes (i.e. cross-sectional dimensions), length of members, weight of structural member, roof slope etc. This design generation is an automated process and it is assumed that the proposed design generated from the software has passed the sway and stability tests as these checks are part of the automated processes in FASTRAK. The structural engineer negotiates over this new design change with other members of design team. In the context of MAS, the *negotiation* would be between the agents on a peer-to-peer basis over the design space. The ensuing sections describe the negotiation objects.

The steel fabricator re-evaluates the constructability requirements. This involves checking ease of transportation of the members specified by the structural engineer. He could request for splicing of the sections if necessary. He also checks the overall cost implications of the design changes. He makes suggestions that would improve design and construction of the frame structure, so as to minimise the design changes later after construction has commenced. A fragment of the information flow in the business process and workflow diagram associated with the steel fabrication process is shown in Figure 5.2.

The building services engineer reviews the services requirements for building, and could for instance increase the size of the plant supported by the structural frame. This triggers re-evaluation of the structural design by the structural engineer to ensure structural safety. The safety advisor checks if the roof material specified is acceptable. If fragile material has been specified, the safety advisor rejects it.

The outlined re-designs/re-evaluations are undertaken on behalf of the team members by their specialist agents, which through rounds of negotiation agree on a final design. The specialist agents make reference to various resources such as design standards, and expert knowledge in evaluating and negotiating over design proposals. Hence the negotiation over design changes is a deliberative process and on a peer-to-peer basis. All the agents must reach a mutually agreed design decision before the negotiation terminates. Results of the agreed design (and indeed any negotiation deadlocks) are communicated to the agent's owner using dedicated Graphical User Interfaces (GUI) or other existing communication infrastructure including web-based email systems.

The ADLIB project scenario captures several interesting aspects of collaborative design for agent research investigation. However, this chapter focuses on automated design generation achieved by integrating agents with existing design software (legacy systems) and automated peer-to-peer negotiation in the MAS design space. The specific aspects of agent research, implementation and collaborative working discussed include: agent research methodology, business processes capture and computational modelling, co-ordination, interactions and negotiation in MAS design spaces at

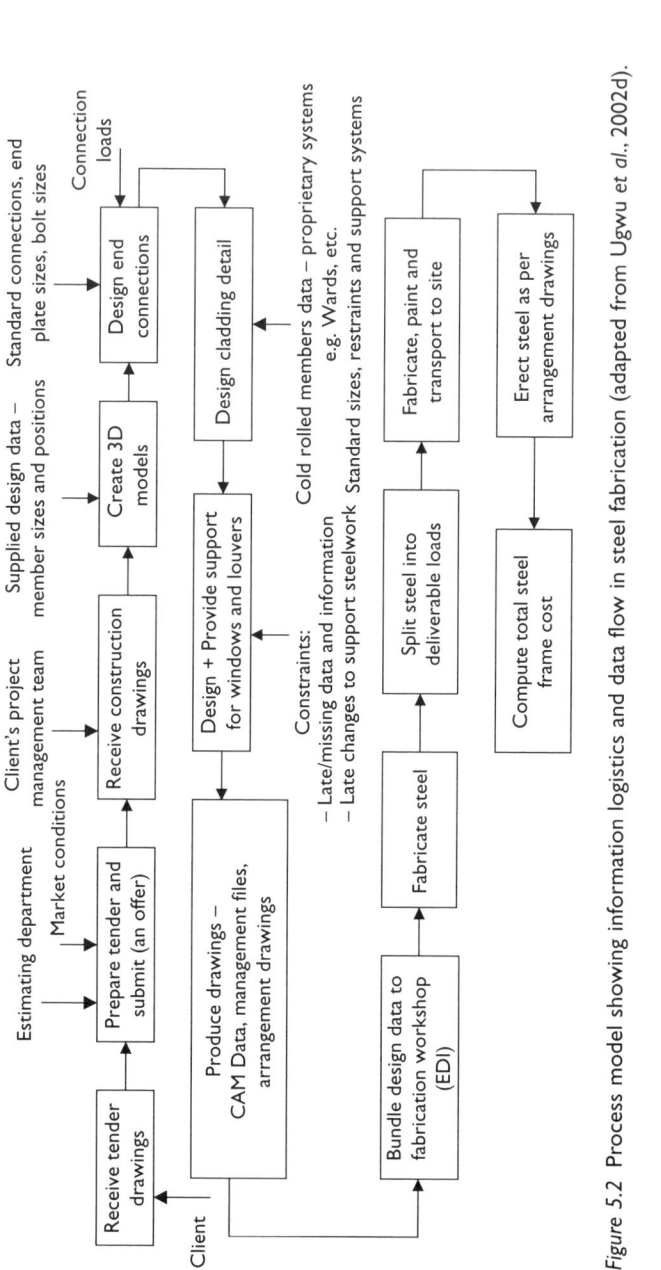

Figure 5.2 Process model showing information logistics and data flow in steel fabrication (adapted from Ugwu et al., 2002d).

agent application levels. The authors have extensively discussed various aspect of the ADLIB project research in related publications (Ugwu *et al.*, 1999a,b, 2000a,b,c, 2001, 2002b; Ugwu and Anumba, 2000; Anumba *et al.*, 2001a,b, 2002).

5.4.3 Research methodology

This section discusses a methodology for developing such agent-based systems and its application in the ADLIB research project, with a focus on the knowledge content of a MAS environment. It identifies the following key stages in a MAS development process in the context of the ADLIB project domain, such as requirements gathering, knowledge acquisition, MAS analysis and MAS design, implementation, testing and validation. These are highlighted in Figure 5.3. The ensuing sections discuss these stages.

5.4.3.1 Stage 1 – Requirements capture and analysis

The major activity here involves potential users specifying their requirements (or as more often the case their expectations) in plain simple language. This

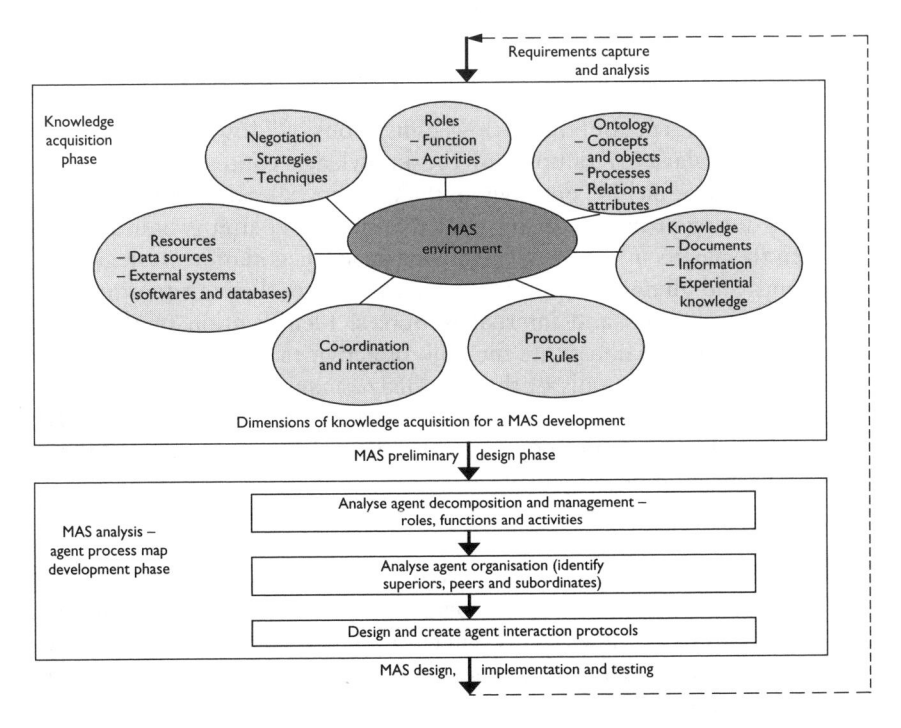

Figure 5.3 A framework for MAS development in collaborative design (adapted from Ugwu *et al.*, 2002a).

is sometimes stated poorly and with some ambiguities. Ugwu (2004) discusses wider issues in user requirements capture in developing information technology systems (including MAS) for decision-support in construction. In any case, it is the responsibility of the MAS development team to gather all the necessary data, clarify the ambiguities, and then translate the sets of requirements into a MAS configuration that incorporates all the technical requirements. Section 5.4 discusses this in detail in the specific context of the ADLIB project.

5.4.3.2 Stage 2 – Knowledge acquisition

The second stage involved deep knowledge acquisition from domain experts. This involved the use of a combination of techniques such as rapid prototyping, workshops and visual modelling using the unified modelling language (UML), (Rational2003-URL) techniques. This was useful in developing the agent knowledge base (k-b), which contains the data structures representing the entities from the experts' application domain such as objects, relation between the objects, actions, processes and procedures. The agent's (k-b) underpins its ability to execute its designated tasks. The next section examines the knowledge content of MAS in detail.

KNOWLEDGE CONTENT OF THE MAS ENVIRONMENT

A MAS environment contains various dimensions of knowledge at different levels of granularity. Examples of the knowledge content are shown in Figure 5.3. These include ontology of the domain, roles and responsibilities of agents during problem solving, co-ordination and patterns of interaction between the agents in the problem-solving space, negotiation strategies and techniques and interaction protocols in resolving conflicting situations, and the required external and internal resources. Identification of the above knowledge segments constitute the bulk of activities during the knowledge acquisition process. Details of the knowledge acquisition process and subsequent ontology development are discussed in Ugwu *et al.* (2001). The next section discusses ontology in greater detail because of its pivotal role in MAS design and development.

ONTOLOGY

The (k-b) of a MAS consists of an ontology that defines the concepts from a domain and a set of rules expressed in terms of these concepts. This approach clearly separates the concepts that are characteristic of a certain domain (the ontology) from the knowledge. As an illustration, in collaboration within the ADLIB domain (i.e. steel structures) the design team would include: structural engineers, steel fabricators, building services engineer, quantity surveyors, health and safety personnel, and constructability (manufacturing and production) experts. Each of these members of the design team has their own perspective on what the important issues and knowledge fragments are for related design problem. For example, all the experts mentioned are often interested in designing facilities/structures that are safe. In the context of light industrial buildings (portal frame structures) which is fairly standardised, the structural engineer is most interested in the relationship between building span, bay-width, building height and design load on the cross section of the structural members (columns, beams, rafters, trusses and bracings).

The steel fabricator, who is also interested in issues of component fabrication and constructability, would be most interested in the technical details of the joint and connections design for the structural members such as column bases, etc. This will relate cross sections of structural members and foundation details to such connection design decisions as: requirement for grout holes, holding down bolts, standard bolt lengths, washer plates, nuts, flats, angles, etc. The concepts in the ontology for such problem would capture the requirements of these experts in a design problem. Consequently, the ontology would be constructed at the knowledge level (Ugwu et al., 2002c).

5.4.3.3 Stage 3 – MAS domain analysis

The purpose of the domain analysis stage is to understand the application domain (i.e. collaborative design). This analysis normally feeds from findings in the knowledge acquisition stage. Some of the activities at this stage include analysis of the agent decomposition, agent organisation (identification of superiors, peers and subordinates agents in the problem-solving space) and the design of appropriate interaction protocols that capture the specific requirements of the problem domain. The analyses were conducted using UML techniques. Ugwu (2004) also discusses the advantages of this visual modelling approach in eliciting user requirements from domain experts. In the ADLIB project and within the specific context of MAS analysis Use Cases implemented as scenarios facilitated the discovery, identification and modelling of the domain problem(s) that the proposed MAS would solve after the implementation. Thus within the specifics of ADLIB prototype system development, Use Cases modelling enabled the project research team to

- communicate with end users and customers to delineate the MAS boundaries;
- define system functional requirements;
- identify the bases for object derivation and user interface design;
- migrate from functional requirements to objects and components structures; and
- define object interaction and object interfaces.

Section 5.4.4 gives detailed descriptions of the UML models for the ADLIB domain.

5.4.3.4 Stage 4 – MAS design, implementation and testing

This stage of the agent development process involves translation of agent role and responsibilities into agent-level problems that they represent and then deriving appropriate solutions. In the ADLIB project, the programmatic solution at this stage was to equip the agent(s) with the domain knowledge using *modal logic*. The knowledge fragments were expressed in terms of actions (primitive tasks), their pre-conditions their effects and then directing the agent's behaviour in a way to achieve the desired goals using the underpinning domain ontology. Section 5.5.3 gives detailed description of MAS development (design and implementation) and highlights some examples of knowledge fragments in the ADLIB domain. The next section discusses the business process modelling.

5.4.4 Business process modelling of collaborative design – user requirements and system functional specification

The MAS specification was outlined using the UML. This section discusses different kinds of system properties that are required for the ADLIB prototype. The specifications include:

- *Agent role specification*: This outlines the various expected roles of the task agents and captures the organisational structure in construction supply chain (Section 5.4.4.1).
- *Functional and behavioural specification*: This outlines the various functions of the participating agents in the context of the collaborative design of structures. The data requirements for each function are also specified in the corresponding Use Cases. The system function and behaviours are captured in the main Use Case diagram. These were extracted from the protocol analysis data of the KA interviews with the domain experts. It also illustrates the interaction between different segments of the system (Section 5.4.4.2).
- *Communication specification*: This shows the communication protocols at the agent level. It also shows the communication at component levels (Section 5.4.4.3).
- *Decomposition specification*: This part of the specification shows different user views of data and information requirements in the problem domain. The information content of each expert is modelled and encapsulated within packages in the UML notation. This underpinned the ADLIB ontology implementation at the application level (Section 5.4.4.4).

The system specification adapted object-based development and UML semantics with appropriate extensions when necessary. Agent modelling is

Agent-support for collaborative design   117

not completely identical with classical object oriented (OO) modelling and the current trend is to adapt extensions of OO models to describe certain properties of agent systems. The specification emphasised domain decomposition (information model), interactions and logistics of communication between agents in MAS (Use Cases, Sequence and Interaction, Activity diagrams). These parts of the specification are discussed in Section 5.4.4.4.

The functional requirements were used to describe the high-level implementation independent requirements of agents in the MAS prototype. These functional requirements are presented at a generic conceptual level, and they capture the needs of members of the project team in a collaborative design. The next section gives a detailed description of the identified agents including their general roles and functions in the MAS environment.

5.4.4.1   ADLIB domain agents

This section discusses the composition of the collaborative design agents (domain agents) in the context of ADLIB research project. The domain agents are the sets of specialist task agents that undertake collaborative design. They have a set of domain-specific ontology, and interaction/communication protocols, that are designed and implemented based on the speech act metaphor (Searle, 1969, 1979). These are summarised in the ensuing sections. Table 5.2 describes the supporting task agents and their roles in the collaborative design space.

THE TASK DESCRIPTION PROCESS AND WORKFLOW FOR THE ADLIB AGENTS

The details of the task description of the agents are described at the application level in terms of the task's preconditions, effects and constraints. Figure 5.4 illustrates the resources produced and consumed in executing tasks such as *DesignPortal* and then *Negotiate*. The top vertical arrows show task decomposition at the micro level, the left horizontal arrows show the resources consumed (essentially data and information), while the right side horizontal arrow shows the product of the resources consumption – data and information that define the conceptual design.

Different workflows are associated with task execution at the micro level, during the resource consumption and production. Figure 5.5 shows an example of typical workflow for the task compute cost, which is used to evaluate the cost of design proposal during the negotiation.

The abstracted workflow for cost evaluation is expressive enough to capture salient features of agent-based collaborative working that need highlighting. The abstraction focuses on design proposal evaluation by a costing agent because it encapsulates different aspects of automated design generation in a collaborative working context. These include: (i) agent–application

Table 5.2 Agents in ADLIB and their responsibilities

Task agent	Responsibility
Interface/Architectural Agent (IAA)	Responsible for rough initial design that meets the specifications
Structural Design Agent (SDA)	Responsible for containing structural design knowledge of steel portal frames, purlins and claddings. It integrates with a commercial design software (FASTRAK), passes design configurations to other peer agents, evaluates proposed designs from the peer agents and negotiates with them suggesting appropriate changes.
Building Services Agent (BSA)	Responsible for thermal load analysis based on building envelop configurations.
Costing and Constructability Agent (CCA)	Responsible for costing all plans. Returns suggestions on the cost and constructability implications of the proposed design (Figure 5.1 illustrates the associated workflow).
Safety Advisory Agent (SAA)	Responsible for broad but not exhaustive safety principles. Returns suggestions to meet minimum standards.

Note
Task agents differ from utility agents as outlined in Section 5.2.1 (Table 5.1). Utility agents are designed and embedded within the AP and provide the services discussed in Section 5.3, while the task agents are designed and constructed to suit the problem-solving situations in given domain.

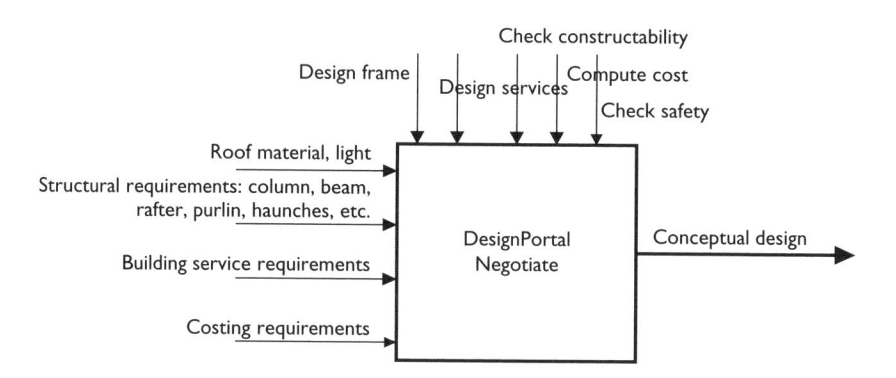

Figure 5.4 Consumption and production of resources (information).

integration (Section 5.5.4), (ii) process automation such as: communication, negotiation and agent–human integration vis-à-vis level of agent autonomy, (iii) component level collaboration (e.g. using Java Objects and available World Wide Web (WWW) infrastructure such as email systems to facilitate

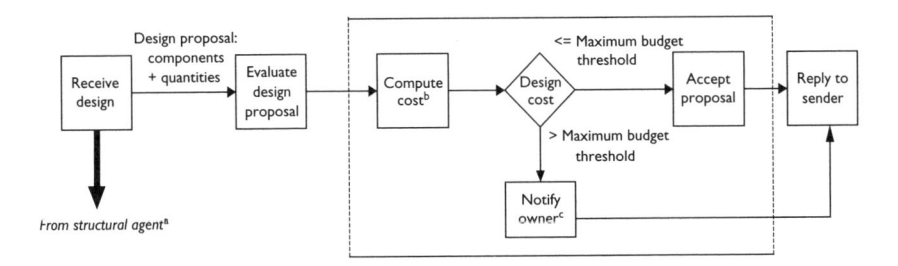

Figure 5.5 Costing agent workflow for the task compute cost.

Notes
a Agent-software integration is at the data level only, achieved using a Wrapper (Parser). Initial design is generated using existing commercial design software and then passed as Neutral Data File to the SDA, for automated negotiation.
b Cost computation is an external resource/service that is encapsulated in, and provided using Java Objects.
c Agent owner notification is an email automation process (achieved by extending the MAS platform (Zeus) functionality).

communication. See Section 5.5.3.4 for detailed discussion at the implementation; and (iv) structuring workflow and action sequencing to achieve desired knowledge-level integration. The next section describes domain knowledge representation of collaborative design processes and how the above outlined issues were addressed. The ensuing sub-sections describe the different parts of the requirement specification in detail.

5.4.4.2 Function and behaviour specification – the Use Case diagram

In order to define the specific requirements and detailed functional specifications, a main Use Case diagram was developed from the results of the knowledge acquisition. In the context of ADLIB, a Use Case is a special sequence of transactions performed by a user of the system (or the user's agent within the MAS) in a dialogue. Only a specific user can initiate a given Use Case and a given user can initiate different Use Cases. The Use Case diagram was verified in a presentation made to the domain experts (ADLIB industry partners). The presentation was part of initial validation of the KA exercises. The domain experts made suggestions, which were incorporated in developing the final models (Figure 5.6).

Within the scope of ADLIB project the user requirement translation in the context of the above validated Use Cases is as outlined in Box 5.1a. Details of some of the system functional specification are outlined in Boxes 5.1b–e. The representative agents for each discipline are shown in brackets.

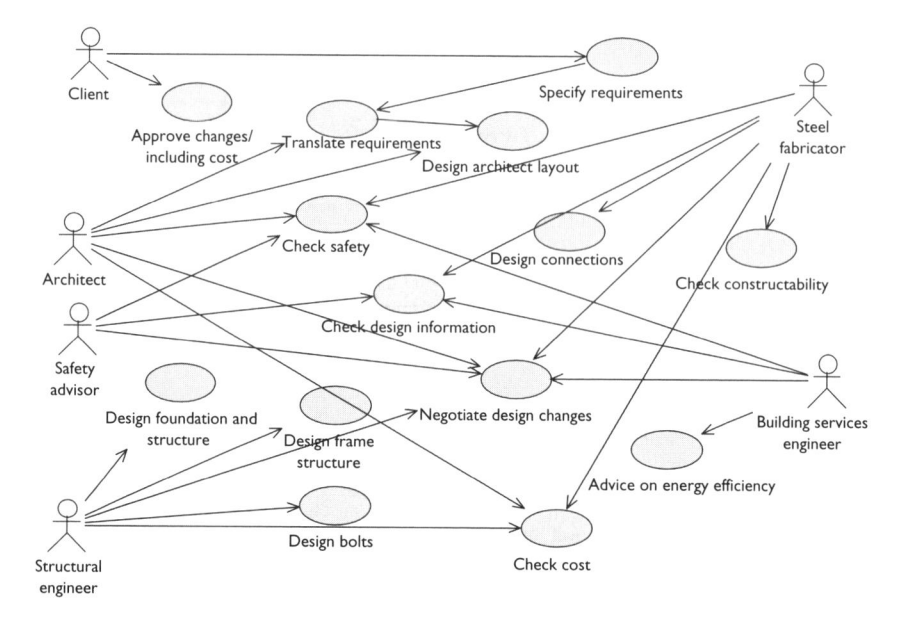

Figure 5.6 ADLIB main Use Case diagram (illustrating roles/functions).

Box 5.1a Informal specification of the ADLIB prototype system (Ugwu et al., 2000d)

The major requirement is for a multi-agent system (MAS) that will facilitate collaborative design of portal frame structures (light industrial buildings), by members of design team. The design team includes the architect, structural engineer, building services engineer, steel fabricator, safety advisor and quantity surveyor (QS). These members constitute the stakeholders and main actors that will interact externally with the MAS and deploy it in the search for design solutions. Each of these actors has different functional requirements from the system. The MAS includes a set of task agents that assist users in design and/or carry out specialist design functions on behalf of the user. The system shall automate some design functions such as negotiations over non-critical design decisions that can be delegated to a user's agent. It will have different functional capabilities to cater for the individual requirements of the different members of the design team (users) engaged in a typical design problem. In addition the agents shall deploy the underlying utilities provided by the supporting AP to carry out such functions as agent name and service registration, peer identification and inter-agent communication. Full details of the MAS prototype specifications are documented in Ugwu *et al.* (2000d).

Box 5.1b Structural Design Agent (SDA) roles

Structural engineer (SDA)

Role: To design the frame structure and communicate relevant design information to appropriate members of the collaborative design team.

Use Case: Design Frame structure.

Description: Use the appropriate design information from the IAA and design elements of the frame structure to satisfy the Ultimate limit states (ULS) and serviceability requirements as specified in the project documents and recommended in appropriate design standards. Carry out appropriate design checks – implemented using FASTRAK.

Reads: Portal frame loading details and other essential design information as extracted and translated from the project documents.

In: Building/Project definition attributes such as ProjectID, Building Grid line, Load, etc.

Out: Structural Design Information, e.g. Member sizes, Fabrication design information, e.g. Member sizes, section, loads, etc.

Assumes: The appropriate complete design information has been extracted from the Architect's project brief and used to populate the IAA.

Results: A design for the portal structure that is structurally safe and OK.

Transactions:

- The CSC Portal32 system automates the structural design process, and passes the result data to CCA, SAA and CA.
- The recipient agents check the design information and negotiate with the SDA as appropriate.

Box 5.1c Costing and Constructability Agent (CCA) roles

Steel fabricator (CCA)

Roles: This agent's additional roles include checking cost and constructability of designs.

- To design connections, joints and bolts, (as well as the frame structure in the case of Design and build projects) and communicate relevant design data information to appropriate members of the team.
- To check that designs are OK from point of view of design and construction.
- To check that budgetary restraints are met.

Use Cases:

- Design connections
- Check constructability
- Check cost.

Description: To utilise the information from the SDA, and design connections, joints, and bolts in the frame structure.

Reads: Member sizes, sections, loading, grade, length, etc.

In: Building/Project definition attributes such as ProjectID, Building Grid line, Load, etc.

Out: Structural Design Information, e.g. Member sizes, Fabrication design information e.g. Member sizes, section, loads, etc.

Assumes: The typical details of the frame design using StruCad will not be covered in the implementation, as is the interaction with the fabrications sites, etc. However, the basic operations/design processes like ensuring that section/members are standard and easily available could be automated. This requirement implies that SFA may need to execute other tasks, such as checking out the availability of steel sections (product). Note that the connection design function can be implemented using the CSC Connect32 software.

Results:

- The fabrication design information for the portal structure is structurally safe and OK.
- The design information for the portal structure is OK from point of view of constructability.
- A safe designs first and economical second.

Transactions:

- The CCA designs the connections, checks the design information and negotiates with the SDA for any alterations as appropriate.
- The CCA checks for the safety of the connection design.
- Because of the dual roles of checking constructabiliy and cost, the CCA interacts with the IAA, BSA, SDA and SAA as appropriate.

Box 5.1d Building Services Agent (BSA) roles

Building services (BSA)

Role: To advice on energy efficiency of the building, U-values, running costs, infiltration and lightening levels.

Use Case: Advice on energy efficiency of the building.

Description: This use case encapsulates the overall functions of the BSA at the conceptual level. At a lower level, this translates to recommending building service requirements based on user-supplied operational data, as extracted from the architect drawings and project specifications.
Reads: Appropriate building operational data, plant loading, etc.
In: Use of building, building type, number of people, usage, section of building and use of space (such as administration, storage, etc).
Out: Recommended building service requirements.
Assumes: The recommendations are based on some sets/sequence of production rules, that the agent uses in making HVAC design decisions.
Results: Satisfactory building services requirements.
Transactions:

- The BSA interacts with IAA, SAA, SDA and CCA as necessary.

Box 5.1e Safety Advisor Agent (SAA) roles

Safety advisor (SAA)

Role: To check general design and construction safety issues, and advice client plus other agents on the health and safety requirements and the necessary plan(s) of action.
Use Case: Check safety.
Description: This is a general use case that checks the overall safety of any design proposal from the point of view of design and construction/erection. Note that by devolving safety responsibility to all members of the design team in the construction, design and management (CDM), this use case can be instantiated by all other agents to check various aspects of safety at the appropriate design interface(s).
Reads: Appropriate safety parameters, dependent on the aspect of safety that is being considered.
In: Design parameters required in evaluating a given aspect/type of safety.
Out: OK or NOT OK safety decision. *There is nothing to negotiate if safety is NOT OK (i.e. when minimum safety is not met).*
Results: A design that is safe from point of view of design and construction.
Transactions:

- The SAA interacts with IAA, SDA, BSA and CCA as when necessary.

5.4.4.3 State charts, interaction (sequence) and activity diagrams

State chart diagrams represent the *behaviour* of various agents (components). It was adapted to object-oriented use. Communication between the agents is represented by collaboration and sequence diagrams. In addition, the UML Activity diagram was used to represent dependencies between agent process activities. The UML diagrams show the interaction diagram (Figure 5.7). The numbering shows the logical sequence of process execution to realise the use case 'Design Frame Structure'. The activity diagram in Figure 5.8a shows the protocol for agent to evaluate design proposal during negotiation and its used to realise the use 'Negotiate Design Changes'. It also shows the communication at the logical component/package levels (Figure 5.8b–d).

The ensuing section describes communication specifications at the component level. Figure 5.8b delineates the logical partitioning of the MAS at implementation level based on the information requirements, Figure 5.8c shows the physical partitioning of the MAS while Figure 5.8d shows the component decomposition view of agent resource requirements in the MAS design space.

The following is a description of some of the logical packages:

- *Domain agent*: ADLIB specific domain agent defined by organisation model of the collaborative design environment.
- *Common shared database*: The general worldview of the domain (i.e. the product features component of the ontology) that is common to all the actors. It is noted that – aecXML can facilitate data transformation into such common/global view.

In Figure 5.8c, Agent Interface and Applet View provides a WWW interface with human users so that design constraints and other project information/data can be input to the appropriate databases. Also the agent interface enables users to interact with their agents and specify negotiation strategies, monitor progress and also intervene as appropriate. This is discussed in detail in Section 5.5.3.4 – handling exceptional cases during negotiation.

5.4.4.4 Decomposition specification – the (object) class diagrams

The class diagrams capture different views of all the stakeholders in the collaborative design solving space. They serve two broad but interrelated purposes: (i) capture data and information requirements of the different

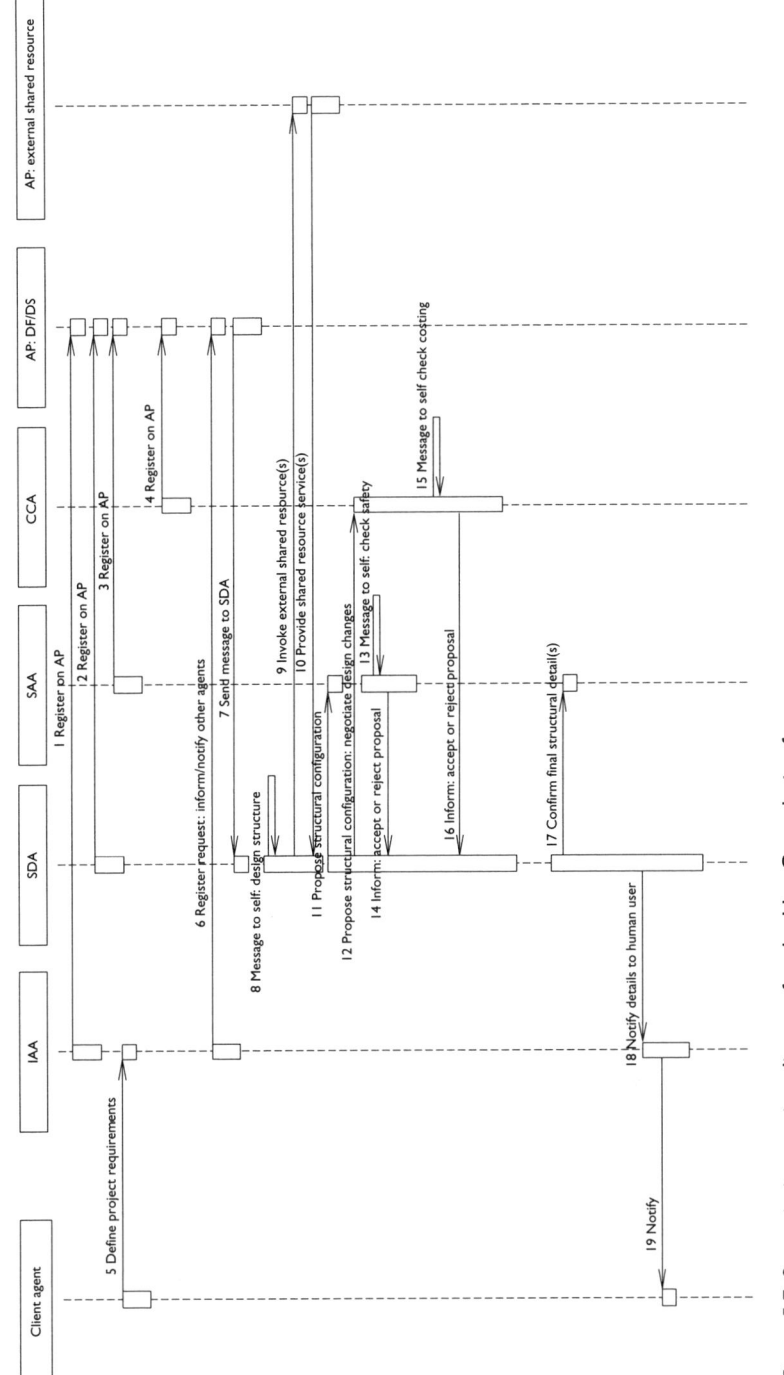

Figure 5.7 Scenario interaction diagram for the Use Case: design frame structure.

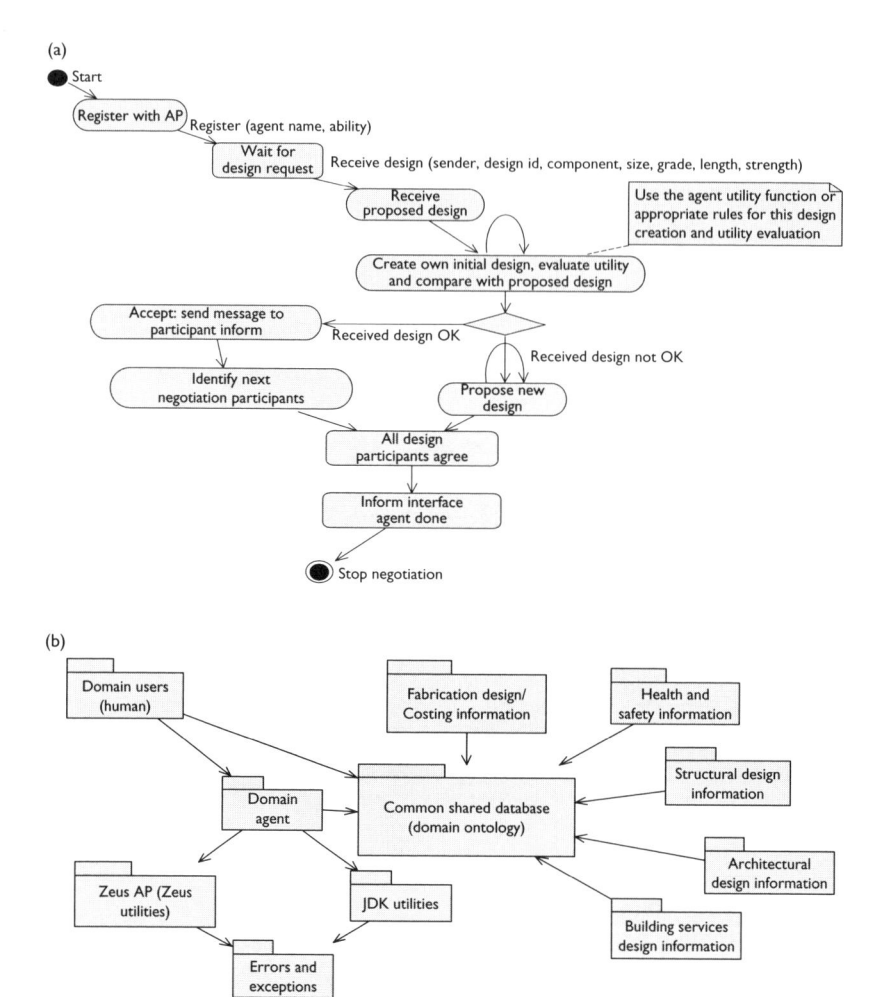

Figure 5.8 (a) Activity diagram for the Use Case: negotiate design changes. (b) Logical packages showing the logical partitioning of the MAS based on information requirement.

Note

The agent designers take implementation decisions on the type of resources the agent may require in task execution and also the appropriate existing system and/or programming language to use in providing such resources(s). Section 5.5.3.4 gives examples of interactions at the package level (i.e. agent invocation of external resources and component interactions at the package levels).

expert and (ii) facilitate explicit representation and understanding of the interconnections between these views at the micro level of problem solving. Thus the static class diagram represents *conceptual decomposition* of the system and the problem domain. This is the main information

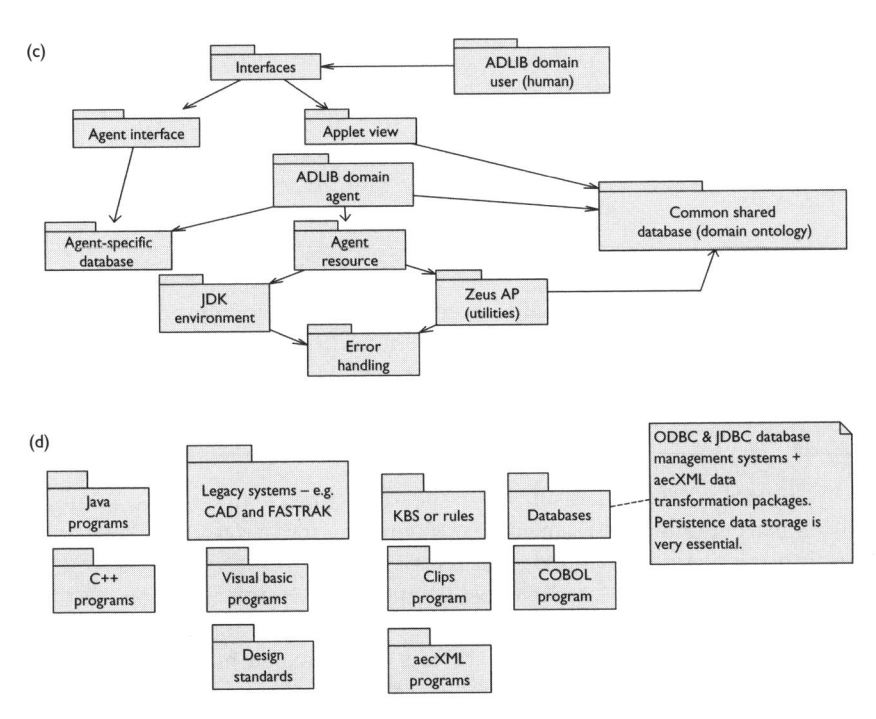

Figure 5.8 (c) Component packages showing the physical partitioning of the MAS.
(d) Component packages showing examples of agent resources.

model showing the various information contents/views of the domain by the different design experts. The data model is object based, which means that the problem domain is represented as a set of discrete recognisable objects such as structural components, columns, rafters, beam, foundation, etc. This underpins the product view of the ADLIB ontology developed from the knowledge acquisition and the verification process plus the comments from the Industrial Partners. Additional resources used in the knowledge acquisition include steel design technical specification documents (SCI 1999), and structural engineering application software such as FASTRAK (CSC2003-URL), publications in journals and other related literature. The classical static structure UML diagram was extended to represent individual ADLIB domain agents (objects) and classes. The ensuing section shows different object views (information model) of the portal frame structure by the experts (and by extension their representative agents) involved in collaborative design in the ADLIB domain. These views are further aggregated to generate a global view of the problem domain.

STRUCTURAL ENGINEER'S VIEWPOINT

The structural engineer is typically interested in the following aspects of the project: the type and use of building, loading, existing environment and its potential impacts on geotechnical properties of the soil, structural layout of the frame, service requirements and loading effects and the roof design. An example of typical detailed structural design information is shown in Figure 5.9.

STEEL FABRICATOR'S VIEWPOINT

The steel fabricator is typically interested in the following aspects of the project information: loading, joint and connection loads, member sections/sizes and the steel grade. Figure 5.10 captures these views.

OTHER COLLABORATIVE DESIGN TASK VIEWS

- *Constructability assessment* – The data object views that relate to constructability assessment are: member cross sectional dimensions, ground conditions and site access.
- *Costing* – Data related to costing include quantity and unit costs of elements and client's budget.
- *Building services* – Building services data include plant size, location and load, u-Values and thermal energy efficiency.
- *Safety advice* – Safety could be considered from point of view of member sections size and weight for lifting mechanisms. Figure 5.11 captures these views.

OBJECT RELATIONS VIEW

Figure 5.12a shows the integrated object view of collaborative designing for the domain (i.e. people, products and process dimensions).

In practical terms, the aggregated views stated earlier are required for joint decision-making as part of collaborative design process during which the agents address the design problems from their individual perspectives in the problem-solving space. For example, the expanded class diagram for costing a design proposal is shown in Figure 5.12b (this also relates to the workflow diagram in Figure 5.5, Section 5.4.4.1). The ensuing Section 5.4.5 discusses collaborative design joint decision-making using a decision tree.

5.4.5 Collaborative design joint decision-making process (decision tree)

The collaborative design negotiation process is illustrated using a decision tree. Figure 5.13 delineates the basis and issues that various task agents that represent domain experts, negotiate on.

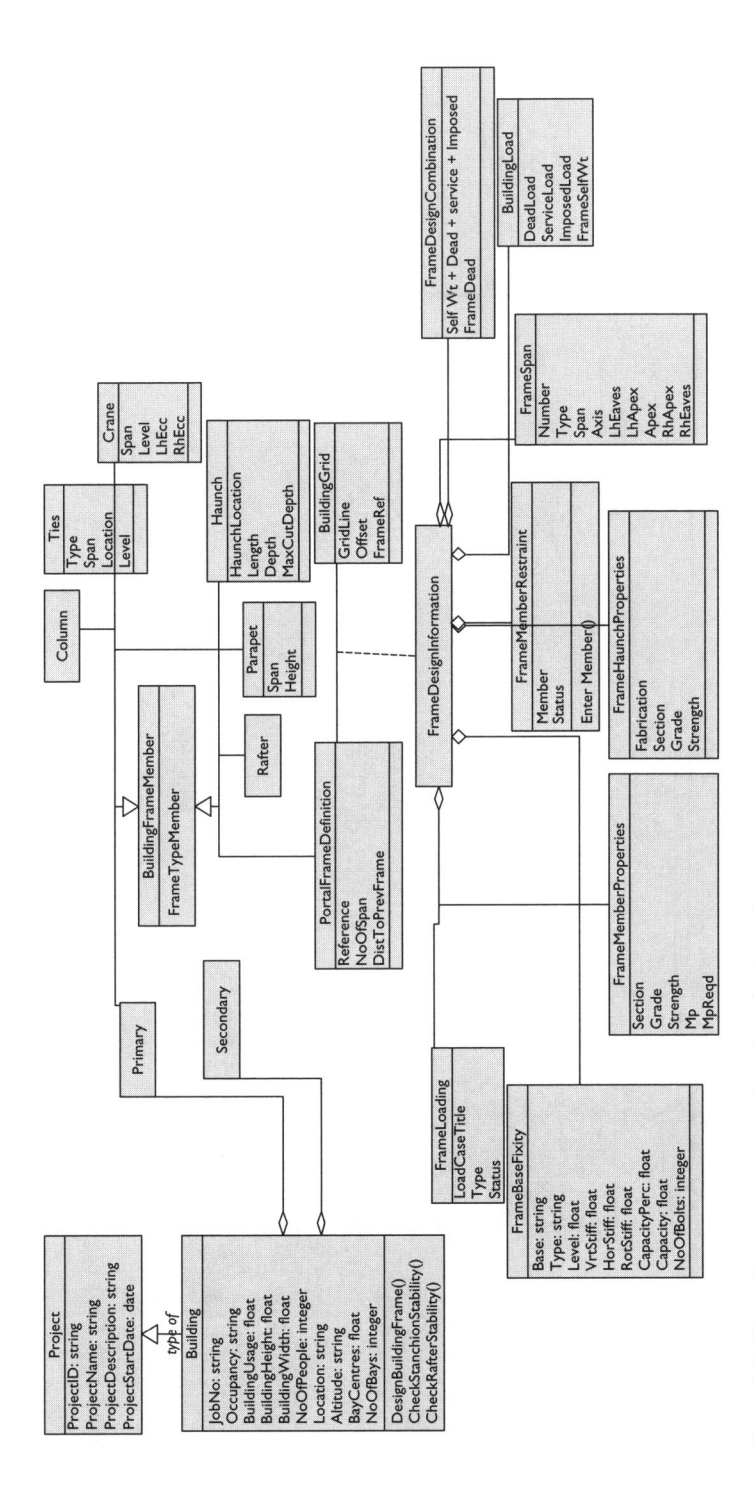

Figure 5.9 Class diagram – structural engineer's viewpoint.

Figure 5.10 Class diagram – steel fabricator's viewpoint.

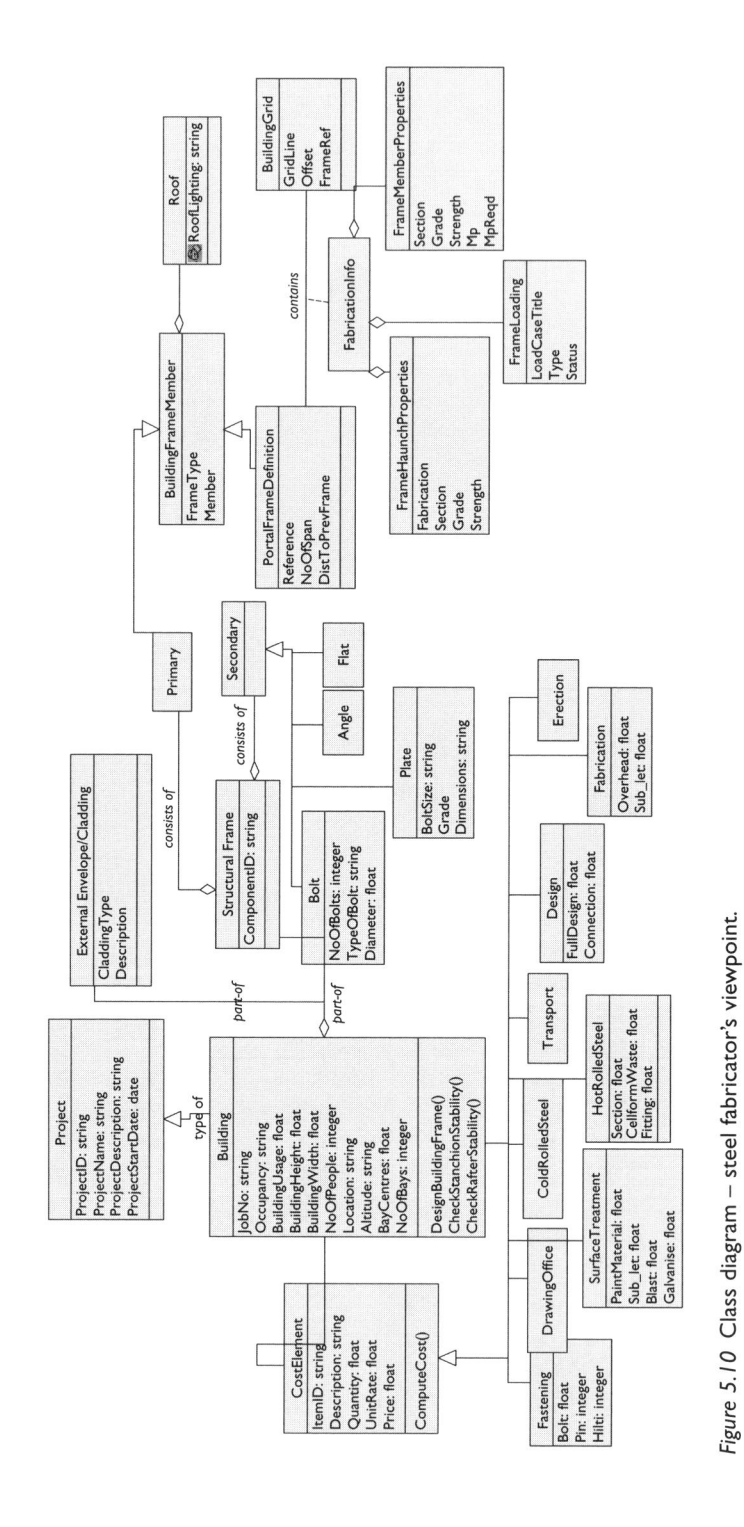

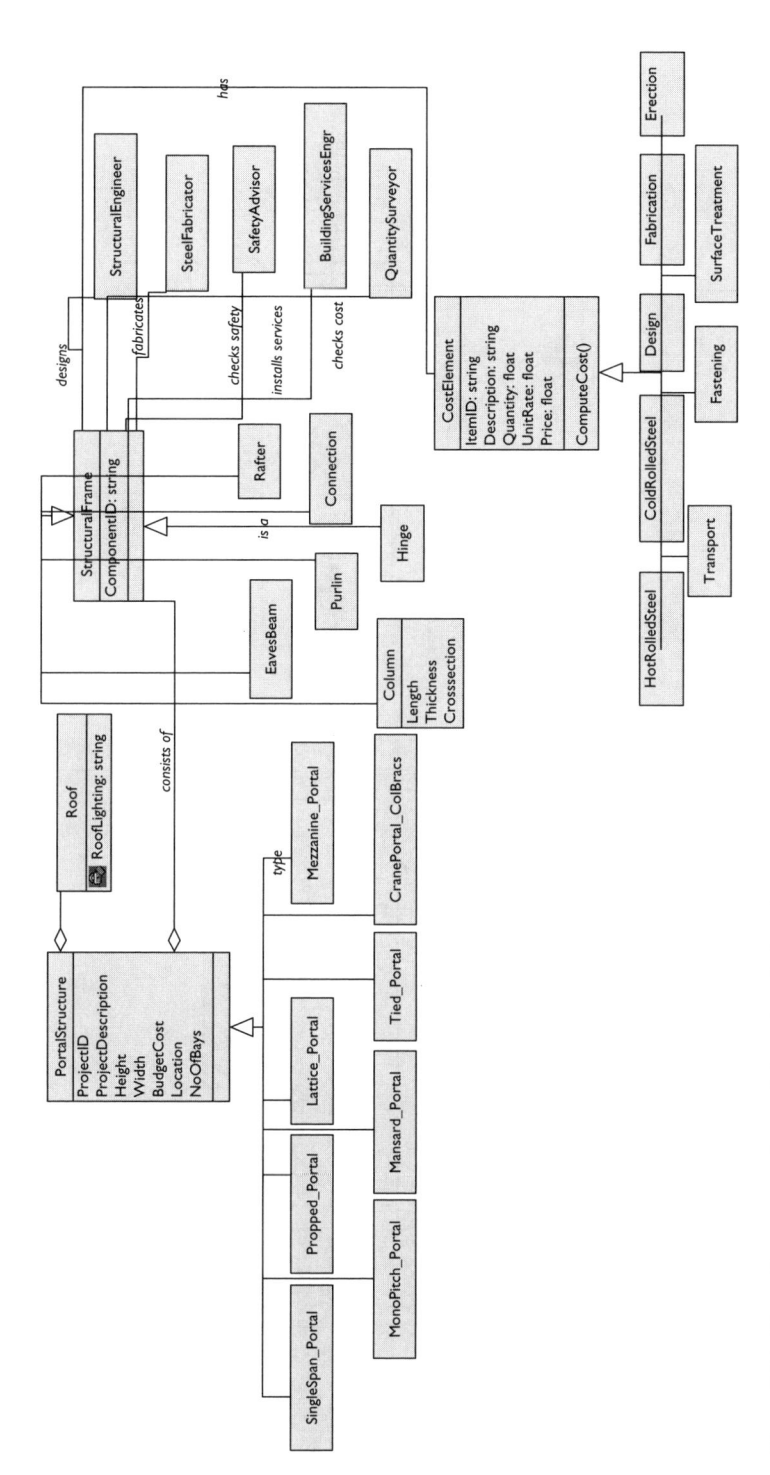

Figure 5.11 Class diagram – other collaborative design views.

(a)

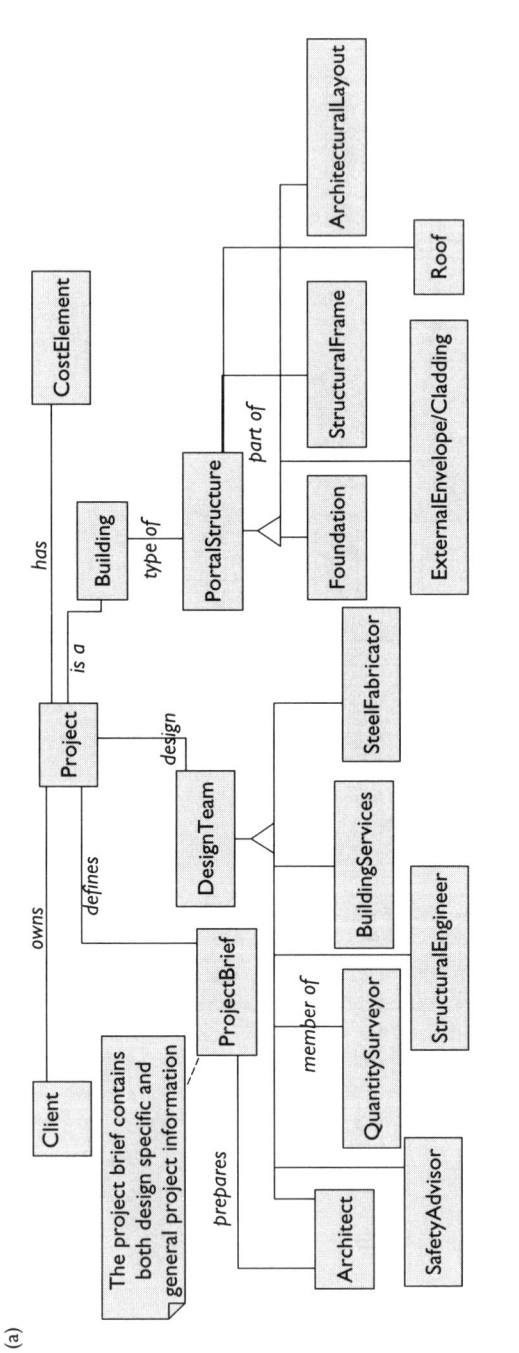

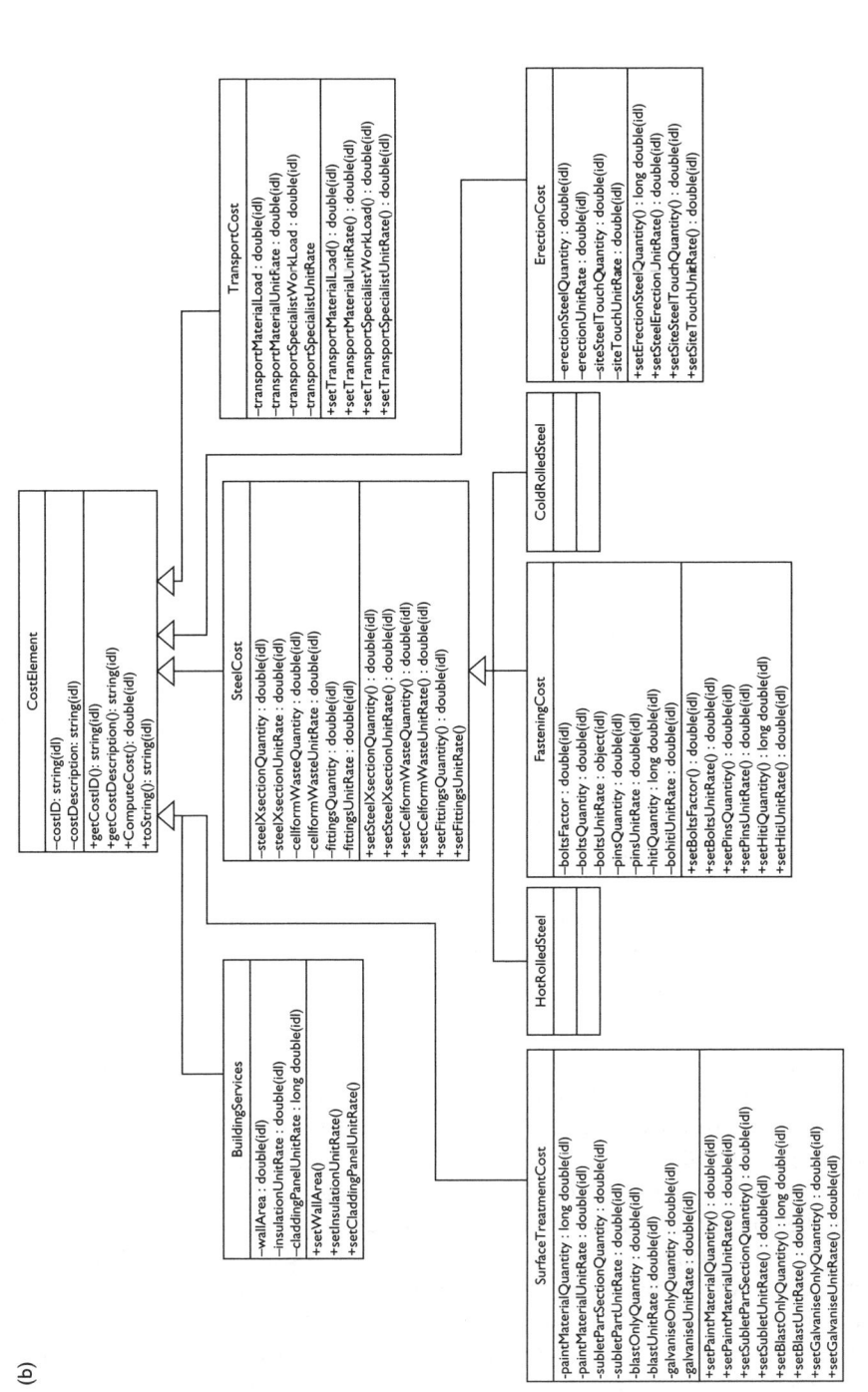

Figure 5.12 (a) Class diagram – object relationships in the ADLIB domain. (b) Class diagram – cost estimation ontology for steel frame structures: fabrication and installation (adapted from Ugwu *et al.*, 2002d).

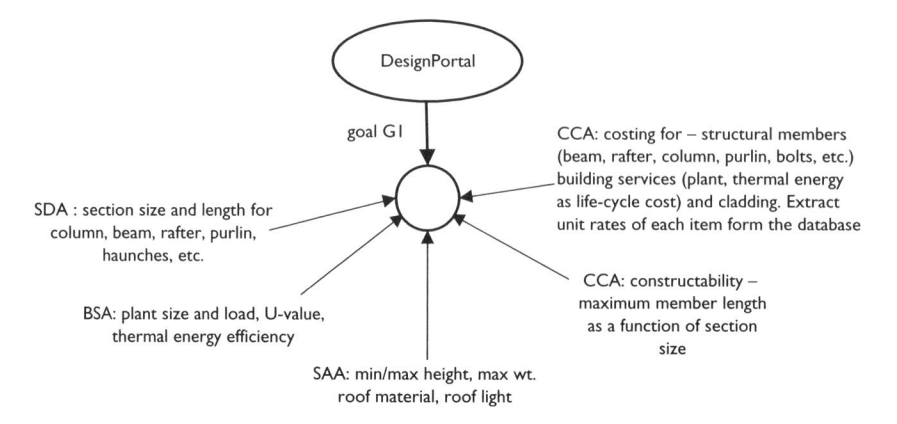

Figure 5.13 Joint/collaborative decision-making process in design space – negotiation issues.

Diagram labels:
- DesignPortal
- goal G1
- SDA : section size and length for column, beam, rafter, purlin, haunches, etc.
- BSA: plant size and load, U-value, thermal energy efficiency
- SAA: min/max height, max wt. roof material, roof light
- CCA: costing for – structural members (beam, rafter, column, purlin, bolts, etc.) building services (plant, thermal energy as life-cycle cost) and cladding. Extract unit rates of each item form the database
- CCA: constructability – maximum member length as a function of section size

The goal G1 is to agree on a design that is mutually satisfying to all the participating agents. This is a desideratum in collaborative working. The ensuing section describes how the previously captured business process features were mapped into MAS design features.

5.5 MAS design and implementation in ADLIB domain

5.5.1 *Knowledge-level representation for collaborative design – modelling product and design processes*

In the case study project, knowledge is modelled as sets of declarative facts, which consists of abstract or concrete concepts that define the domain ontology. For example, the concrete facts define the structural elements of a portal structure such as rafter, beam, column and the associated design attributes (see Figure 5.14). These were encapsulated in the fragmented views shown in Figures 5.9–5.11 in the preceding sections.

In the context of a MAS organisation, the abstract facts model an agent's design and negotiation knowledge including the social behaviour (such as interaction with other agents during negotiation) in the design space. These facts cumulatively define the sequences of negotiation actions, and the design processes/tasks are decomposed to such a discrete level that they are amenable to process automation through a series of symbolic logic manipulations. However, definition of the agent tasks and responsibilities reflect the functional and disciplinary roles and responsibilities of domain experts, in a collaborative design context. As an illustration, the abstract fact *NegotiationOrder* that is encapsulated within the Negotiation protocol is used to define an agent's sequence of negotiations (rules of engagement) with other task agents in the design space. This is an initial resource that the agent is equipped with, and it defines an important social behaviour of the agent in the design space. However, the negotiation order was determined by the logistics of communication and interaction between the domain experts. This very important social behaviour was specifically captured during the knowledge acquisition and user requirements elicitation from the project industrial partners.

5.5.2 Design process knowledge representation – application domain ontology

As discussed in the preceding section, the design and implementation of ADLIB ontology is predicated on two interconnected levels of knowledge representation in agent implementation. These include:

- abstract level that captures symbolic representation of collaborative design processes and business rules and
- concrete level that captures the object concepts that describe the portal frame structure physically.

The abstract facts were used to model and represent the design process in a symbolic way, and they encapsulate the various design processes and knowledge in a collaborative context (Figure 5.14). This is designed to facilitate rule generation and symbolic logic manipulation. Figures 5.15 and 5.16 show these two hierarchies of the ADLIB ontology.

The following section describes the various abstract facts are as they relate to agent knowledge modelling and MAS organisation in the collaborative design process.

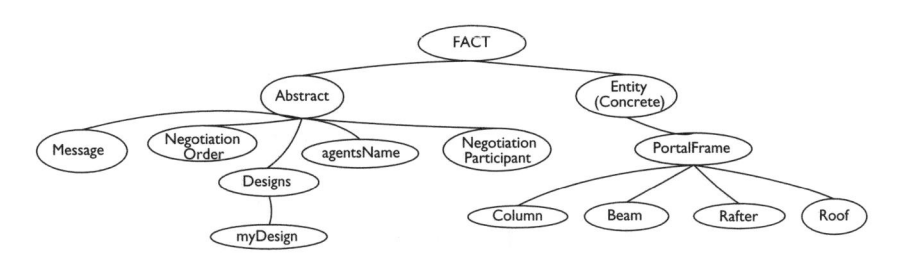

Figure 5.14 Knowledge-level ontology for agent-based collaborative design.

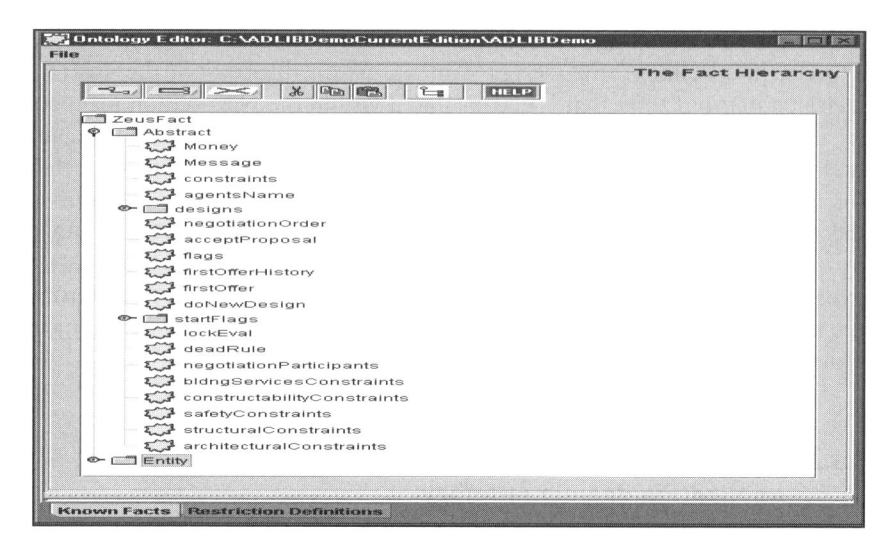

Figure 5.15 Symbolic representation of collaborative design processes: abstract facts hierarchy.

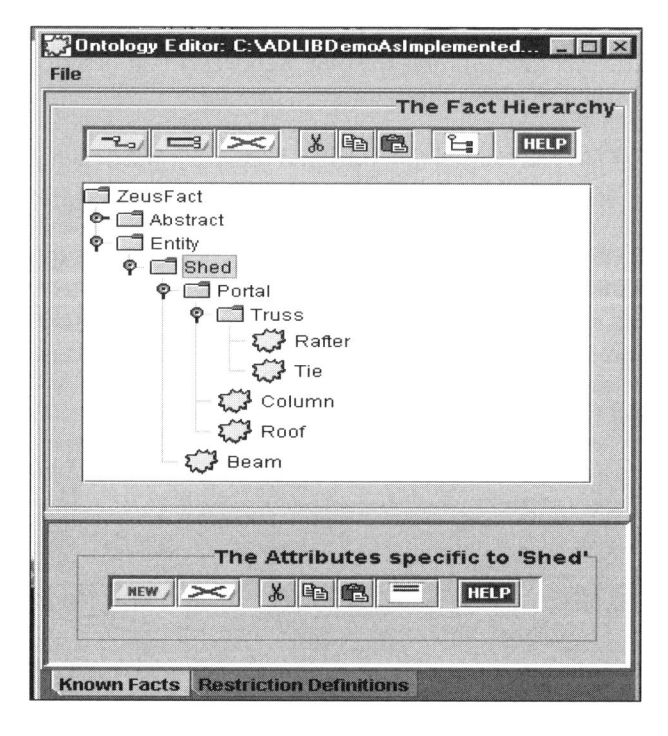

Figure 5.16 ADLIB ontology – concrete level domain concepts.

5.5.3 Design of agent tasks and knowledge fragments

5.5.3.1 Co-ordination, communication, messaging and social behaviour in MAS environment

Table 5.3 summarises the various components of the ontology (i.e. abstract and concrete) as they relate to co-ordination, messaging and communication in the MAS design space.

There are two types of task: Rulebase and Primitive. The Zeus Rule Engine activates the Rulebase tasks, while the Primitive tasks are essentially implemented as Java codes (agent external resources) and hooked up to the agents using the ZeusTask, in a polymorphic behaviour. Details of some of the tasks are described in the following section.

AGENT REGISTRATION AND IDENTIFICATION

Agents register with a unique identity to facilitate identification by peers in the MAS design space. The agents communicate by sending messages to each other. The message syntax, which is FIPA compliant, is given in the following Box 5.2. Box 5.3 shows the rule for agent registration with the interface.

PUBLISHING DESIGN PROPOSAL

A Rulebase task defines the rules for the Interface agent to distribute the design proposals to the respective participating agents. The knowledge fragment is shown in Box 5.4.

5.5.3.2 Negotiation protocols – modelling negotiation in MAS design space

The ensuing section describes communication loop and message passing in the MAS design space. The protocol is the set of rules of interaction that the agents will follow to converge on a solution in design space. A common method is the Monotonic Concession Protocol (MCP) (Rosenchein, 1994). For two agents the standard version is as follows. Agents start by simultaneously proposing a deal (in this case a design). Agreement is reached if one agent matches or exceeds what the other has asked for (in terms of utility). If an agreement is not reached the negotiation proceeds to another round. An agent can only propose deals that have a greater or equal utility for the other agent, that is, concede or do nothing. If neither agent concedes then negotiation ends as conflict is reached, otherwise negotiation continues. It is a protocol particularly suited to task-oriented domains (TODs) such as the collaborative working in construction, as it ensures convergence to a deal (where a deal is possible). In the ADLIB prototype, the protocol involves the following sequences: (i) an agent registers with the interface agent supplying its name as

Table 5.3 Components of the ADLIB ontology

Fact name	Description and attributes
FACT	A root level ontology that is inherited from the MAS platform Zeus, [BT, 2002], developed by British Telecom Plc, UK.
Money	This is an abstract class inherited from Zeus and is designed to cater for transactions in which actual cost is incurred and payment made for services by agents. In ADLIB we assume an agent in executing its tasks incurs no monetary costs.
Message	Agents communicate by sending messages to each other and this is how an external stimulus is used to initiate action from an agent. It is inherited from the MAS platform (Zeus) and has been designed to follow the FIPA specifications for messaging [FIPA, 1997a]. It has the following attributes: *sender, receiver, content, reply-with, in-reply-to, ontology, protocol, conversation-id*.
agentsName	Defines the name of an agent and offers the agent unique identity in the MAS environment. Used for agent registration and subsequent discovery by its peers in the design space.
Constraints	This is a cumulative definition of the constraints that a design must meet in a collaborative context. This is a translation of project specification documents.
Designs	Holds the objects to design and captures a symbolic object-attribute representation of the portal structure (i.e. Structural members such as Column, Beam, Rafter, etc.).
myDesign	A child fact to Design. Instantiates a given design owned by an agent. Typical content of message sent from one agent to another.
NegotiationOrder	An abstract fact that defines the social behaviour of an agent in a MAS space. Values are defined for a given agent and used to equip it as an initial resource during agent design. Also because not all agents need to interact with every agent, the peer-to-peer interaction is decomposed and controlled in a particular chain, and the interaction decomposition is controlled using *NegotiationOrder*.
NegotiationParticipant	Used to define the agents participating in design and/or negotiation process, and whether they are *initiator* or *respondent*.
acceptProposal	This abstract fact is used to take decision on whether to accept or reject a design proposal. It has two attributes: accept (Boolean – false by default), and name (String – that defines the agent taking the decision within the design space).
flags	This abstract fact is used to initiate a given design and agent process/action depending on the current value of the attributes.

Box 5.2 Messaging syntax for ADLIB agents

```
Message <sender, receiver, content, reply-with,
in-reply to, ontology, conversation-id>
content <design>
design <portal frame design attributes{Beam, Column,
Rafter} etc>
```

As an illustration, the syntax for Agent Registration (see Box 5.3 for the associated rule) is: *registerWithInterface* <agentName, Ability>*
Uses the MAS Platform (ZEUS) utilities.

Box 5.3 Knowledge fragment for agent registration – *registerWithInterface*

```
if aN <- (agentsName (name ?var148))
and ?flg <- (flags (registrationDone false))
    then (send_message (receiver Interface) (content ?aN)
    (type inform))
    (modify ?flg (registrationDone true))
```

The knowledge fragment specifies that the abstract fact agentsName contains the agent name description (agentsName (name ?var148)) and the condition is that trigger flag for registration done is false (flags (registrationDone false)). The action is that the agent should register with the interface by sending a message of performative type inform (send_message (receiver Interface) (content ?aN) (type inform)), and modify the trigger registration flag to true thereby signing off the registration process – (modify ?flg (registrationDone true)).

the parameter, (ii) after it receives a design proposal an activation trigger starts negotiation, (iii) it first evaluates the utility of a design proposal and (iv) it takes appropriate decision and actions based on the utility of the proposal. In the case of a costing agent, the utility would be measured by comparing the computed cost of the proposal with the budget threshold (see Figure 5.5 – workflow of the costing agent). The activity diagram in Figure 5.8d captured this protocol. Table 5.4 describes the corresponding production rules and performative acts. Searle (1969, 1979) describes the philosophical underpinnings of speech acts in language and communication. Box 5.5 describes the knowledge fragments associated with an agent starting negotiation.

5.5.3.3 Negotiation strategy

In the real world, designers often adapt different strategies in negotiating with other members of the project team. The ADLIB project agents adhere

Box 5.4 Knowledge fragment for publishing the design proposal to all participating agents that are registered with the Interface Agent – *SendToAllAgents*

```
if ?cons <- (constraints (hasBeenSent false))
and ?scid <- (startCreateInitialDesign (start true))
and ?np <- (negotiationParticipants (initiator ?ini))
and ?scnp <- (startChangeNegotiationParticipants)
    then (send_message (receiver ?ini) (content ?scnp)
    (type inform))
    (send_message (receiver HandS) (content ?cons)
    (type inform))
    (send_message (receiver Structural) (content ?cons)
    (type inform))
    (send_message (receiver Costing) (content ?cons)
    (type inform))
    (send_message (receiver HandS) (content ?scid)
    (type inform))
    (send_message (receiver Structural) (content ?scid)
    (type inform))
    (send_message (receiver Costing) (content ?scid)
    (type inform))
    (send_message (receiver HandS) (content ?np)
    (type inform))
    (send_message (receiver Structural) (content ?np)
    (type inform))
    (send_message (receiver Costing) (content ?np)
    (type inform))
    (modify ?cons (hasBeenSent true))
```

The knowledge fragment specifies that the first condition is that design proposal has not been published by the Interface agent [cons <- (constraints (hasBeenSent false))], the second condition is that the trigger to start creating initial design has been activated and the value set to true [scid <- (startCreateInitialDesign (start true))], the abstract fact indicates that the interface is playing the role of an initiator [?np <- (negotiationParticipants (initiator ?ini))], and has a trigger to activate (start changing) the negotiation participant [scnp <- (startChangeNegotiationParticipants)]. The action is that the interface agent should publish the design proposal by sending a message of the type inform to all the participating agents who are now playing the role of receivers [(send_message (receiver ?ini) (content ?scnp) (type inform))], each participating agent being uniquely identified with its name e.g. Costing agent [(send_message (receiver Costing) (content ?cons) (type inform))], and also a message to start creating its own design [(send_message (receiver Costing) (content ?scid) (type inform))]. The final action is to sign off that design proposal has been published to all the agents and modify the relevant trigger to true [(modify ?cons (hasBeenSent true))].

Table 5.4 Performative acts in ADLIB domain

Performative	Description
StartNegotiation	Used to activate the negotiation process by an agent
ProposalEvaluateUtility	Trigger for an agent to evaluate the utility of a design proposal
RespondToProposal	Agent responds to the proposal
SendAcceptanceNotice	Agent sends acceptance (or rejection) notice
SendProposalBack	Agent sends alternative design proposal if applicabe
RespondentSetUpNegotiationPair	Agent sets up the next peer in the negotiation
MoveToNextPair	Moves to the next participant in the negotiation loop
StopNegotiation	Trigger for Interface agent to stop negotiation once agreement is reached

Box 5.5 Knowledge fragment for negotiation protocol – *StartNegotiation*

```
if ?ssn <- (startStartNegotiation (start true))
and ?nps <- (negotiationParticipants (initiator ?an))
(agentsName (name ?an))
and ?prop <- (myDesign (status ?cs))
and ?flg <- (flags (stop false))
(negotiationOrder (next ?nxt))
and ?scnp <- (startChangeNegotiationParticipants)
and ?speu <- (startProposalEvaluateUtility)
       then (modify ?ssn (start false))
       (modify ?scnp (initiator ?an) (respondent ?nxt)
       (start true))
       (send_message (receiver ?nxt) (content ?scnp)
       (type inform))
       (retract ?scnp)
       (modify ?nps (respondent ?nxt))
       (modify ?prop (owner ?an) (status proposal)
       (toBeEvaluatedBy ?nxt))
       (modify ?flg (negotiationInProgress ?nxt))
       (send_message (receiver ?nxt) (content ?prop)
       (type inform))
       (modify ?prop (status mine))
       (modify ?speu (start true))
```

The knowledge fragment specifies that the conditions are that the trigger to start negotiation is activated ssn <- (startStartNegotiation (start true)). The agent's state changes to that of an initiator for the negotiation [?nps <- (negotiationParticipants (initiator ?an))], and the abstract facts include agents unique identity – name [(agentsName (name ?an))],

> proposed design [?prop <- (myDesign (status ?cs))], the flag to proceed with negotiation remains active [?flg <- (flags (stop false))]. The abstract facts also include negotiation order [(negotiationOrder (next ?nxt))], and triggers to; (i) start changing the negotiation participant [?scnp <- (startChangeNegotiationParticipants)] and (ii) begin evaluating the utility of the design proposal [?speu <- (startProposalEvaluateUtility)]. The action after the agent has evaluated the utility of the design proposal and taken appropriate decisions are to: (i) modify its start negotiation sate to false [(modify ?ssn (start false))], (ii) modify its position in the negotiation from *initiator* to *respondent*, and send the message to the next pair and retract the contents of the knowledge base [(send_message (receiver ?nxt) (content ?scnp) (type inform))(retract ?scnp)(modify ?nps (respondent ?nxt))]. Other facts manipulated include respondent to the proposal, claiming ownership of the design and identifying that it be evaluated by the next respondent [modify ?nps (respondent ?nxt))(modify ?prop (owner ?an) (status proposal) (toBeEvaluatedBy ?nxt))(modify ?flg (negotiationInProgress ?nxt))], sending message performative type inform to the next peer agent and bundling all these as contents of the message of [(send_message (receiver ?nxt) (content ?prop) (type inform))] and then modifying the status to proposal to that of the agent [(modify ?prop (status mine))] and activating the trigger for negotiation process [(modify ?speu (start true))].

to the above negotiation protocol in proposing alternative deals during negotiation. The negotiation strategy is based on a simple gradient decent algorithm. In this case, an agent makes an offer that will decrease its utility by the least amount with the minimum step size being governed by other considerations such as the time cost. At the application level the algorithm is provided as primitive task (Java object codes) and dynamically bound to the agent during task execution. Figure 5.17 shows the opening offer for negotiation in design space. By a process of rounds involving concession making, each individual agent will move from its (in its own terms) most optimal design (represented by black dots in Figure 5.17) to a final compromise design. The agents would reach a final design (even if it were a *satisficing* solution). This deal or final design will be somewhere within the interaction of all three circles in Figure 5.17.

5.5.3.4 Handling exceptional cases during negotiation

There could be instances of exceptional cases during the negotiation process, and it is necessary that an agent be able to handle such situations. As an illustration, the costing agent on evaluating the utility of a design proposal may observe that the cost far outweighs the client's budgetary threshold, and is

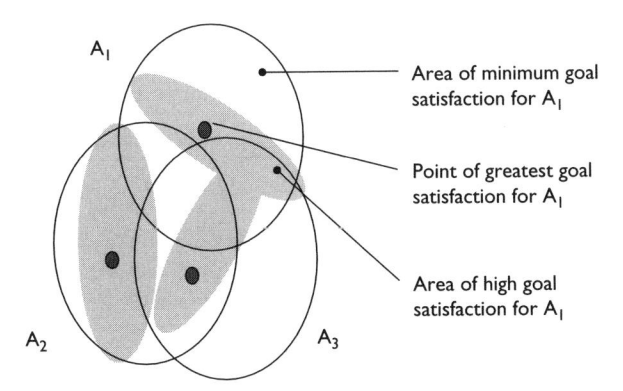

Figure 5.17 Gradient decent algorithm.

therefore unable to proceed with the negotiation. This situation would require human (i.e. agent owner) intervention. A solution could be to provide the agent with a resource such as WWW email infrastructure so that it could notify/alert the owner for appropriate intervention. This solution captures the representative business practice in real world where such critical consultations become inevitable at project design interfaces (Ugwu *et al.*, 2003). Figures 5.8c and 5.8d captured this at the conceptual system requirements modelling level (Section 5.4.4.3). However, in ADLIB project this implementation required extending existing functionality of the agent development platform – Zeus. Figure 5.22 in Section 5.5.5 (prototype system validation) illustrates such email alert scenario in the ADLIB prototype. The typical code fragments of this functionality at the application level is shown in Boxes 5.6a and b.

5.5.4 Agent–system integration

The task agents in the MAS framework executes using a defined workflow sequence illustrated in Figure 5.18. At the initial stage an agent has an empty execution plan until it receives an event-driven stimulus that presents it with a design goal. This goal presentation is activated at the user-interface level by clicking the Design button (Figure 5.19). The collaborative design process plan generation begins by the agents invoking the primitive and Rulebase tasks as discussed in the preceding sections. The structural agent receives as the initial resource, design generated using FASTRAK and this is mapped into the Zeus resource base ontology using the Parser Translator. It is on completion of this sequence that the peer-to-peer negotiation process begins (Figure 5.18), with the structural agent as the Initiator.

The neutral data parser is also provided as a package-level resource to the structural agent. The next section describes example outputs from the MAS prototype (Anumba *et al.*, 2002).

Box 5.6a Code fragments for CCA to handle exceptional cases during negotiation

```
import zeus.extension.*;
import zeus.extension.MMX; - (see the package resource -
MMX in Box 5.6b)
import zeus.adlibagentresources.FasteningCost;
import zeus.adlibagentresources.ColdRolledSteelCost;
import zeus.adlibagentresources.HotRolledSteelCost;
import zeus.adlibagentresources.SurfaceTreatmentCost;
import zeus.adlibagentresources.BuildingServicesCost;
import zeus.adlibagentresources.ErectionCost;
import zeus.adlibagentresources.TransportCost;
{
if (designCost > maxAllowableCost) // Automated
Messaging to handle negotiation deadlock
{MMX mmx = new MMX();
   mmx.setServer("NetworkServerAddress");
   mmx.setRecipient("ReceiverEmailAddress");
   mmx.setSender("SenderEmailAddress");
   mmx.addMsg("SenderEmailAddress.");
   mmx.send();
     }
   }
```

Some notes on the simulated results: user-defined values and other attributes – see output in Figure 5.22.

1. designCost is dynamically computed by the agent based on the design proposal configuration
2. maxAllowableCost is defined by the user as a budget threshold constraint, at the appropriate agent-user graphical user interface
3. NetworkServerAddress = staff-mailout.lboro.ac.uk
4. ReceiverEmailAddress = joeugwu@hotmail.com
5. SenderEmailAddress = O.O.Ugwu@lboro.ac.uk (must be an authorised network user in the email server
6. UserDefinedMessage = Testing Message Automation. This is an automated messagage from the ADLIB Design Agent – CCA. The clients maximum allowable buget "+maxAllowableCost+" has been exceeded. Design cannot proceed Negotiation has terminated. Please relax the contraint for negoiation to proceed.

Similar simulations were successfully tested for message delivery to mobile phones using SMS technology.

Box 5.6b Package level resource definition for integrating CCA and WWW infrastructure (email)

MMX package resource – Zeus extension: MMX: send mail

Usage:

1. `MMX mmx = new MMX();`
2. `mmx.setServer(String mailServer)`
   2a. to read the server name from database (if defined), do:
   `mmx.setServer("");`
   2b. to read and online verify the server name from database, do:
   `mmx.setServer(".");`
3. `mmx.setRecipient(String emailAddress)`
4. `mmx.setSender(String senderEmailAddress) // optional`
5. `mmx.addMsg(String message)`
6. `mmx.send();`

Note:
This extension package has an online facility to help you set up the mail server name database. Do: java zeus.extension.SetMailServer() in your unix or DOS window.

Remark:
For window users, your mail server name can be found by tools/account in outlook express.

For unix users, usually you can specify "mailhost" as your server. Otherwise, consult your administrator.
*/

```
package zeus.extension;
import java.util.*;
import java.io.*;
import java.awt.*;
import java.net.*;
import zeus.util.SystemProps;
public class MMX {
String sender = "Zeus-MMX-Extension";
String recipient = null;
String server = null;
Vector msg = new Vector();
boolean readServerFile = true;
boolean debug = true;
```

```java
public MMX ()
{
}
public void setServer (String server)
{
   if (server !=null)
   if (server.equals("") && readServerFile)
   // quick read
     new SetMailServer().display(this, true);
   else
     if (server.equals(".") && readServerFile)
     // check database
       new SetMailServer().display(this, false);
     else
     this.server = server;
   else
     if (readServerFile)
       new SetMailServer().display(this, true);
   if (debug) System.err.println("Server is set as" +
   this.server);
}
public void setRecipient (String emailAddress)
{
   if (emailAddress !=null) this.recipient =
   emailAddress;
   if (debug) System.err.println("recipient is set as"
   + this.recipient);
}
public void setSender (String senderEmailAddress)
{
   if (senderEmailAddress !=null &&
   !senderEmailAddress.equals(""))
     this.sender = senderEmailAddress;
   if (debug) System.err.println("sender is set as" +
   this.sender);
}
public void addMsg(String message)
{
   if (message !=null) msg.addElement(message);
}
public void send() {
   if (server ==null || recipient ==null ||
   sender ==null ||
     msg ==null || msg.size() ==0)
     return;
   Socket mailSession;
```

```java
// establish connection
try {
    mailSession = new Socket(server, 25);
    BufferedReader in =
      new BufferedReader(new
        InputStreamReader(mailSession.
        getInputStream()));
    DataOutputStream out =
      new DataOutputStream(mailSession.
      getOutputStream());
// check dialogue from SMTP
    String resp = getResponse(in);
    if (resp.charAt(0) != '2')
      throw new IOException("Bad SMTP dialogue I\n");
    resp = mailCommand("HELO dummy", in, out);
    if (resp.charAt(0) != '2')
      throw new IOException("Bad SMTP dialogue II\n");
    resp = mailCommand("MAIL FROM:" + sender, in, out);
    if (resp.charAt(0) != '2')
      throw new IOException("Bad SMTP dialogue III\n");
    resp = mailCommand("RCPT TO:" + recipient, in, out);
    if (resp.charAt(0) != '2')
      throw new IOException("Bad SMTP dialogue IV\n");
    resp = mailCommand("DATA", in, out);
    if (resp.charAt(0) != '3')
      throw new IOException("Bad SMTP dialogue V\n");
    for (int i = 0; i<msg.size(); i++)
      out.writeBytes(((String)msg.elementAt(i))+"\n");
    resp = mailCommand(".", in, out);
    if (resp.charAt(0) != '2')
      throw new IOException("Bad SMTP dialogue VI\n");
// close connection
    mailSession.close();
    if (debug) System.out.println("Msg has been sent successfully");
    }
    catch (IOException e)
    {
      System.err.println("Unable to open SMTP connection.\n");
      System.err.println(e);
    }
}
public String mailCommand(String s, BufferedReader in,
DataOutputStream out) {
    try {
      out.writeBytes(s + "\n");
```

```java
            return getResponse(in);
        }
        catch (IOException e)
        {System.err.println("Unable to send command to
        SMTP server.\n");}
        return null;
    }
}
public String getResponse(BufferedReader in) {
String resp = "";
for (;;) {
    try {
      String line = in.readLine();
      if (line == null)
         throw new IOException("Bad SMTP response\n");
      if (line.length() < 3)
         throw new IOException("Bad SMTP response\n");
      resp += line + "\n";
      if (line.length() == 3 || line.charAt(3) != '-')
         return resp;
      }
    catch (IOException e) {}
    }
}
public static void tutorial()
{
    System.out.println("*** Tutorial ***\n");
    System.out.println("To use mail extension, you need
    four pieces of information:");
    System.out.println(" (1): your mail server name");
    System.out.println(" (2): your recipient's email
    address");
    System.out.println(" (3): your email address
    (optional)");
    System.out.println(" (4): the message to be sent\n");
    System.out.println("Note:");
    System.out.println("If you are using window and have
    Outlook Express as your mail client, then the server
    name can be found in <tools>/<account>. \n");
    System.out.println("If you are using UNIX, then
    usually \"mailhost\" can be used as the subitute
    for the server name.");
    System.out.println("If it does not work, consult
    your UNIX administrator.\n");
    System.out.println("To call the mail extension from
    a Java program, do the following:");
    System.out.println(" import zeus.extension.*;\n");
    System.out.println("At the point in your Java
    program you want to send an email, do:");
```

```java
        System.out.println("(1): MMX mmx = new MMX();");
        System.out.println("(2): mmx.setServer(String
        mailServer);");
        System.out.println("(3): mmx.setRecipient(String
        emailAddress);");
        System.out.println("(4): mmx.setSender(String
        senderEmailAddress); // optional");
        System.out.println("(5): mmx.addMsg(String
        message);");
        System.out.println("(6): mmx.send();\n");
        System.out.println("Step 4 is optional.");
        System.out.println("You can do step 5 repeatedly to
        write a paragraph. <New Line> will be automatically
        added.\n");
        System.out.println("Now start trying...\n");
    }
    public static String getInput()
    {
        int msg;
        StringBuffer buffer = new StringBuffer();
        byte hibyte = 00;
        try {
        while ((msg = System.in.read()) != 10) {
          char c = (char) (((hibyte & 0xff) << 8) | (msg &
          0xff));
          buffer.append(c);
         }
        } catch (Exception e) {}
        return buffer.toString();
    }
    public static void main(String arg[]) {
        tutorial();
        MMX mmx = new MMX();
        System.out.print("Specify your server [enter: get
        from database, \".\": check database]:");
        mmx.setServer(getInput());
        System.out.print("Specify your recipient's email
        address:");
        mmx.setRecipient(getInput());
        System.out.print("Specify your email address
        [enter: default]:");
        mmx.setSender(getInput());
        String m;
        System.out.println("Write message [\".\":
        terminate]: \n");
        while (!(m = getInput()).equals("."))
        mmx.addMsg(m);
        mmx.send();
    }
```

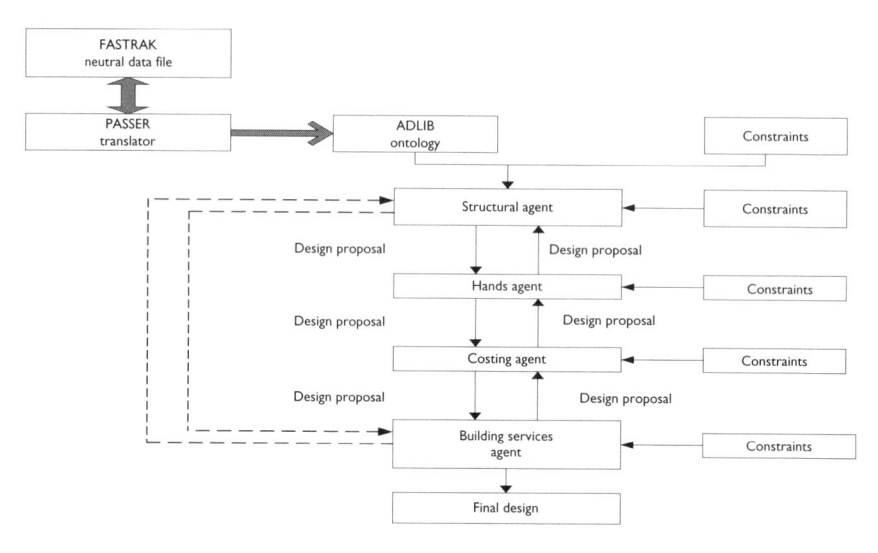

Figure 5.18 Flow chart of the ADLIB demo.

5.5.5 ADLIB prototype outputs

Figures 5.19–5.21 below show some sample outputs from the ADLIB prototype system. Figure 5.19 shows the GUI that enables the agent owner to input specific constraints and design data, based on his perspective in the collaborative problem-solving environment. In Figure 5.20, the agents register themselves in the Agent Name Server (ANS), and are able to locate and communicate with each other during design. For instance, the illustration captures the interaction between the CCA and SAA agents. After the registration process, the Interface Agent distributes the project design constraints (as extracted from the project specification document and used to generate initial design using FASTRAK. The interface agent then sends trigger to the SDA to begin design. This message exchange from the Interface Agent to SDA acts as an external stimulus for the SDA to initiate the design process and subsequently begin negotiation with the negotiation participants. On completion of its design, the structural agent identifies and enters into negotiation with the first negotiation participant in the chain SAA, proposing the structural configuration that maximises its own utility (Figure 5.21a).

In Figure 5.21a, each agent proposes a design that maximises its utility; for example, the SDA proposes the biggest structure possible while the CCA proposes the cheapest structure possible. The abstract concepts *Negotiation-Participants*, and *NegotiationOrder* (discussed in the Section 5.5.3.1 and

Figure 5.19 Sample agent user interfaces.

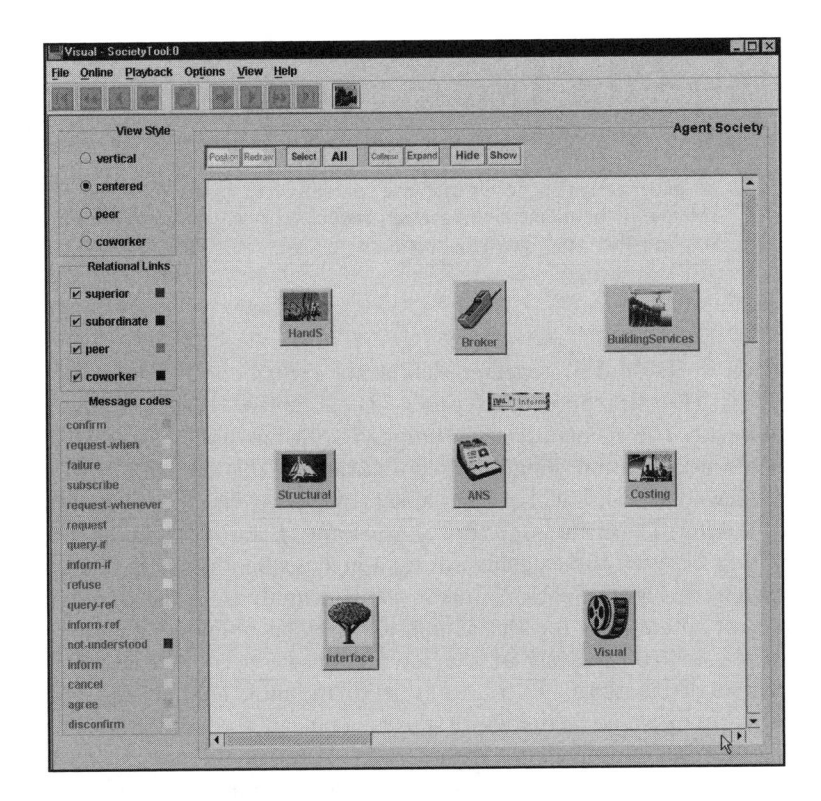

Figure 5.20 Simulated collaborative MAS design space.

152   O. O. Ugwu et al.

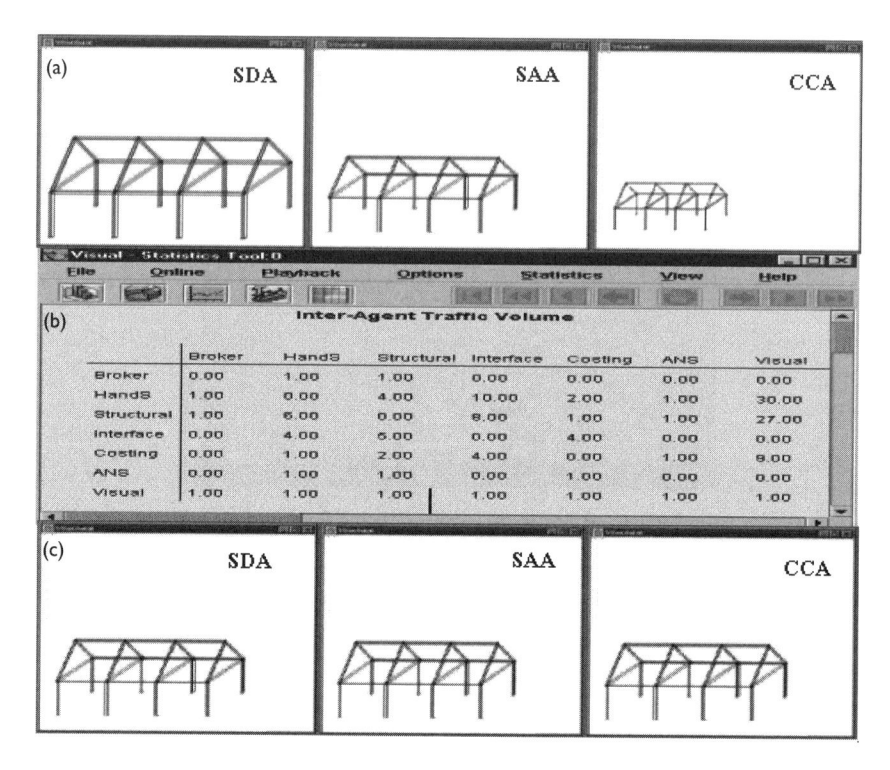

Figure 5.21 (a) Initial design proposals by three specialist task agents. (b) A typical traffic volume in communication between agents. (c) Agents converge on a design solution after automated negotiation.

described in Table 5.3) together define an agent's social behaviour with regard to its interactions with other agents during design in the MAS environment. The agents are also equipped with this piece of knowledge as initial resources during design and configuration. This interaction sequence can be viewed in real time, but Zeus also provides the visual tools that enable both the sequence of message transactions and a statistical analysis of the interaction between these agents during a given collaborative design session to be viewed (Figure 5.21b). Thus a detailed analysis of the frequency of inter-agent interaction for the simulation can be obtained. Figure 5.21b shows the matrix analysis of the interaction between the agents in the automated design space. Users are able to monitor, analyse and evaluate the performance and activities of their agents in the network using the visual/reporting tools in Zeus. The interaction matrix in Figure 5.21b shows all the agents in the MAS environment including the utility and task agents.

The utility agents include: ANS that provides a directory service for agent registration and peer identification, the Broker agent provides directory services for agent-discovery, while the Visual agent provides services for visualisation of the agents' behaviours in real time. (The task agents were described in Section 5.4.4.2 and Table 5.2.) Figure 5.21c represents the final design agreed after several rounds of negotiation.

The thread of conversation/transactions between the SDA and other task agents during the negotiation is summarised in Table 5.5. The complete scenario consists of the parametric design of the following structural elements in a portal frame: beam, column, rafter and purlin. However, some of the structural elements and attributes have been suppressed due to space constraints, but negotiations and decisions on the parametric design are simply extensions of the inter-agent transactions demonstrated in the Table 5.5. Only a portal frame structural element (beam) is shown, and the table only shows the attributes that were negotiated and altered in the process before agreement was reached. This includes length and depth of the beam. The depth is a function of the cross-sectional dimensions. The agents communicate by sending a design proposal (*myDesign* – refer to the knowledge level ontology shown in Figure 5.14 in Section 5.5.1) as the main content in a message body. The abstract concept *message* is inherited from the agent development platform (Zeus) and the syntax and protocol are FIPA-compliant. All the messages shown in the table are of type *inform*.

The next session describes the sequence of actions in the inter-agent interaction. The negotiations trail follows these sequences. There is a progressive increase in the time, which is generated from the system:

1. Interface passes design constraints to all the task agents.
2. Interface sends trigger for SDA to start design, as well as the other task agent.
3. Identifies the negotiation participants, default value is none.
4. ANS responds to SDA's request for the address/location of SAA, responds with: *Structural1649* as the conversation ID in the 'In-Reply-To' field of the message.
5. SDA proposes a design (*myDesign*) and sends it for SAA to evaluate.
6. Counter-proposal sent to SDA: 1st alteration is on Beam Depth.
7. Counter-proposal from SAA: 2nd alteration is on Beam depth, and length.
8. Structural informs Interface that proposal from HandS has been rejected, proposes an alternative design and notifies HandS that respondents should use *NewDesignEvaluateUtility* as the conversation ID in the Reply-With field.
9. Counter-proposal: 3rd alteration is on Beam depth.
10. Counter-proposal: 4th alteration is on Beam depth.

Table 5.5 Summary of agent transaction report: trail of conversation/negotiation between SDA and other task agents

S/N	Time	From/To	Content
1	8.95	>From: IAA	data (:type constraints :attributes ((maxHeight 16) (maxCost 100000) (maxWidth 15) (minHeight 6) (minWidth 5) (tgtCost 6000) (maxDepth 16) (minDepth 6)))
2	8.959	From: IAA	data (:type startCreateInitialDesign:attributes ((start true)))
3	8.970	From: IAA	data (:type negotiationParticipants :attributes ((initiator Structural) (respondent notYetKnown)))
4	11.302	From: ANS	(:name HandS :host "158.125.72.17" :port 6703 : type Agent)
5	14.261	To: SAA	data (:type myDesign :attributes ((beam (:type Beam :attributes (Depth 0.30) (Length 12.0) (owner HandS) (status proposal) (toBeEvaluatedBy Structural)))
6	17.264	From: SAA	data (:type myDesign :attributes ((beam (:type Beam :attributes ((Depth 0.4994) (Length 12.0) (owner Structural) (status proposal) (toBeEvaluatedBy HandS) (completed false)))
7	20.337	From: SAA	data (:type myDesign :attributes ((beam (:type Beam :attributes (Depth 0.3599) (Length 9.6) (owner HandS) (status proposal) (toBeEvaluatedBy Structural)))
8	21.238	To: IAA	data (:type acceptProposal :attributes ((accept false) (name none)))
9	22.283	To: IAA	data (:type myDesign:attributes ((beam (:type Beam :attributes ((Depth 0.49552) (Length 9.6) (utility 1.0) (owner HandS) (status mine) (toBeEvaluatedBy HandS) (completed false)))
10	23.265	To: SAA	data (:type myDesign :attributes ((beam (:type Beam (Depth 0.4992) (Length 9.6) (owner HandS) (status proposal) (toBeEvaluatedBy HandS) (completed false)))
11	24.259	From: SAA	data (:type acceptProposal :attributes ((accept true) (name HandS)))
12	25.293	From: CCA	data (:type negotiationParticipants :attributes ((initiator Costing) (respondent Structural) (start true)))
13	25.295	From: CCA	data (:type myDesign :attributes ((beam (:type Beam :attributes ((Depth 0.4992) (Length 9.6) (owner Costing) (status proposal) (toBeEvaluatedBy Structural) (completed false)))
14	26.285	To: CCA	data (:type acceptProposal :attributes ((accept true) (name Structural)))
15	26.312	To: IAA	data (:type myDesign :attributes ((beam (:type Beam :attributes ((Depth 0.4992) (Length 9.6) (owner Costing) (status mine) (toBeEvaluatedBy Structural) (completed false)))
16	26.312	To: SAA	data (:type startChangeNegotiationParticipant :attributes ((initiator Structural) (respondent HandS) (start true)))
17	26.344	From: IAA	data (:type firstOfferHistory :attributes ((HandS true) (Structural true) (Costing true)))

11 HandS accepts the proposal and enters negotiation with Costing (trigger not shown in the table).
12 CCA initiates negotiation with SDA.
13 CCA sends a proposal. There is no change from that agreed upon with SDA, as CCA is also satisfied with the proposal.
14 SDA confirms acceptance to CCA.
15 Sends confirmation of accepted design from CCA to the Interface and includes *SendAcceptNotice* as the conversation ID in the 'Reply-With field' of the message.
16 Informs HandS that proposal from costing has been accepted.
17 The Accepted Final Design Configuration is: Beam (Depth 0.4992, Length 9.6) @ 26.3125.
18 Interface certifies that all three agents have agreed and sends trigger to all the Agents to stop the negotiation (see Figure 5.21c).

The scenario transaction described earlier holds for situations where there is no deadlock during the peer-to-peer negotiation among the agents. However, in real life that does not hold true in every situation. Figure 5.22 below captures a scenario in which there is deadlock and the agent notifies its owner through automated messaging using the WWW infrastructure (refer to Section 5.5.3.4 – handling exceptional cases during negotiation).

This agent–human communication was also successfully simulated in a mobile phone using Short Messaging Service (SMS) technology. It demonstrates that agents can seamlessly integrate with evolving wireless technology.

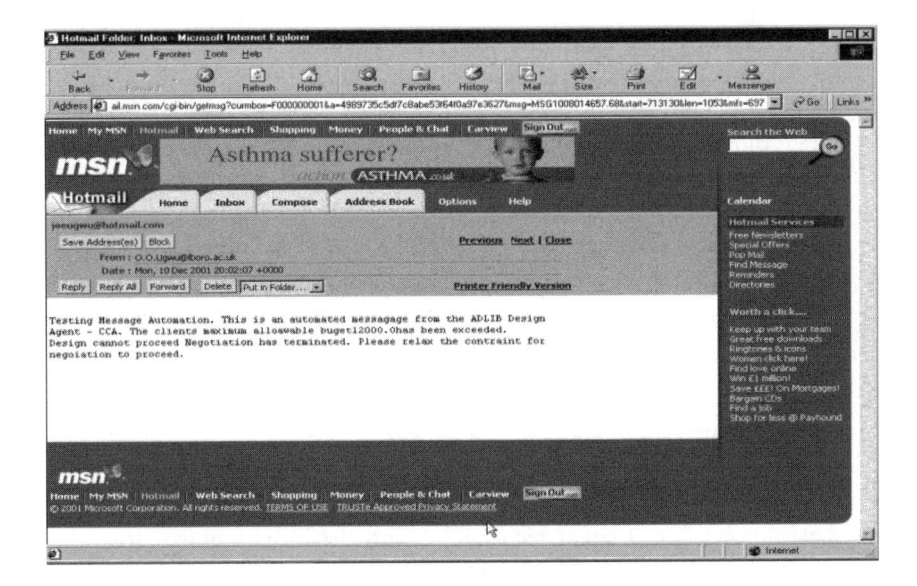

Figure 5.22 Typical email automated message in a negotiation deadlock.

Another feasible option to handle the negotiation deadlock (based on discussion with the Zeus team from the BT Plc UK) would be for the agent to dial-up the owner's contact telephone number. This functionality was not implemented because of specific constraints in telephone network resource requirements. However, BT had already implemented such level of agent functionality in its network services management.

These research explorations indicate the next possible levels of agent and MAS application in construction. For example, it is essential to consider the possibility of the agent-owner responding to the automated message alert and then remotely starting the agent after necessary intervention, so that negotiation could proceed (either from the deadlock position or afresh with of course consequential network overhead costs). Again this protocol replicates what often obtains in real-life work situations. For example, an administrator/secretary would normally consult his/her boss (through either face to face, telephone, fax or email) when confronted with an issue that he/she is unable to take immediate decision on depending on its potential impact on the organisation's business operations. Thus, imbuing agents with such functionality would add human face(s) to MAS applications in construction (Ugwu *et al.*, 2003).

5.5.6 Prototype system validation with ADLIB industrial partners

The ADLIB prototype development was an iterative process as shown in the framework diagram (Figure 5.3, Section 4.3). The test data for the validation was generated using FASTRAK. The generated data was in the neutral data format that FASTRAK normally generates after design. The final ADLIB prototype was demonstrated to the collaborating industrial partners, showing various systems level functionality that encapsulate the user requirements identified during the knowledge acquisition and user requirements elicitation and analysis stages of the project. The industry collaborators gave the following comments and feedback:

- Agent technology is well ahead of present times, but the concept is innovative. The ADLIB project has proved the concepts of MAS to distributed collaborative design. The project shows where next to go while most of the issues raised and findings from the project are also the sort of issues that the International Alliance for Interoperability (IAI) is addressing. The project provided guidance on where current thinking is going, and future trends.
- The ADLIB project has identified the requirements and defined the basic principles of agent technology, and the ideas required for more complex negotiation. More intelligent search engines combined with

agent-based negotiation proposed in the ADLIB project will be good practical application of the technology.
- The automated email messaging is an important user requirement to account for deadlock and facilitate human intervention when necessary. This is of potential practical use in deploying agent systems in construction (vis-à-vis integration with evolving web services and other mobile/wireless technologies).
- The project identified problems related to agent integration with existing applications such as FASTRAK.

The visualisation of the negotiation is good and facilitates appreciation of what is going on between the agents.

5.6 Summary

This chapter discussed the application of agents for decision-support in collaborative design. It described various dimensions of MAS research in construction with a focus on the development process and agent implementation at the application level. The chapter highlights the need to understand the processes, problem-solving methods, organisational structures and social contexts within which collaborative working occur in the problem-solving space. It used the ADLIB project as a case study to highlight methodological aspects of agent development (design and implementation) in solving design problems. The chapter demonstrates that the application of agents in collaborative working, demands a good understanding of working practices (workflow) in order to be able to replicate these practices and automate some of the processes associated with task execution. Some of the predominant features of, and research issues that need to be addressed in agent-based collaborative working include: organisational structures, co-ordination, communication, interaction, negotiation, cooperation, competition and product and process views of the problem domain. Other aspects of MAS development explored include integration of MAS with external (existing) systems, symbolic knowledge modelling and representation of the design process (ontology) amongst others.

Agent-based technologies provide a natural metaphor to integrate distributed collaborative working environments. This chapter has demonstrated that agent-based construction integration would be underpinned by the underlying domain ontology. The ADLIB project case study also shows that the pivots of actions in agent research remain the encapsulation of business processes and rules within appropriate task agents. This demands a clear understanding of the work processes, information logistics and data flows as well as the data-process interactions across the various decision points in the construction supply chain.

From the work reported, it is concluded that collaborative design agents that can interact using available agent technologies have numerous potential applications and advantages. In the context of the AEC sector such agents can:

a   Support information filtering and reductions in information overload on users, so that a user receives only the information that is essential for decision-making.
b   Support effective negotiation between different members of a project team. Such negotiations are often complex and project participants downstream of the project management need to be involved at appropriate decision points. This will reduce incidence of change orders and associated downtime costs as project commences.
c   Support the integration of existing software and (k-b) systems technology, thereby facilitating the construction of more robust decision-support systems (through efficient knowledge exchange). Such systems have potentials for application in training within academia and industry.

The case study research project also shows that developing ubiquitous agents and MAS requires a fusion of different technologies. These include WWW infrastructures (e.g. emails, Portals and Web services) and computational techniques at the implementation level, including the use of programming constructs such as Rulebases, and object oriented programming. The case study simulation results show that agent paradigm provides the necessary ubiquity, computational power and intelligence that users require for decision-support using the next generation IT systems in construction (Anumba *et al.*, 2003, 2004; Ugwu, 2004). At the same time the chapter demonstrates that the agent/multi-agent paradigm is robust enough to handle the complexities engendered by the emerging distributed collaborative work practices.

Acknowledgements

The ADLIB project was funded by the Engineering and Physical Sciences Research Council (EPSRC), UK under its Innovative Manufacturing Initiative (IMI). The industrial collaborators were: Building Research Establishment (BRE) Ltd, Steel Construction Institute (SCI), British Telecommunications (BT) Plc, Curtins, WS Atkins, Health and Safety Executive (HSE), Wescol Glosford, Ferguson McIlveen Architects and CSC Ltd. They provided the domain knowledge and practical industry perspectives on agent applications in the construction sector.

References

Anumba, C. J., Newnham, L., Ugwu, O. O. and Ren, Z. (2001a) 'Intelligent agent applications in construction engineering', in A. Singh (ed.) *Creative Systems in Structural and Construction Engineering*, Balkema, Rotterdam, pp. 161–6.

Anumba, C. J., Ugwu, O. O., Newnham, L. and Thorpe, A. (2001b) 'A multi-agent system for distributed collaborative design', *Journal of Logistics Information Management*, 14(5&6): 355–66 (Special Issue on Softwares and Methods for Improving Decision-Making in Civil Engineering).

Anumba, C. J., Ugwu, O. O., Newnham, L. and Thorpe, A. (2002) 'Collaborative design of portal frame structures using intelligent agents', *Automation in Construction*, 11(1): 89–103.

Anumba, C. J., Ren, Z., Thorpe, A., Ugwu, O. O. and Newnham, L. (2003) 'Negotiation within a multi-agent system for the collaborative design of light industrial buildings', *Advances in Engineering Software*, 34(7): 389–401, Elsevier Science, Oxford.

Anumba, C. J., Aziz, Z. and Ruikar, D. (2004) 'Enabling technologies for next generation collaboration systems', *Designing, Managing and Supporting Construction Projects Through Innovation and IT Solutions*, in P. Brandon, H. Li, N. Shaffii and Q. Shen (eds) INCITE 2004 Conference, Langkawi, Malaysia, 18–21 February, pp. 85–96.

BT (2002) *Zeus Development Toolkit*, Zeus Tool Kit Documentation, British Telecommunications Plc, UK.

CIRC (2001) *Construct for Excellence*, Report of the Construction Industry Review Committee, HKSAR.

Construction Task Force UK (1998) *Rethinking Construction*.

CSC2003-URL: CSC http://www.cscworld.com

Foundation for Intelligent Physical Agents (FIPA) (1997a) *Agent Communication Language*, FIPA 97 Specification, Version 2.0 Part 2, Geneva, Switzerland.

Foundation for Intelligent Physical Agents (FIPA) (1997b) *Agent Software Integration*, FIPA 97 Specification Part 3, Geneva, Switzerland.

Foundation for Intelligent Physical Agents (FIPA) (1997c) *Personal Travel Assistance*, FIPA 97 Draft Specification Part 4, Geneva, Switzerland.

Foundation for Intelligent Physical Agents (FIPA) (1999) *FIPA Architectural Overview*, FIPA Technical Committee A, San Francisco, CA.

International Standards Organisation (ISO) (1994) 'Classification of information in the construction Industry', ISO TR 14177.

Latham, M. (1994) *Constructing the Team*, Final Report on Joint Review of Procurement and Contractual Arrangements in the UK Construction Industry, HMSO.

M4I-URL http://www.m4i.org.uk/m4i/, Last accessed on 15 July, 2003.

Rational2003-URL: Rational Rose http://www.rationa.com

Rosenschein, J. S. and Zlotkin G. (1994) 'Rules of encounter', MIT Press, Cambridge, MA.

SCI (1999) 'Design of single-span steel portal frames', *Draft specification SCI Publication 252*, The Steel Construction Institute (SCI), UK.

Searle, J. R. (1969) *Speech Acts – an Essay in the Philosophy of Language*, Cambridge University Press, New York.

Searle, J. R. (1979) *Expressions and Meaning – Studies in the Theory of Speech Acts*, Cambridge University Press, New York.

Ugwu, O. O. (2004) 'Developing next generation decision-support systems in construction: a perspective on user requirements capture', in P. Brandon, H. Li, N. Shaffi and Q. Shen (eds) International Conference on Construction Information Technology (INCITE 2004): WORLD IT for Design & Construction –

Designing, Managing and Supporting Construction Projects through Innovation and IT Solutions, Langkawi, Malaysia, 18–20 February 2004.

Ugwu, O. O. and Anumba, C. J. (2000) 'An e-commerce framework for integrating product design, manufacture, and procurement in construction,' in *Proceedings of Second International Conference on Decision Making in Civil and Urban Engineering*, Lyon, France, 20–22 November, 2000, pp. 933–44.

Ugwu, O. O., Anumba, C. J., Newnham, L. and Thorpe, A. (1999a) 'Agent-based collaborative design of constructed facilities', in *Artificial Intelligence in Structural Engineering – Information Technology for Design, Manufacturing, Maintenance, and Monitoring*, in Proceedings of the Sixth EG-SEA-AI Workshop, Adam Borkowski (ed.) Wierzba 1999, pp. 199–208.

Ugwu, O. O., Anumba, C. J., Newnham, L. and Thorpe, A. (1999b) 'Agent-based decision support for collaborative design and project management', *The International Journal of Construction Information Technology*, 7(2): 1–16.

Ugwu, O. O., Anumba, C. J., Newnham, L. and Thorpe, A. (2000a) 'Agent-oriented collaborative design of industrial buildings', in R. Fruchter, F. Pena-Mora and W. M. K. Roddis (eds) *Proceedings of the Eighth International Conference of Computing in Civil & Building Engineering, ICCCBE-VIII*, Stanford University, USA, 14–17 August 2000, pp. 333–40.

Ugwu, O. O., Anumba, C. J., Newnham, L. and Thorpe, A. (2000b) 'Towards a common ontology for the collaborative design of portal framed structures', *Second International Conference on Decision Making in Civil and Urban Engineering*, Lyon France, 20–22 November 2000, pp. 789–800.

Ugwu, O. O., Anumba, C. J., Newnham, L. and Thorpe, A. (2000c) 'The application of DAI in the construction industry', in G. Gudnason (ed.) *Proceedings of International Conference on Construction Information Technology – CIT2000*, Icelandic Building Research Institute, Reykjavik, Iceland, 28–30 June 2000, 2: 950–70.

Ugwu, O. O., Anumba, C. J. and Thorpe, A. (2000d) 'Specifications and requirements for a multi-agent system for the collaborative design of light industrial buildings', *ADLIB Research Report* No. ADLIB/03, October 2000, Interim Working Document for the ADLIB Project, Loughborough University, Loughborough, UK.

Ugwu, O. O., Anumba, C. J. and Thorpe, A. (2001) 'Ontology development for agent-based collaborative design', *Engineering, Construction, and Architectural Management*, Blackwell Science, Oxford, 8(3): 211–24.

Ugwu, O. O., Anumba, C. J. and Thorpe, A. (2002a) 'Developing multi-agent systems for collaborative design – towards a methodology', *Advances in Intelligent Computing in Engineering*, in M. Schnellenbach-Held and H. Denk (eds) Proceedings of Ninth International Workshop of the European Group of Intelligent Computing in Engineering (EG-ICE), Technische Universitat Darmstadt, Germany, 1–3 August 2002, pp. 61–70.

Ugwu, O. O., Anumba, C. J., Thorpe, A. and Arciszewski, T. (2002b) 'Building knowledge level ontology for the collaborative design of steel frame structures', *Advances in Intelligent Computing in Engineering*, in M. Schnellenbach-Held and H. Denk (eds) *Proceedings of Ninth International Workshop of the European Group of Intelligent Computing in Engineering (EG-ICE)*, Technische Universitat Darmstadt, Germany, 1–3 August 2002, pp. 71–8.

Ugwu, O. O., Arciszweski, T. and Anumba, C. J. (2002c) 'Teaching agents how to solve design problems – a mixed initiative learning strategy', ASCE International Workshop, *IT in Civil Engineering*, 2–3 November 2002, pp. 11–24.

Ugwu, O. O., Butterworth, J., Anumba, C. J. and Thorpe, A. (2002d) 'Development and application of integrated product and process models for the cost management of steel frame structures', in *Proceedings of the Third International Conference on Decision-Making in Urban and Civil Engineering (DMinUCE)*, November 2002, London, UK.

Ugwu, O. O., Kumaraswamy, M. M., Ng, T. S. and Lee, P. K. K. (2003) 'Agent-based collaborative working in construction: understanding and modelling design knowledge, construction management practice and activities for process automation', *The Hong Kong Institution of Engineers (HKIE) Transactions, Tenth Anniversary Issue, Emerging Technology in the 21st Century*, 10(4): 81–7.

Chapter 6

MASCOT
A multi-agent system for construction claims negotiation

Z. Ren and C. J. Anumba

6.1 Introduction

In the construction industry, negotiation is preferred for the settlement of claims (Powell-Smith and Stephenson, 1993). Negotiation plays an important role in resolving claims, preventing disputes and keeping a harmonious relationship between project participants. However, claims negotiations are commonly inefficient due to the diversity of intellectual backgrounds, many variables involved, complex interactions and inadequate negotiation knowledge of construction participants. Most project managers consider negotiation as the most time and energy-consuming activity in claims management (Hu, 1997). Inefficiencies in negotiation make claims resolution much more difficult and adversarial and may delay resolution or even lead to expensive litigation (Zack, 1994). Thus, it is necessary to develop an approach to facilitate claims negotiation to reduce the tremendous time and human resources invested.

Unlike most of the former studies in construction claims management, which seek to improve negotiation efficiency through the improvement of fundamental theories and principles, this study tries to look for solutions through the improvement of technologies, which can facilitate the negotiation process. Although several computer-aided negotiation support systems (NSS) have been developed, these systems focus solely on the analytical negotiation activities based on expert systems. None provides direct assistance for the entire interaction bargaining process (Anson and Jelassi, 1990). This study is concerned with the development of a high-level multi-agent system architecture for construction claims negotiation (MASCOT) in which autonomous agents, acting on behalf of project participants, can directly negotiate with each other to resolve construction claims. By applying such a system, the current problems in claims negotiation such as inefficiency, late involvement of the client and the influence of unhealthy human factors can be reduced.

The development of MASCOT involves five key stages: review of the major negotiation theories and models to provide a sound basis for the development of agent interaction within MASCOT; study of multi-agent systems (MAS) and the associated theoretical underpinning; examination of the nature and important characteristics of construction claims negotiation to identify its peculiar requirements; the development of an appropriate construction claims negotiation model; and the implementation and evaluation of the developed model. This chapter describes the MASCOT system development process. An example is used to demonstrate the workings of the model.

6.2 Multi-agent systems in negotiation

Inefficiency has been widely recognised as a common problem in negotiation (Kraus, 1996). Many research projects (Anson and Jelassi, 1990; Rosenschein and Zlotkin, 1994; Shell, 1995) have sought to improve negotiation efficiency by using computer technology. MAS technology provides an opportunity to overcome the problem of inefficiency. Agents in MAS act collectively as a society and they collaborate (or compete) to achieve their own individual goals as well as the common goals (Anumba and Newnham, 2000). This co-operative and competitive feature matches the essential nature of negotiation. Moreover, the essential characteristics of agents (i.e. social ability, reactivity and proactiveness) make them ideal for supporting negotiations.

Unlike other negotiation support systems, such as expert systems and decision analysis systems which can only provide suggestions to human negotiators, agents in MAS, on behalf of their owners, can directly negotiate with each other about items to reach an agreement within a specified time frame. This constitutes a promising approach to solving complicated negotiation problems in a natural and flexible way (Ferber, 1998). Rosenschein and Zlotkin (1994) highlight that negotiation is a subject of central interest in MAS.

An efficient negotiation mechanism is the basis of a successful MAS negotiation system. Many different negotiation mechanisms have been developed. Most of them (Rosenschein and Zlotkin, 1994; Kraus, 1996; Sandholm, 1996) are rooted in the game/economic theory approach, since there is a particular match between these theoretical approaches and MAS. One of the major assumptions of these theories is that players are rational. When they are applied to analyse human negotiation, the problem regularly faced is that human beings do not always act rationally and frequently do not have consistent preferences between alternatives. On the other hand, agents, being pre-programmed in their behaviour, make concrete the notion of 'strategy' which plays a central role in game/economic theory – the idea that a player adopts rules of behaviour before starting to play a given game and that these rules entirely control his responses during the game (Rosenschein and Zlotkin, 1994).

However, the focus of classic game-theory approaches, such as Nash (1950, 1953) is on the prediction of outcomes, under certain assumptions about the player and the outcomes themselves. They do not really specify the negotiation process (Bacharach and Lawler, 1981). On the other hand, economic theoretical models, such as Zeuthen (1975) and Cross (1977) provide a plausible approach for the actual bargaining process (Harsanyi, 1956). Since this study intends to build automated negotiators, the economic model is adopted for the development of the MASCOT negotiation mechanism. However, as shown later in the chapter, cognisance is taken of the importance of behaviour theory in the development of a MAS negotiation system because it reflects the essential human aspect of claims negotiation.

6.3 Characteristics of construction claims negotiation

To improve the efficiency of claims negotiation, the MASCOT system was developed based on a thorough analysis of the major characteristics and the problems of construction claims negotiation is described in the following sections.

Contractually obliged self-interested relationship. Unlike the negotiations in other businesses where one party may simply leave a negotiation if the negotiation falls into a deadlock, nobody can easily walk away from construction claims negotiation. First, claims negotiation participants are legally obliged by the project contract. Negotiations are conducted within the framework of the contract. Second, if a negotiation ends in conflict, the negotiating parties may be forced into an arbitration or litigation that they can hardly afford. Thus, both parties will try to avoid a conflict outcome. On the other hand, project participants are from different organisations. Each participant will try to maximise his own benefit as long as it does not break the co-operative relationship.

Role-dependent information. Due to their different roles, participants have different perspectives on a project. The client knows clearly the final functional requirements, budget and the financial status of the project while the engineer understands well the client's requirements, contract documents as well as the contractor's financial situation, progress and quality of the work. The contractor has detailed information about schedule, progress and the circumstances that led to a claim. Furthermore, each party also has different expertise. Since construction claims negotiation is evidence-oriented, evidence plays a critical role. Thus, each party will try to use his specific information and expertise to explain, argue and persuade the other party to accept his offer.

Strategy-influenced process. Incomplete information and different strategies will influence the payoffs of negotiation (Pena-Mora and Wang, 1998). Given the contractual bounded self-interested nature and the role-dependent information, the contractor and the engineer often adopt a number of strategies in an attempt to draw the settlement point from the middle towards their expected outcomes. In construction claims, either or both parties may inflate the opening demands; misrepresent their positions or interests; withhold sensitive or potentially damaging information; use threatening behaviour; or adopt an intransigent stance until the other side is ready to move (Pickavance, 1997).

Time – an important factor. Claims negotiation is a time-consuming process that may often change the negotiation consequences. Before negotiation, documents need to be specially presented, negotiators need to be gathered, and there is often a substantial delay between the submission of documents and negotiation meetings. During negotiation, a party may adopt a time-consuming strategy, expecting to benefit from the opponent's time pressure or emotional exhaustion. Moreover, claims often can be settled only through several meetings which allow both parties to make offers and counter-offers. Many standard contracts, such as *Fédération Internationale des Ingénieurs-Conseils* (FIDIC), require the contractor to complete revised work first regardless of possible claims or disputes. Negotiations for compensation are conducted long after the work to avoid delays caused by potentially long arguments.

Role Definition and the Client Involvement. The roles of the engineer as an independent professional expert, an agent of the client and an independent organisation place him in a conflicting position in claims negotiation. Such conflicting roles often become a major contributing factor to the inefficiency of negotiation. For example, the engineer tends to discourage those claims resulting from his defaults. Moreover, the current involvement of the client is very low in claims management (Vidogah and Ndekugri, 1997). In such cases, the engineer is likely to take advantage of the low client intervention, which may finally increase the difficulty of the negotiations and the possibility of disputes (ECI, 1992). Therefore, the European Construction Institute (ECI) recommends earlier and greater client involvement in claims management and negotiation, which is often difficult to achieve practically.

6.4 Design of MASCOT

In designing the MASCOT model, the important aspects that needed to be addressed included the key assumptions of the model, the choice of a negotiation model, the development of the negotiation protocol and strategies, and the implementation and evaluation of the model. The development of these aspects takes into consideration the existing negotiation theories, the nature of MAS and the characteristics of construction claims negotiation, discussed earlier.

6.4.1 Assumptions

A number of simplifying assumptions are made to lay the foundations for the MASCOT model. These address both the claims negotiation and the multi-agent aspects of the system:

- *Quantitative negotiation* – the contractor's entitlement to the claim being negotiated has been established. Negotiation is only concerned with the amount of compensation in the claim.
- *Rationality* – agents are rational and willing to maximise their utilities or willing to reduce the risk of loss caused by a conflict deal. Agents prefer to reach an agreement which may not bring them the maximum benefits rather than have a conflict deal.
- *Fixed utility function* – agents' utility function will not change during the negotiation process. Therefore, agents' risk attitudes are consistent during negotiation.
- *Incomplete information* – agents have incomplete information about each other. However, an agent can estimate another's key negotiation features and update its beliefs through a learning mechanism.
- *Isolated negotiation* – no further change or information comes to agents during the negotiation. Moreover, each agent stands alone and an agent cannot commit itself to any future negotiation.

6.4.2 Choice of negotiation model

Although the economic theoretical approach provides a general framework for modelling MAS negotiation, few essential issues need to be addressed for its applications in claims negotiation. One of them is to choose a strategic bargaining model which is applicable for the specific MAS negotiation problem and to match the MAS scenarios with the economic theoretic definition of the chosen model (Kraus, 1997). Considering the characteristics of claims negotiation, this study adopted Zeuthen's model as the essential negotiation model (concession mechanism). In this model, the two players' bargaining problem is considered as a one-player decision process. First, players calculate the maximum probability of conflict they would be willing to accept in preference to acquiescing on the opponent's current offer and then the player who has a lower maximum risk acceptability will make the next concession (Young, 1975).

Zeuthen's risk evaluation model specifically addresses the participants' risk perception in conflict avoidance, which matches the contractually obliged self-interested nature of claims negotiation. Moreover, Harsanyi (1977) has demonstrated that Zeuthen's solution is identical to Nash's method (1950), that is, parties will settle at the point that maximises the product of the difference between what bargainers get from conflict and what each gets from the settlement point. Meanwhile, this study also adopted the concepts of time accounting and negotiation learning to make Zeuthen's model work in the claims negotiation circumstance. The former addresses the important time issues in claims negotiation and helps to keep the negotiation stable. The latter overcomes the incomplete information assumption on which Zeuthen's model is based. Aspects of the behaviour-oriented approach are incorporated by the provision of a learning mechanism for the MASCOT agents.

6.4.3 Negotiation protocol

The design of the MASCOT negotiation mechanism covers two distinct aspects: the definition of an acceptable protocol for agent interaction and the formulation of strategies for the agents participating in the negotiation. The protocol specifies the kinds of deals that the agents can make, as well as the sequence of offers and counter-offers that are allowed (Rosenschein and Zlotkin, 1994).

Since claims negotiation, in most cases, is characterised as bargaining and involves alternate offers between both parties, a specific negotiation protocol based on the Monotonic Concession Protocol (MCP) was developed for MASCOT. In the MCP, the agents start by simultaneously proposing a deal from the space of possible deals. An agreement is immediately reached if one agent matches (or exceeds) what the other one has asked for. The protocol continues to go on for another round if neither agent matches or exceeds the other's demand. An agent is not allowed to offer the other agent less than what it did in the previous round. If neither agent concedes at a certain step, then the negotiation ends and the protocol specifies that a deadlock has been reached. In this protocol, the agents cannot backtrack, nor can they both simultaneously stand still in the negotiation more than once.

The advantage of the MCP is that it ensures convergence or puts a stop to the negotiation promptly when convergence is not occurring. To satisfy the rules in a standard MCP, the agents need to know each other's utility functions. However, this complete information assumption is not true in claims negotiation. To make the MCP work in claims negotiation, several important modifications are made as stated in the following sections.

Conflict deal. In the MCP, if neither agent concedes at the same step, then the negotiation ends with a conflict deal. This restriction is relaxed in MASCOT. Negotiations will not necessarily fall into a conflict deal even if agents stand still for more than one encounter. Three factors reduce the possibility of conflict deals, which are: the involvement of the client agent; the expanded solution searching process; and the information enquiry from the mediator agent.

Incomplete information. In the MCP, each agent needs to know the other's full information to monitor whether the other has offered it more (or the same) at each step of negotiation. It is mutually verifiable whether the other agent has followed the rules of the protocol. In claims negotiation, the agents cannot have full information about the others' final goals, utility functions and risk perceptions. Thus, it is impossible for an agent to know whether the other is following the rules of the MCP. To overcome this problem, the Bayesian learning mechanism is adopted, which allows an agent to estimate its opponent's key negotiation features and update the estimation during the negotiation. Thus, negotiation can be conducted based on the estimated information rather than the full information assumption.

Time penalty. MASCOT takes, the passage of time during the negotiation process, into account in facilitating convergence. The time taken by each agent during negotiation is considered as a penalty element in its utility function. Each agent has a different cost $C > 0$ for a *time unit* which it spends during negotiation. The value of C depends on each agent's time sensitivity. The time penalty can be a fixed amount (e.g. \$5 per iteration) or a fixed rate of the agent's utility (e.g. a deduction of 1% of utility per iteration). For example, if an agreement is reached in time period t (or t iterations), the agent's time penalty will be Ct. Thus, time penalty becomes a driving force for agents, especially those with a high time penalty, conceding to reach an agreement.

Involvement of the client-agent. In this study, the client-agent will get involved in a negotiation when it finds that there is no chance for the contractor-agent and the engineer-agent to reach an agreement. The client-agent can identify this situation directly by observing both parties' offers or indirectly by observing the time penalties of both parties. In some other cases, the client-agent may be invited to get involved in the negotiation by one of the parties because a conflict deal is very harmful to that party. The involvement of the client-agent will not change the negotiation relationship between the contractor-agent and the engineer-agent. That is, the client-agent and the engineer-agent will work together as one negotiator to negotiate with the contractor-agent. The contractor-agent still receives offers or counter-offers from the engineer-agent (Figure 6.1).

6.4.4 Negotiation strategies

Given the above protocol, a negotiation strategy specifies precisely each agent's reaction to every possible course of events. The selection of

```
                    ┌─────────────────────────┐
                    │   ┌──────────────┐      │
            ────────┼──▶│ Client-agent │      │
            ┌──────────┐│└──────────────┘     │
            │Contractor││        ▲            │
            │  -agent  ││        │            │
            └──────────┘│        ▼            │
                 ▲  ────┼──▶┌──────────────┐  │
                 │      │   │Engineer-agent│  │
                 └──────┼───└──────────────┘  │
                        └─────────────────────┘
```

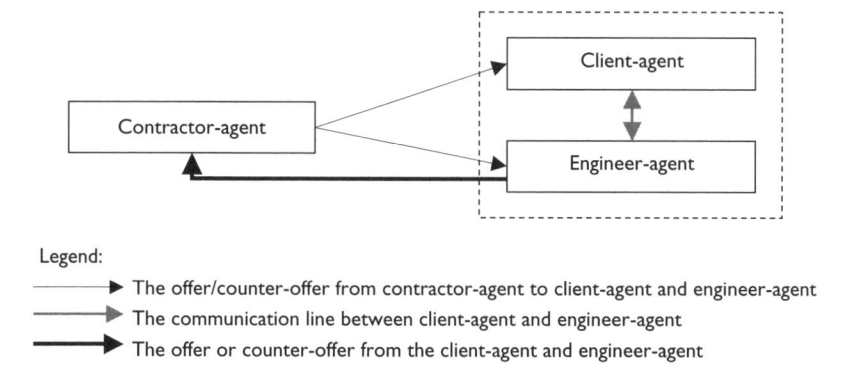

Legend:
──────▶ The offer/counter-offer from contractor-agent to client-agent and engineer-agent
──────▶ The communication line between client-agent and engineer-agent
──────▶ The offer or counter-offer from the client-agent and engineer-agent

Figure 6.1 Tripartite negotiation.

strategies depends on how the agents evaluate a possible deal and their position in such a deal. In a co-operative-competitive negotiation, the agents evaluate a deal from both individual and group perspectives. Decisions are made based on the principle that they maximise both personal and group utilities. In MASCOT, the agents are self-interested. Each agent is concerned only about its own utility without considering the utility of the group or the other agents. On the other hand, the agents are bounded by the project contract and the willingness to not to break the negotiation. Thus, the agents in MASCOT adopt Zeuthen's negotiation strategy.

6.4.4.1 Concession mechanism – conflict avoidance

Unlike the common utility maximisation approach, this study adopts Zeuthen's strategy in negotiation. This is because the risk avoidance principle in Zeuthen's model reflects an important characteristic of claims negotiation (i.e. both parties try to avoid the loss caused by a conflict deal). Furthermore, this model is identical to Nash's method, that is, parties settle at the point that maximises the product of the difference between what bargainers get from the conflict and what each gets from the settlement point (Harsanyi, 1977).

According to Zeuthen's strategy (1975), an agent makes its decision of concession based on how much it has to lose by running into conflict at that time. If an agent has already made many concessions, it will have less to lose from a conflict and will be less willing to concede. Thus, it has a high acceptability to risk conflict. If each agent's willingness to risk conflict can be measured, the agent with less willingness to risk will make a concession. The criteria for risk evaluation can be formulated into the following equations (Rosenschein and Zlotkin, 1994):

$$\text{Risk}_1^t = \frac{\text{the utility Agent 1 loses by conceding and accepting Agent 2's offer}}{\text{the utility Agent 1 loses by not conceding and causing a conflict}}$$

or

$$P_{cmax} = \frac{U_{cc}^t - U_{ce}^t}{U_{cc}^t - U_c(C)} \qquad P_{emax} = \frac{U_{ee}^t - U_{ec}^t}{U_{ee}^t - U_e(C)} \qquad (6.1)$$

where P_{cmax} (P_{emax}) is the contractor's (*engineer*) maximum likelihood of risk acceptability; U_{cc}^t (U_{ee}^t), the contractor (*engineer*) agent's utility based on its offer in iteration t; U_{ce}^t, the contractor-agent's utility based on the engineer-agent's offer in iteration t; U_{ec}^t, the engineer-agent's utility based on contractor-agent's offer in iteration t; and $U_c(C)$ ($U_e(C)$), the contractor (*engineer*) agent's utility for a conflict deal.

At every step, each agent calculates and compares the Risk_i^t (or P_{max}) for itself and its opponent. If Agent 1's Risk_i^t (or P_{max}) is higher than that of Agent 2, Agent 1 will have less to lose from a conflict and will be less willing to concede and risk reaching a conflict. Therefore, Agent 2 (with smaller risk acceptability) will make the next concession. The concession rate should be the minimum, sufficient to make its opponent's maximum risk acceptability (P_{max}) smaller or equal than its own. Otherwise, the agent will offer the same deal as the previous one (Rosenschein and Zlotkin, 1994). By following this approach, the agents will concede alternately until the maximum risk of conflict for both parties is zero. However, it is quite conceivable that an agent will have to make more than one concession before the opponent's P_{max} becomes lower than its own. Thus, Zeuthen does not suggest that Agent 1's concession will invariably be followed by Agent 2's concession, but only that this will occur if Agent 1's concession reduces Agent 2's P_{max} below that of Agent 1. This study adopts the simple alternate concession approach suggested by Rosenschein and Zlotkin (1994).

6.4.4.2 Bayesian learning mechanism

The application of Zeuthen's model is limited by the perfect information assumption. A learning mechanism, Bayesian learning approach, is introduced for the agents to estimate their opponents' key negotiation data. Bayesian Inference has a long history being used as a simple and powerful learning approach (Iversen, 1984; Jordan, 1992). Various Bayesian learning approaches have been developed in artificial intelligence (AI) research, such as: Bayesian learning (Zeng and Sycara, 1998), Bayesian network learning (Sahin, 1999) and Bayesian classifier learning (Bui et al., 1996). This study focuses on the Bayesian learning mechanism.

Based on the Bayesian learning mechanism, when an agent receives an offer (or counter-offer) from its opponent, the agent can analyse the offer, modify its beliefs about the opponent and make a counter-offer accordingly. The updated belief then becomes the agent's prior knowledge in the next updating process. An agent can finally get a relatively accurate belief about the opponent even if its initial domain knowledge is not so accurate (Figure 6.2).

Such beliefs are normally about the opponent's key negotiation features, such as: reservation value, risk attitude, payoff functions or negotiation strategy. This study focuses on the 'reservation value' which is the maximum amount that the engineer-agent can offer to the contractor-agent and vice versa (Figure 6.3).

Since reservation values are private information, it is impossible for an agent to know its opponent's exact reservation value. Nevertheless, an agent can update its beliefs about its opponent's reservation value based on its interactions with the opponent and its domain knowledge by using the Bayesian Inference. Therefore, an agent can have more accurate expectation about its opponent's utility and maximum risk acceptability and make a counter-offer based on the information available at this stage. The following section discusses how the contractor-agent updates its beliefs about the engineer-agent's reservation value. Some important terms used in

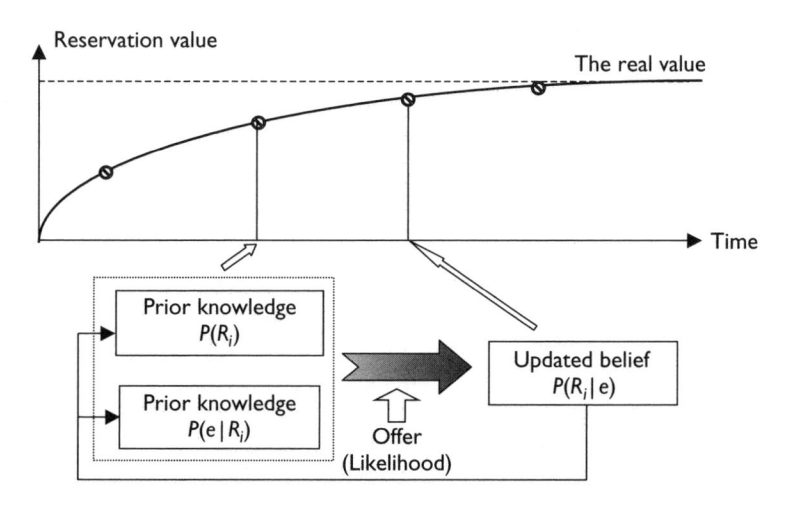

Figure 6.2 Bayesian updating mechanism.

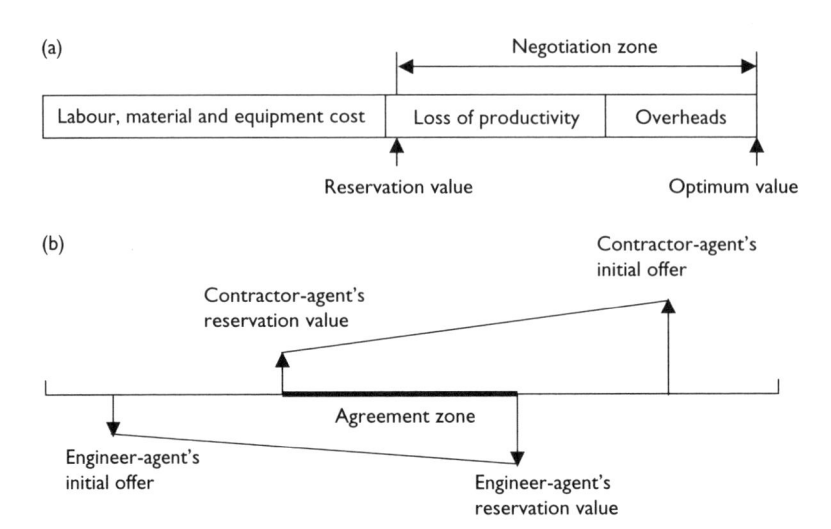

Figure 6.3 (a) An example of the contractor's negotiation zone. (b) An example of the agreement zone between agents.

MASCOT are defined as follows:

R	the engineer-agent's reservation value;
e_i	the engineer-agent's offer at encounter i;
R_i	a set of the contractor-agent's partial beliefs (hypotheses) about the engineer-agent's reservation value R, for example, $R_1 = 100$; $R_2 = 150, \ldots (i = 1, 2, \ldots, n)$;
$P(R_i)$	the probabilistic evaluation over the set of hypotheses $\{R_i\}$, which are the contractor-agent's prior knowledge, for example, $P(R_1) = 0.75$, $P(R_2) = 0.60, \ldots (i = 1, 2, \ldots, n)$;
$\sum R_i^* P(R_i)$	the current estimate of R can be calculated as a mean.

The Bayesian learning mechanism is applied when the contractor-agent receives a new offer from the engineer-agent. Based on its prior knowledge about the engineer-agent, a new offer enables the contractor-agent to acquire new insights about the engineer-agent's reservation value in the form of posterior subjective evaluation over R_i. The contractor-agent's prior knowledge about the engineer-agent's strategy can be expressed as 'usually the engineer-agent will offer an amount, which is 20% lower than its reservation value' (Zeng and Sycara, 1998). Such relationship can be represented by a set of conditional statements, for example: $P(e_2 | R_2) = 0.95$, where

e_2 and R_2 represent the events that offer$_{engineer} = 120$ and the contractor's partial belief reservation value$_{engineer} = 150$.

Given the encoded contractor-agent's domain knowledge in the form of conditional statements and the engineer-agent's offer, the contractor-agent can use the standard Bayesian rule to revise its beliefs about the engineer-agent's reservation value R:

$$P(R_i|e) = \frac{P(R_i)P(e|R_i)}{\sum_{k=1}^{n} P(e|R_k)P(R_k)} = \frac{P(R_i)P(e|R_i)}{P(e)} \quad (6.2)$$

where $P(R_i|e)$ is the probability that the engineer-agent's reservation value is R_i under the condition that his offer is e; $P(R_i)$, the probability that the engineer-agent's reservation value is a certain R_i; $P(e|R_i)$, the probability that the engineer-agent's offer is a certain e under the certain reservation value R_i; $P(e)$, the probability that the engineer-agent's offer is e.

6.4.4.3 Other strategies

Besides the above standard concession strategies, MASCOT also allows agents to adopt two other strategies to avoid a conflict deal. They are the involvement of the client-agent and expanded solution searching. The involvement of the client-agent facilitates claims negotiation in two ways: first, it facilitates the communications between the contractor-agent and the client-agent. The client-agent can monitor the negotiation process and make its judgement independently. The problems caused by the engineer-agent's attempt to screen the contractor-agent's information can be avoided. Second, if the negotiation between the contractor-agent and the engineer-agent falls into a deadlock and there is a possible deal within the client-agent's negotiation zone that can be accepted by the contractor-agent to reach an agreement, the client-agent will get involved in the negotiation and request the engineer-agent to adopt the client-agent's reservation value. Thus, the engineer-agent's negotiation zone is expanded and the conflict is avoided.

The expanded solution searching strategies are highly related to the innovation ability of agents and their owners. These strategies can resolve the conflicts which the autonomous agents themselves cannot solve by maximising the utility of a single negotiation item. The strategies include trade-offs between negotiation items in the same claim; trade-offs between different claims; and relaxation of the negotiation constraints by the agents' owners. Figure 6.4 illustrates the MASCOT process model represented using the IDEF0 notation along with the key stages, sub-stages, actors, inputs, outputs and controls clearly shown.

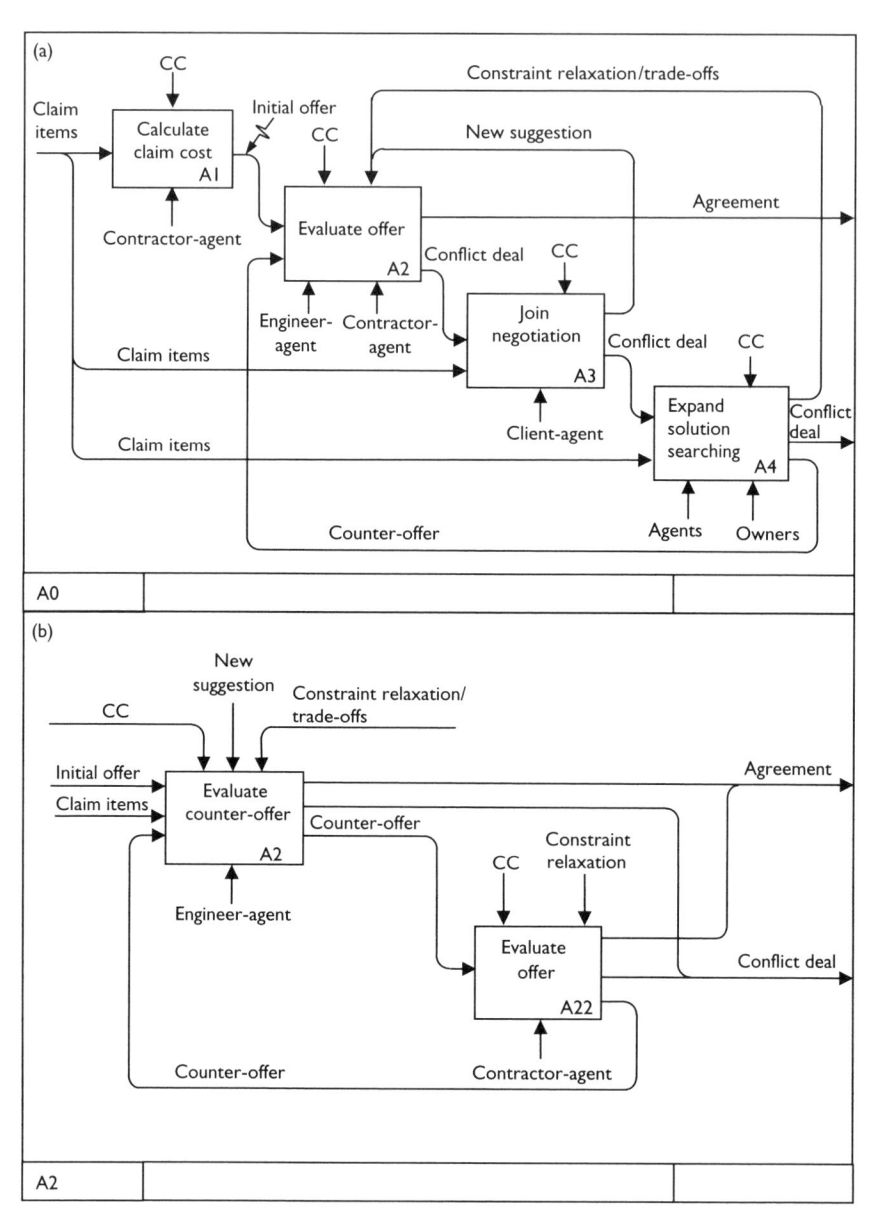

Figure 6.4 The MASCOT process model represented using the IDEF0 negotiation. (a) Negotiate claims. (b) Evaluate offer.

Note
CC contract and conditions of contract.

6.5 Example

The following example illustrates the application of the MASCOT in a real claim case for the piling work in a water supply project. According to the design, the length of pile was 8 m with 7.5 m to be driven into the river bed. During the construction, most of the piles could only be driven about 3.5 m into the river bed. The engineer insisted that the problem was caused by the contractor's old piling machine. Although the contractor tried all the possible piling methods, only three piles could be driven 4.5 m into the river bed before the work was completed. The work was delayed by 60 days. Later, borehole testing at the site showed that a dense gravel-sand layer lay underneath the river bed, which was shown as soft clay in the original drawing. Consequently, the contractor submitted a claim including time extension, new pile rate, pile cutting and removal, overheads and loss of productivity. Since similar problem may occur at other sites, both parties were very cautious about their claim.

The real case took more than four months to settle. The most difficult negotiation item was loss of productivity and with more than 10 meetings held at different levels. In this example, MASCOT will be adopted to resolve the claim for loss of productivity. The focus is on the contractor-agent's offer of mechanism, and how it enables to include and update its beliefs which are used to determine the level of concession. The negotiation strategies such as 'the involvement of the client-agent' and 'expanded solution searching' are not demonstrated.

6.5.1 Utility function

Uncertainty about the opponent's utility function is a critical feature of construction claims negotiation. In this example, it is assumed that all the participants' utility functions are linear, which can be determined by two points: the optimum point and the reservation point. Here, the utility of the optimum value is assumed as 1 and 0.6 at the reservation point (Figure 6.5). Each agent can estimate the opponent's utility function based on these two critical points. Meanwhile, it is reasonable to assume that each agent's initial offer is its optimum amount. Thus, an agent can know the opponent's utility function if it can estimate the opponent's reservation value through the learning mechanism.

6.5.2 Negotiation preparation

Before negotiation, the contractor estimates that his real loss of productivity is $9,000. Considering his current situation and the importance of the claim, the contractor decides that his reservation value is $9,000 and his

(a)

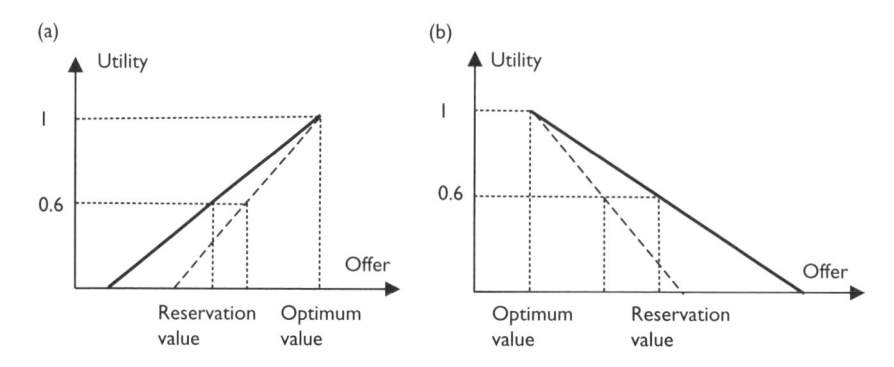

(b)

Figure 6.5 (a) The contractor-agent's utility function. (b) The engineer-agent's utility function.

Table 6.1 The contractor's prior knowledge of the engineer's reservation value

Hypothesis	R_1 $7,000	R_2 $8,000	R_3 $8,500	R_4 $9,000	R_5 $10,000	R_6 $11,000
Probability $P(R_i)$	0	0	0.25	0.25	0.25	0.25

Table 6.2 The conditional probabilities of the engineer's offer given the contractor's hypothesis

Hypothesis	e_0 £6,000	e_1 £7,000	e_2 £8,000	e_3 £8,500	e_4 £9,000	e_5 £9,500	e_6 £10,000	e_7 £11,000
£7,000	0.35	0.35	0.25	0.04	0.01	0	0	0
£8,000	0.10	0.40	0.35	0.12	0.03	0	0	0
£8,500	0	0.14	0.50	0.30	0.05	0.01	0	0
£9,000	0	0.10	0.15	0.45	0.25	0.05	0	0
£10,000	0	0.03	0.10	0.20	0.40	0.22	0.05	0
£11,000	0	0	0	0.05	0.10	0.25	0.40	0.20

optimum value is $11,000. Meanwhile, the contractor also tries to estimate the engineer's reservation value based on his domain knowledge. Table 6.1 shows the contractor's estimate about the possible distribution of the engineer's reservation value. Table 6.2 shows the contractor's estimate of the conditional probabilities of the engineer's offers given his hypothesis, which are encoded from the contractor's negotiation domain knowledge with the engineer.

6.5.3 Negotiation process

THE INITIAL OFFER AND COUNTER-OFFER

The contractor-agent makes its initial offer of £11,000, which is assumed to be its optimum claim amount. After receiving the contractor-agent's initial offer, the engineer-agent makes a counter-offer of £7,000 to the contractor-agent, which is also assumed to be the engineer-agent's optimum amount.

THE CONTRACTOR-AGENT'S OFFER IN THE 2ND ITERATION

Updating the belief of the probability of the engineer-agent's reservation amount – Based on the engineer-agent's counter-offer and the contractor's prior knowledge about the engineer-agent (Tables 6.2 and 6.3), the contractor-agent updates its belief about the probability of the engineer-agent's reservation amount being R according to the Bayesian rule (Equation 6.2). In this case, the engineer's offer is £7,000 ($e_1 = 7,000$), thus,

$$P(R_3|e_1) = \frac{P(R_3)P(e_1|R_3)}{\sum_{k=1}^{6} P(e_1|R_k)P(R_k)}$$

$$= \frac{0.25 * 0.14}{(0.25 * 0.14) + (0.25 * 0.1) + (0.25 * 0.03)} = 0.518$$

$$P(R_4|e_1) = \frac{P(R_4)P(e_1|R_4)}{\sum_{k=1}^{6} P(e_1|R_k)P(R_k)}$$

$$= \frac{0.25 * 0.1}{(0.25 * 0.14) + (0.25 * 0.1) + (0.25 * 0.03)} = 0.370$$

$$P(R_5|e_1) = \frac{P(R_5)P(e_1|R_5)}{\sum_{k=1}^{6} P(e_1|R_k)P(R_k)}$$

$$= \frac{0.25 * 0.03}{(0.25 * 0.14) + (0.25 * 0.1) + (0.25 * 0.03)} = 0.111$$

From Table 6.2:

$P(R_1|e_1) = 0$, $P(R_2|e_1) = 0$ because $P(R_1) = P(R_2) = 0$

$P(R_6|e_1) = 0$ because $P(e_1|R_6) = 0$

Estimating the engineer-agent's reservation amount – Prior to receiving the engineer-agent's offer (£7,000), the contractor-agent's belief about the engineer-agent's reservation amount is:

$$R = \sum P(R_i) * R_i = 0.25 * 8,500 + 0.25 * 9,000 + 0.25 * 10,000 + 0.25 * 11,000 = 9,625$$

After receiving the counter-offer, the contractor-agent's estimation of the engineer-agent's reservation amount is updated as follows:

$$R = \sum P(R_i) * R_i = 0.518 * 8,500 + 0.37 * 9,000 + 0.111 * 10,000 = 8,843$$

Utility functions

(a) *The contractor-agent's utility function* – in this example, agents' utility functions are assumed to be linear, which can be expressed as $u_c = kx + b$. Also, the two key points in the utility functions are known, that is, optimum point: (11,000, 1) and reservation point: (9,000, 0.6). Thus, the contractor-agent's utility function can be calculated to be: $u_c = 2 \times 10^{-4}x - 1.2$.

(b) *The contractor-agent's estimate of the engineer-agent's utility function* – can be expressed as $u_e = kx + b$, where the contractor-agent knows two points along this line based on its updated beliefs: optimum point: (7,000, 1) and reservation point: (8,707, 0.6). Thus, the contractor-agent estimates the engineer-agent's utility function to be: $u_e = -2.3 \times 10^{-4}x + 2.64$.

(c) *The combined utility function of the contractor-agent's and the engineer-agent's are* – $u_c = 2 \times 10^{-4}x - 1.2$ and $u_e = -2.3 \times 10^{-4}x + 2.64$, the correlation between the contractor-agent and the engineer-agent's utility functions can be calculated as: $u_e = -1.17u_c + 1.234$.

Risk evaluation – According to Zeuthen's model, the maximum likelihood of risk acceptable to the contractor-agent (P_{cmax}) and the engineer-agent (P_{emax}) can be calculated as:

$$P_{cmax} = \frac{U_{cc}^t - U_{ce}^t}{U_{cc}^t - U_c(C)} \qquad P_{emax} = \frac{U_{ee}^t - U_{ec}^t}{U_{ee}^t - U_e(C)} \tag{6.1}$$

In this iteration, the offer of the contractor-agent and the engineer-agent are (11,000, 7,000). Thus, the maximum likelihood of risk acceptability to the contractor and the engineer are:

$$P_{cmax} = \frac{U_{cc}^t - U_{ce}^t}{U_{cc}^t - U_c(C)} = \frac{1 - 0.2}{1} = 0.8 \tag{6.3}$$

$$P_{emax} = \frac{U_{ee}^t - U_{ec}^t}{U_{ee}^t - U_e(C)} = \frac{1 - 0.11}{1} = 0.89 \qquad (6.4)$$

Concession – Since the $P_{cmax} < P_{emax}$ (i.e. the contractor's maximum risk acceptability is less than that of the engineer-agent), the contractor-agent knows that it should make a concession in the next iteration. In this example, a simple concession approach is adopted to calculate the concession rate (i.e. the contractor-agent will make the minimum concession sufficient to make the engineer-agent's maximum acceptable risk smaller than its own in the next iteration). The concession step can be calculated as:

$$P_{emax} = \frac{U(w_e) - U(D_c)}{U(w_e) - U(e)} = \frac{|1 - u_{ec}|}{1 - 0} = 0.8$$

$u_{ec} = 0.2 \Rightarrow u_c = 0.8837 \Rightarrow x = 10{,}418$

Thus, the contractor-agent's new offer will be equal to or lower than £10,418.

THE ENGINEER-AGENT'S COUNTER-OFFER IN THE 2ND ITERATION

By following the same procedure as outlined earlier for the contractor-agent while the engineer-agent is required to make a concession and makes a counter-offer of £8,508.

THE CONTRACTOR-AGENT'S OFFER IN THE 3RD ITERATION

By receiving the engineer-agent's new counter-offer (£8,508) while the contractor-agent repeats the evaluation process, as in the 2nd iteration, in order to decide its concession amount.

Updating the belief of the probability of the engineer-agent's reservation amount – After the last iteration, the contractor-agent updates its belief about the probability of the engineer-agent's reservation amount as given in Table 6.3. Based on this prior knowledge (at £7,000) and the engineer-agent's new counter-offer £8,838, the contractor-agent updates its belief as follows:

$$P(R_i | e_1, e_2, e_3) = \frac{P(R_i | e_1) P(e_1, e_2, e_3 | R_i)}{\sum_{k=1}^{6} P(e_1, e_2, e_3 | R_k) P(R_k)}$$

Table 6.3 The contractor agent's belief of the probability after the 2nd iteration

Hypothesis	1 £7,000	2 £8,000	3 £8,500	4 £9,000	5 £10,000	6 £11,000
Probability $P(R_i|e_1)$	0	0	0.518	0.370	0.111	0

where,

$$P(e_1, e_2, \ldots, e_n | R_i) = P(e_1 | R_i) * P(e_2 | R_i) * \cdots * P(e_n | R_i)$$
$$P(e_1, e_2, e_3 | R_i) = P(e_1 | R_i) * P(e_2 | R_i) * P(e_3 | R_i)$$
$$= (0.14, 0.1, 0.03) * (0.30, 0.45, 0.20)$$
$$= (0.042, 0.045, 0.006)$$

The given equations are based on the assumption that event e_1 and e_2 are independent. In this iteration, the contractor-agent's new prior conditional probabilities become:

$$P(R_1 | e_1, e_2, e_3) = \frac{P(R_1 | e_1) P(e_1, e_2, e_3 | R_1)}{\sum_{k=1}^{6} P(e_1, e_2, e_3 | R_1) P(R_1)}$$

$$= \frac{0.518 * 0.042}{(0.518 * 0.042) + (0.37 * 0.045) + (0.111 * 0.006)}$$

$$= 0.556$$

$$P(R_2 | e_1, e_2, e_3) = \frac{P(R_2 | e_1) P(e_1, e_2, e_3 | R_2)}{\sum_{k=1}^{6} P(e_1, e_2, e_3 | R_2) P(R_2)}$$

$$= \frac{0.37 * 0.045}{(0.518 * 0.042) + (0.37 * 0.045) + (0.111 * 0.006)}$$

$$= 0.426$$

$$P(R_3 | e_1, e_2, e_3) = \frac{P(R_3 | e_1) P(e_1, e_2, e_3 | R_3)}{\sum_{k=1}^{6} P(e_1, e_2, e_3 | R_3) P(R_3)}$$

$$= \frac{0.111 * 0.006}{(0.518 * 0.042) + (0.37 * 0.045) + (0.111 * 0.006)}$$

$$= 0.017$$

Thus, after receiving the engineer-agent's two offers, the contractor agent's current estimation of the engineer-agent's reservation amount is:

$$R = 0.556 * 8{,}500 + 0.426 * 9{,}000 + 0.017 * 10{,}000 = 8{,}738.$$

Making concession according to Zeuthen's strategy – Through the process of estimating the opponent's utility function, maximum risk acceptability and concession amount based on Zeuthen's strategy, the contractor-agent identifies that the new offer should be equal to or lower than £10,073.

Table 6.4 The negotiation process

Negotiation iteration	0	1st	2nd	3rd	4th	5th	6th
The contractor-agent's offer	—	11,000	10,418	10,073	9,814	9,624	9,548
The contractor-agent's estimate of the engineer's reservation value	9,625	8,843	8,738	8,952	9,394	9,576	—
The engineer-agent's offer	—	7,000	8,508	8,818	9,232	9,429	9,500
The engineer-agent's estimate of the contractor's reservation value	8,600	9,464	9,520	9,528	9,532	9,496	—

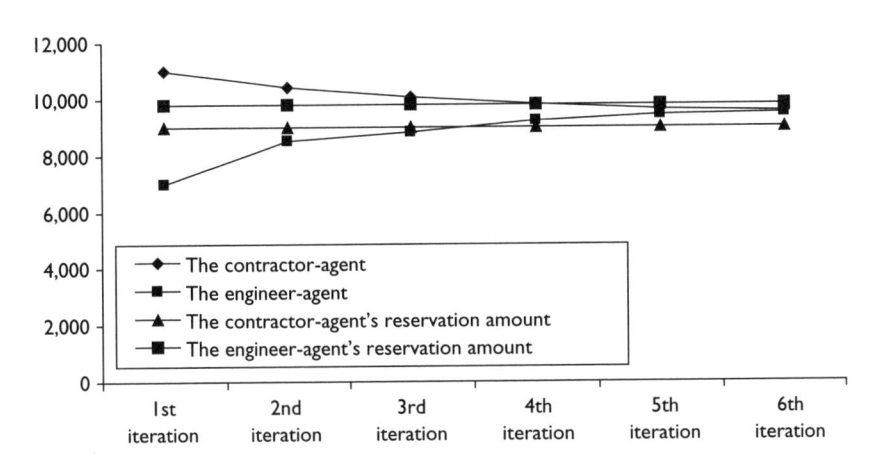

Figure 6.6 The negotiation process (negotiation converged at the 8th/9th iteration).

Due to space constraints, this example only presents the first two iterations. Further iterations of negotiation will follow the same procedure. Table 6.4 and Figure 6.6 show the outcomes of the full negotiation process in each iteration.

6.6 Implementation

The MASCOT model has been implemented in a software system using the ZEUS agent building tool kit. Since the MASCOT agent negotiation mechanism is rather complex, the existing negotiation protocol and strategies provided by ZEUS cannot satisfy the MASCOT requirements. As a result, the MASCOT negotiation protocol and strategies are developed with JAVA

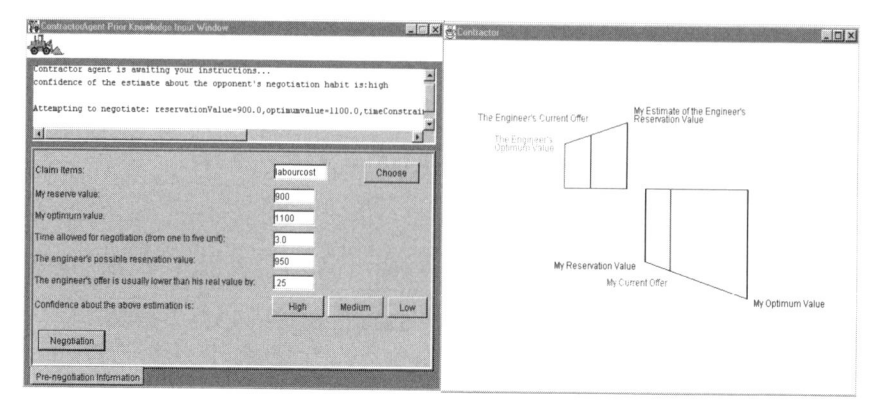

Figure 6.7 The input and negotiation display windows of MASCOT.

and embedded in the ZEUS environment (for details see Ren *et al.*, 2001). This approach is helpful for any complex and large system which cannot be simply implemented by using the existing tool kit. Also, it provides a basis for integrating industrial legacy systems and building user's specified interfaces. Figure 6.7 shows the input window and negotiation process display window design for MASCOT.

Overall, the MASCOT prototype worked very well with industrial examples, but two problems were observed. First, since one condition of Bayesian updating is that an agent should have sufficient prior knowledge about its opponent; the model may not work properly (e.g. an agent may make too much concession or stand still at the beginning of the negotiation due to its wrong estimate of the opponent's reservation value) if an agent's estimate of the opponent's negotiation habit is very different from the opponent's real value, but the agent still believes that its confidence in the estimate is high. To overcome this problem, two extra rules were set, which limit the agents' concession speed so that they can make reasonable concessions even if their estimate of the opponent and the opponent's real value are very different.

Second, the current expression of the agents' domain knowledge 'usually the engineer-agent will make an offer, which is 20% lower than its reservation value' needs to be improved further. Unlike Zeng and Sycara's (1998) straightforward concession mechanism, an agent's concession amount in MASCOT is determined by its estimate of the opponent's reservation value according to Zeuthen's model. Thus, it is possible for an agent to stop conceding to make a very small concession even if its current offer is still far from its reservation value. This is caused by the problem that an agent always makes the same exaggerated amount irrespective of whether it is at the beginning or the end of negotiation. This is often not true in the real world. It might be more realistic to explain the agent's domain knowledge as 'usually the engineer-agent will make an initial offer which is 20% lower than its reservation value, but about 15% lower at the end of the negotiation', because it is quite possible that negotiators reduce their exaggerated amount as negotiation goes on.

6.7 Conclusions and further study

The development of the MASCOT model not only improves the efficiency of construction claims negotiation, but also provides an opportunity to overcome some major problems of claims negotiation caused by the industrial organisational structure. The major advantages of the MASCOT negotiation mechanism are twofold.

First, the MASCOT negotiation mechanism matches the nature and characteristics of construction claims negotiation. Based on a detailed analysis, this study has outlined the nature and major characteristics of construction claims negotiation. The MASCOT has been developed to match these characteristics and to resolve the highlighted problems.

- Based on a contractual bounded self-interested model, MASCOT adopts Zeuthen's model as the essential concession model in which the participants try to avoid a conflict. Meanwhile, they can also maximise their utilities if the negotiation can reach an agreement.
- The application of the Bayesian learning mechanism provides a simple and effective approach to resolving the problems of incomplete information assumption in construction claims negotiation.
- The introduction of a time penalty reflects the different time pressures of construction claims negotiation participants.
- The support of the early involvement of the client-agent resolves the problems posed by the conflicting roles of the engineer in claims negotiation.

Second, the MASCOT develops a rational negotiation system. A rational outcome is an essential requirement for all negotiation systems and yet, it is difficult to achieve. For example, an agent may benefit from an unreasonably high initial offer if a binary divisive procedure is adopted. In MASCOT, the learning mechanism and the risk avoidance model make negotiation not only efficient but also rational. The problem in binary divisive procedure is reduced because an agent may have prior knowledge about its opponent's negotiation method. Thus, no matter how high an offer the opponent makes, the agent will make a reasonable counter-offer based on its domain knowledge. The high frequency of claims negotiation in construction projects makes it possible for a party to get enough such domain knowledge. On the other hand, even if an agent does not have sufficient prior knowledge of its opponent (this often happens at the beginning of a project), negotiations can still converge and reach a reasonable result since the agent can make decisions based on the opponent's offers during negotiation.

Since the studies on construction claims negotiation and MAS are relatively new, many problems still need to be addressed. For example, further investigations need to be conducted into the level of empowerment of the MASCOT agents, the encoding of claim participants' domain knowledge and the qualitative claims negotiation. To make the system more efficient, further development of the negotiation mechanism, such as the involvement of a mediator agent, the application of a dual responsiveness model and negotiation protocol in which offers are made with the supporting arguments, could be investigated.

References

Anson, R. G. and Jelassi, M. T. (1990) 'A development framework for computer-supported conflict resolution', *European Journal of Operational Research*, 46: 181–99.

Anumba, C. J. and Newnham, L. N. (2000) 'Computer-based collaborative building design: conceptual model', *International Journal of Construction Information Technology*, 8(1): 1–14.

Bacharach, S. B. and Lawler, E. J. (1981) *Bargaining: Power, Tactics and Outcomes*, Jossey-Bass Publishers, San Franciso, CA.

Bui, H. H., Venkatesh, S. and Kieronska, D. (1996) 'Learning other agents' preferences in multi-agent negotiation using the Bayesian classifier', *AAAI*, 96: 114–19.

Cross, J. G. (1977) 'Negotiation as a learning process', in I. W. Zartman (ed.) *The Negotiation Process: Theories and Application*, Sage Publications, London, pp. 29–54.

ECI (1992) *Client Management and its Role in the Limitation of Contentious Claims*, European Construction Institute Report No. TF 003/3, Loughborough University.

Ferber, J. (1998) *Multi-agent Systems: An Introduction to Distributed Artificial Intelligence*, Addison–Wesley, Harlow.

Harsanyi, J. C. (1956) 'Bargaining and conflict situations in the light of a new approach to game theory', *The American Economic Review*, 55: 447–57.

Harsanyi, J. C. (1977) *Rational Behaviour and Bargaining Equilibrium in Games and Social Situation*, Cambridge University Press, Cambridge, UK.

Hu, J. X. (1997) *Chinese Oversea Construction Contractor's Claim Management* (Chinese Version), China Construction Publisher, Beijing.

Iversen, G. R. (1984) *Bayesian Statistical Inference*, Sage University Paper, Beverly.

Jordan, J. S. (1992) 'The exponential covergence of Bayesian learning in normal form games', *Games and Economic Behaviour*, 4: 202–17.

Kraus, S. (1996) 'Beliefs, time and incomplete information in multiple encounter negotiations among autonomous agents', *Annals of Mathematics and Artificial Intelligence*, 20(1–4): 111–59.

Kraus, S. (1997) 'Negotiation and cooperation in multi-agent environments', *Artificial Intelligence Journal, Special Issue on Economic Principles of Multi-Agent Systems*, 94(1–2): 79–98.

Nash, J. F. (1950) 'The bargaining problem', *Econometrica*, 28: 155–62.
Nash, J. F. (1953) 'Two-person cooperative games', *Econometrica*, 21: 128–40.
Pena-Mora, F. and Wang, C. (1998) 'Computer-Supported collaborative negotiation methodology', *Journal of Computing in Civil Engineering*, 12(2): 64–81.
Pickavance, K. (1997) *Delay and Disruption in Construction Contracts*, LLP Reference Publishing, London.
Powell-Smith, V. and Stephenson, D. (1993) *Civil Engineering Claims* (2nd edn), Blackwell Scientific, Oxford.
Ren, Z., Anumba, C. J. and Ugwu, O. O. (2001) 'The implementation of a multi-agent system for construction claims negotiation', *Proceedings Civil-Comp 2001 and AI-Civil-Comp 2001 Conference*, Vienna, Austria.
Rosenschein, J. and Zlotkin, G. (1994) *Rules of Encounter*, MIT Press Cambridge, MA.
Sahin, F. (1999) 'A Bayesian network approach to the self-organisation and learning in intelligent agents', URL: http://www.ee.vt.edu/sferat/proposal_def/
Sandholm, T. W. (1996) 'Negotiation among self-interested computationally limited agents', dissertation (PhD), University of Massachusetts, Amherst, MA.
Shell, G. R. (1995) 'Computer-assisted negotiation and mediation: where we are and where we are going', *Negotiation Journal*, 11(2): 117–22.
Vidogah, W. and Ndekugri, I. (1997) 'Improving management of claims: contractors' prospective', *Journal of Management in Engineering*, 13(5): 37–44.
Young, O. R. (ed.) (1975) *Bargaining: Formal Theories of Negotiation*, University of Illinois Press, Urbana, IL.
Zack, J. G. (1994) 'The negotiation of settlements – a team sport', *Cost Engineering*, 36(8): 24–30.
Zeng, D. and Sycara, K. (1998) 'Bayesian learning in negotiation', *International Journal of Human-Computer Studies*, 48: 125–41.
Zeuthen, F. (1975) 'Economic warfare', in O. R. Young (ed.) *Bargaining: Formal Theories of Negotiation*, University of Illinois Press, Urbana, IL, pp. 145–63.

Chapter 7

Specification and procurement of construction products using agents

E. O. Obonyo, C. J. Anumba and A. Thorpe

7.1 Introduction

7.1.1 Background

The expansion in the use of the Web and its exponential growth are well-known facts nowadays. However, a problem has emerged from the very core of the success of the Web. The Web has evolved into a very large, unstructured but ubiquitous database holding textual data in the order of one terabyte (Baeza-Yates, 1998). Such information overload on the Web has resulted in substantial losses (KPMG, 2000). It has in fact been established in the US that employees spend 8 hours a week on average retrieving external information.

At the individual company level, there have been great expectations that the use of Internet-enabled systems would improve productivity and reduce cycle times in various operations by collecting and providing the right information. A significant proportion of the forecasted gains is yet to be realised partly due to the lack of interoperability between systems (Madhusudan, 2001; Samtani, 2003). Collaboration has subsequently become more important for any company seeking to make the most out of e-business (Deloitte Research, 2002).

Clearly, there is a need for Internet-based technologies that address the availability of information and the ability to exchange seamlessly and process it across different applications in different organisational units. Consequently, the e-business world has been growing at an unprecedented pace and new market realities, such as the software agent paradigm, have emerged. This chapter is based on a project that explores the role of agents in Construction Industry-specific e-business. The project focuses on developing an agent-based prototype system for use in the specification and procurement of construction products on the Internet.

7.1.2 The motivation and context

Significant research effort has gone into achieving the Web described by Berners-Lee (1989). Work being carried out focuses on providing links to information contained in the documents displayed on the Web. The W3C's Web of the future that will hold machine-processable information has been very broadly defined as the Semantic Web. The Semantic Web will allow people to find, share and combine information more easily (Hendler *et al.*, 2002). W3C has set out to define and link the Web in a way that it can be used for more effective discovery, automation, integration and reuse across applications. The enabling technologies for the Semantic Web include:

- *eXtensible Mark-up Language (XML)* – adds arbitrary structure;
- *RDF* – provides common framework for representing meta-data across many applications;
- *ontologies* – store formal definitions of relations among terms; and
- *software agents* – automate tasks.

The potential contribution of software agents to the success of e-business initiatives lies in their ability to collect Web content from diverse sources, process the information and exchange the results with other programs (Berners-Lee *et al.*, 2001). Sun (2004, Chapter 9) discusses the use of agents for information search and retrieval. Further, information on the rationale behind the use of agent-based system in business operations in the digital era has been provided by Blake and Gini (2002), Blake (2002) and Samtani (2003).

The APRON project emerged from a partnership between Loughborough University and Building Information Warehouse (BIW) Technologies Plc, UK. BIW is a construction industry-specific e-business portal. The company's core product is the BIW Information Channel, which creates a central repository of information about the built asset, and allows users to access information exactly tailored to their needs.

At the start of the APRON project, BIW Technologies had already developed some customised 'intelligent' components in AutoCAD that could be used to execute parametric searches for construction products. The implemented application scenario emerged from the need to provide an interface between these components in AutoCAD and the heterogeneous repositories provided by construction products' manufacturers.

7.1.3 Defining software agents and agent road map

It is not easy to define the term 'agents'. Nwana (1996) provides a number of explanations for this difficulty: it is a common term in every day conversion; it encompasses a broad area; it is a meta-term and researchers in this area have come up with such synonyms as 'bots', 'spiders' and 'crawlers'. Software agents in this project have been explored from the viewpoint of leading researchers. Various researchers such as Brustolini (1991), Ferber (1999), FIPA Architecture Board (2001), Jennings and Wooldridge (1998), Jennings *et al.* (1998), Lieberman (1997) and Maes (1994) have defined the term 'agent' in various ways depending on their interests. It is, however, possible to extract some common attributes of agents from these definitions. There is a general consensus that software agents exist in an environment. They can sense the conditions in the environment and such senses may affect how they act in future. Software agents are also perceived to be adaptive and capable of learning. They are proactive, exhibiting goal-directed behaviour and the execution of tasks occurs autonomously (without human intervention). In short, agents are systems capable of autonomous, purposeful action in the real world.

There are a number of closely related paradigms that need to be distinguished from agents. These include 'ordinary' programs, whose output has no effect on what is sensed in future. Such programs also lack temporal continuity (Franklin and Graesser, 1996). Process control systems and software demons exhibit some features of agent-based systems but they lack intelligence and flexibility. Agent technology bears close resemblance to other paradigms such as artificial intelligence, systemics, distributed systems, expert systems, remote programming and object-oriented programming. Such techniques provide the base on which agent technology is built (Aylett *et al.*, 1997; Kashyap, 1997; Fingar, 1998; Jennings and Wooldridge, 1998; Mahapatra and Mishra, 2000).

Software agents are growing in importance and will gradually become indispensable components of the Web of the future. There are several significant research efforts both across Europe and on the global arena. The European Union (EU) has, for example, made significant investments in large-scale schemes such as Agentcities (URL1) and AgentLink (URL2).

Despite such significant research interest in agent technology, there are, however, no matching large-scale implementations of agent-based applications in commercial scenarios. Agent technology is still maturing and it will take at least another decade before there is a perfect match between the great expectations and implemented systems. Luck *et al.* (2003) have predicted that agent technology will go through four major phases in its transition to maturity. These phases are depicted in Figure 7.1 while Table 7.1 highlights some of the inherent characteristics of these phases.

The work done in the APRON project builds on previous work that resulted in an e-procurement framework for construction products

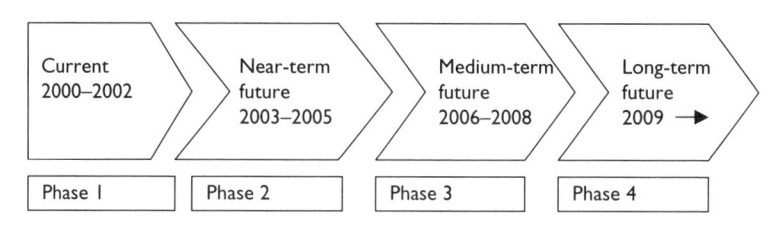

Figure 7.1 The agent road map.

Table 7.1 Characteristics of agent technology phases

Period	Characteristics
Phase 1	Development by a single design team
	Deployment in a single corporate environment
	Common high-level agent goals in a single domain
Phase 2	Development by a single design team
	Deployment across corporate boundaries
Phase 3	Development by different design teams
	Deployment of heterogeneous agent systems
	Deployment in a single application domain
	Having bridging agents to criss-cross domains
Phase 4	Open MAS crossing multiple domains
	Development by diverse design teams

Source: Adapted from Luck *et al.*, 2003.

(Ugwu *et al.*, 1999, 2002; Ugwu and Anumba, 2000). Direct 'agentification' of this system would result in an agent-based application that conforms to applications under Phase 1 in the agent technology road map. The APRON prototype advances the technology used on this electronic procurement system and seeks to deliver a system with some functionalities of future systems defined in Phase 2 and Phase 3. The prototype system has been developed without consultation with the construction product manufacturers to reflect the fact that the Web of the future will have so many potential links that it would not be possible to involve all the information providers actively in the system development.

7.2 The specification and procurement of construction products

The collaboration challenges in the selected domain are similar to those outlined in the general business context in the introductory section of this chapter. The processes in Internet-enabled specification and procurement of construction products involve retrieving information from autonomous units. Firms in the construction industry are under great pressure to improve their ability and capacity for co-operating with their business associates. This can only be realised when co-operating firms have interfaces to the applications used by the team members across corporate boundaries.

APRON can be perceived as an e-business support system that would be hosted by a construction industry-specific information portal. It comprises components that execute search functions, content management, enable collaboration and manage the processes involved in the specification and procurement of construction products. The first basic assumption in the project is that product manufacturers have made the required information on products available on the Internet. It is further assumed that the participating parties operate autonomously as far as their information systems are concerned. In the implemented prototype, the agent-based framework allows the final end-users to have direct access to information published by manufacturers without any changes being made to their existing information system. The very essence of APRON is facilitating the reuse of legacy components in their existing form.

7.3 Related work

The projects highlighted in this section have similarities with APRON in two ways. Some of the projects address a similar problem but utilise different approaches in the solution. The other projects address a generally similar theme with significant variations in the selected domain but the modelled solution is based on the use of the software agent-paradigm.

The fundamental challenge being addressed in the APRON project is similar to that addressed in the Global Engineering Network Intelligent Access Libraries (GENIAL) project: the digital 'anarchy' (Radeke, 1999). The present use of the Internet in construction is characterised by closed markets that cannot utilise each other's services and incompatible applications that cannot interoperate or build upon each other. The objective of the GENIAL project was to define an open architecture and to establish a common semantic infrastructure. The solution adopted in the GENIAL project was largely based on the use of standards. The output of the GENIAL project was extended in the eConstruct project. The focus of the eConstruct project was to develop an XML vocabulary (bcXML) for the European building and construction industry (Stephens *et al.*, 2002). A second project that emerged from the GENIAL project was a collaborative project between Taylor Woodrow and Loughborough University, whose objective was to extend the discrete product search in GENIAL into a product schedule that could be used to perform product comparisons across different suppliers (Ugwu *et al.*, 2002). The APRON project takes this effort to the next level: it utilises the software agent-paradigm to resolve the digital 'anarchy'. This digital 'anarchy' was also the fundamental subject of the e-bip project, which resulted in a business-to-business Broker Service for the Small and Medium-sized Enterprises (SMEs) in the construction tile supply chain (Thiels *et al.*, 2002).

Other researchers have investigated the subject of extracting data from semi-structured sources by using different approaches. Hammer *et al.* (1997) implemented a Web extractor prototype using the Python programming language. In the SEEK project (O'Brien, 2003), Data Reverse Engineering (DRE) and Schema Matching (SM) processes were used to deploy a source wrapper for a legacy information system.

There have been a number of projects focusing on the use of agent to enhance various aspects of design. For brevity, only a few examples of agent-based implementations in domains that are close to the specification and procurement of construction products are cited here. The Agent-based Support for the Collaborative Design of Light Industrial Buildings (ADLIB) focused on developing a multi-agent system (MAS) framework for the representation of the activities and processes involved in the collaborative design of light industrial buildings (Ugwu *et al.*, 2000a, see Chapter 5). In another project a multi-agent architecture for the integration of design, manufacturing and shop floor control activities was developed (Balasubramanian *et al.*, 1996). Two other closely related agent-based projects are ProcessLink and Responsible Agents for Product-Process Integrated Design (RAPPID). ProcessLink was a project aimed at providing a technical infrastructure and methodology for integrating spatially distributed engineers, designers and their heterogeneous tools (Petrie, 1997). In the RAPPID project agents were used to resolve conflicts amongst designers (Parunak, 1996).

7.4 The APRON architecture

In this section a high-level architectural view of the APRON prototype is presented. The functional capabilities of the prototype in the business context outlined at the outset of the chapter are described in the following section.

7.4.1 The APRON functional components

The APRON prototype is modelled using the established mediation/wrapper methodology, that was used in, for example, the InfoSleuth prototype (Bayardo *et al.*, 1996) and the SEEK prototype (O'Brien *et al.*, 2003). In this approach, the developed solution constitutes a software middle layer between the semi-structured repositories of construction products and the end-user applications utilised in specification and procurement. This is depicted in Figure 7.2.

There are three distinct, intercommunicating layers. The construction product manufacturers are at the top level in the architecture. Manufacturers display details of the various product offerings as semi-structured data on the Internet. The kernel of the architecture is the e-marketplace of an Information Provider, who uses the Internet to ensure that requisite project information is available to all the key players in the construction project supply chain. This e-marketplace hosts the APRON solution, which consists of the Download, Extraction, Structuring, Database, Search and Procurement Modules. The e-marketplace also holds a repository of relevant standards that can be used as

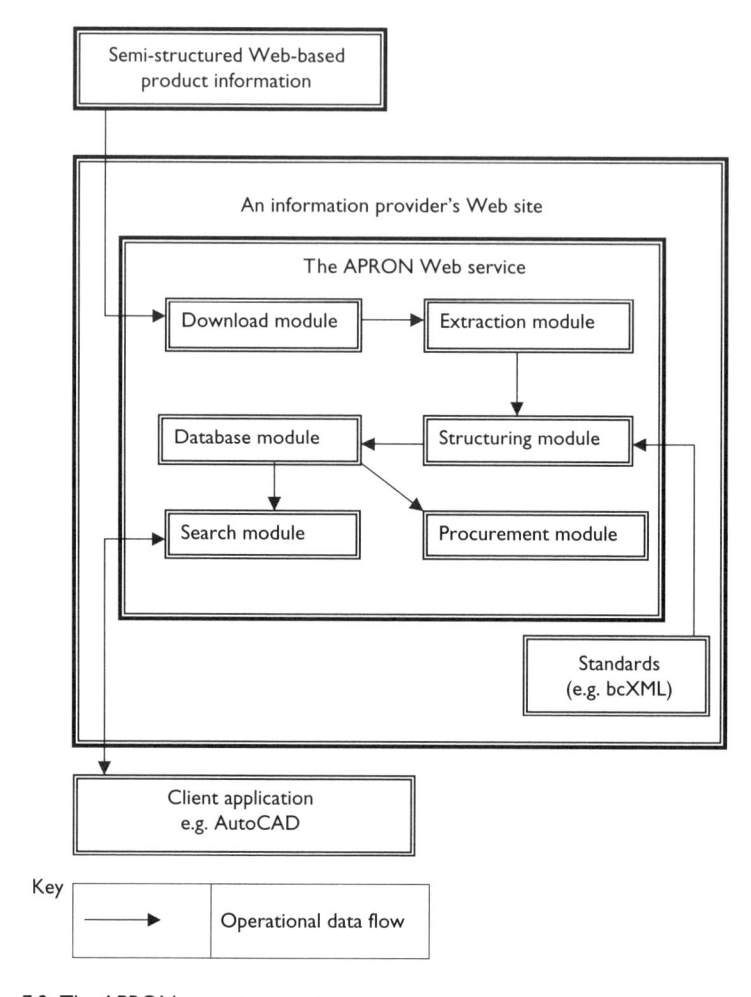

Figure 7.2 The APRON prototype system.

XML schemas for structuring the extracted information. The final layer comprises the end-user firms in the specification and procurement of construction products. The focus in these firms is providing an automated interface to the computing applications used by specifiers and procurers. A client application has been deployed to allow such end-users to communicate with the APRON Web Service.

The APRON system provides a link between the Web site holding product information and the applications used in the specification and procurement of construction products. The Download Module maintains real-time access with these Web sites. The text is then extracted from the downloaded file, and using previously defined XML schemas, structured into a context-specific format. Industry standards, such as IFCXML and bcXML, can be easily adopted and used to create an XML template for structuring the extracted information. The APRON system also stores the relevant information in a database.

Information for the specification and procurement of construction products is obtained from this database. Specifically, APRON, provides a Web-based search engine, which can be used to execute context-specific queries for construction products. The APRON solution also offers a framework for automating the procurement of specified products. The Procurement Module has two types of agents: a buyer agent and a seller agent representing product procurers and product sellers respectively. The two agents exchange requisite information and automate the transactions involved in the procurement of construction products.

7.4.2 Deployment scenario

7.4.2.1 The target source

The example implemented in the APRON prototype focuses on processing product information for the specification and procurement of light

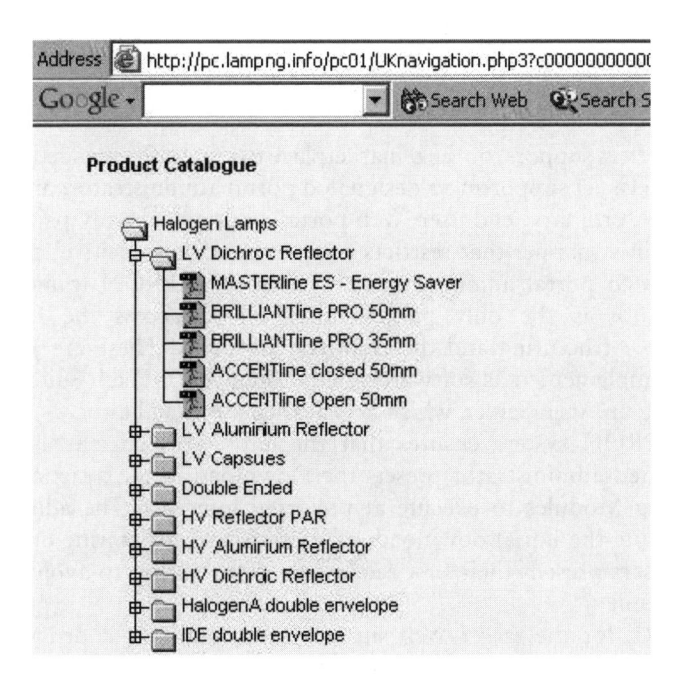

Figure 7.3 Philips Lighting Web site.

Table 7.2 A snapshot of the lighting bulb information source

Commercial product name

Type	Wattage (W)	Cap/base	Voltage (V)
MASTERline ES	20	GU5.3	12
MASTERline ES	20	GU5.3	12
MASTERline ES	30	GU5.3	12
MASTERline ES	30	GU5.3	12
MASTERline ES	30	GU5.3	12
MASTERline ES	30	GU5.3	12
MASTERline ES	35	GU5.3	12

bulbs from the Philips Lighting Web site (Figure 7.3) (URL3). The site hosts close to 200 catalogues in Adobe Acrobat PDF format. The information that would be of interest to an end-user such as wattage, cap size and voltage is presented in a semi-structured format.

The Web site does not have a search facility that would support guided navigation based on, for example, attributes such as wattage and voltage. It is also not possible to query the information directly from another application. Furthermore, relevant data has to be re-keyed for reuse elsewhere. It is therefore necessary to educe the contents of the light bulb table (Table 7.2) from the underlying Web page.

7.4.2.2 APRON functionalities

APRON offers support for the e-marketplace to create an extended enterprise at two levels: (1) supporting a designated portal administrator; and (2) supporting an 'ordinary' end-user. Web portal solutions already package their solutions in a manner that restricts access to certain 'sensitive' capabilities to designated portal administrators. Within the APRON framework, an administrator is the only party authorised to access the Download, Extraction, Structuring and the Database Modules. These are Java-based modules implemented as software agents using the X-Fetch Suite (URL 4), comprising an AgentServer, which acts as mediator/ facilitator.

The APRON system ensures that the data is as current as possible. A designated administrator presets the Download, the Extraction and the Structuring Modules to execute at preferred intervals. The administrator must execute the initial download, extraction and structuring cycle before the end-users present their first request for information to avoid delays in response time.

The URL for the target Web site has been manually coded within the Download Module. It is assumed that many specifiers/procurers of construction products have preferred suppliers with whom they have established a working business relationship over time. Should there be a desire to access information from other Web sites, an administrator can easily broaden the relevant agent's to roaming habits on the Internet.

The Download Module specifically handles product data displayed in the Adobe Acrobatic PDF format. It was established through an informal review of various Web sites that the PDF format is the most widely used format for displaying information on the Web by construction product manufacturers. The scope of APRON can be easily broadened to accommodate many other formats such as HTML.

The Extraction Module is an optional module that was implemented in the implemented prototype because the target PDF files on the Philips Lighting Company's Web site were not in ASCII format. The Extraction Module educes all the text from a source that is not in ASCII format. This is then fed into the Structuring Module, which utilises the capabilities of the X-Fetch Suite to structure the extracted information into XML format.

The Extraction Module parses the source file based on the specification file shown in Box 7.1. This file uses an XML template, which qualifies the input text to be considered. When the program is executed, the identified parameters are used to excerpt the various attributes of light bulbs. The Extraction Module uses the Data Extraction Language (DEL). The first 10 lines in Box 7.1 extract the column titles. These are registered and stored as element names or tags to be used in the structured XML document.

The code listed in Box 7.2 extracts all the relevant product data from the source. The Structuring Module generates an XML file using the extracted column titles as element tags and the product details as values. An excerpt of the resulting XML file is depicted in Box 7.3. Evoking the agent responsible for the Structuring Module a second (and a subsequent) time, updates the original XML output file only if the product manufacturer has made modifications on the source file.

It is important to note that the resulting XML file would have been created regardless of the source data format. The DEL reasons on different file formats in the same way. However, for this code to work on a different source file, in a different format and from a different manufacturer, it is imperative to have consensus on the key attributes that define a selected construction product. Such consensus can be arrived at through the existing standardisation efforts. However, the existing standards are still developing and are not yet robust enough (Froese, 2003).

Once product data has been presented in a structured format, the APRON system provides a component that allows administrators to create permanent data stores of the information. The information that is downloaded from manufacturers' Web sites is only temporarily stored on the server. Designated administrators have authorisation to create and update databases for the structured data. MS Access was used for APRON's Database Module. This component can be easily adjusted to accommodate other types of databases.

Box 7.1 Extracting column titles

```
<mark action="go"/>
<extract type="over" expression="Commercial product
name"/>
<extract type="re_upto" expression="[a-zA-Z]"/>
<extract type="upto" expression=" "
register="title_1"/>
<extract type="re_upto" expression="[a-zA-Z]"/>
<extract type="upto" expression=" "
register="title_2"/>
<extract type="re_upto" expression="[a-zA-Z]"/>
<extract type="upto" expression=" "
register="title_3"/>
<extract type="re_upto" expression="[a-zA-Z]"/>
<extract type="upto" expression=" "
register="title_4"/>
<extract type="re_upto" expression="[a-zA-Z]"/>
<extract type="upto" expression=" "
register="title_5"/>

<convert conversionset="toXML">
   <register name="title_1"/>
   <register name="title_2"/>
   <register name="title_3"/>
   <register name="title_4"/>
   <register name="title_5"/>
</convert>
<repeat>
```

Box 7.2 Extracting product attribute values

```
<extract type="re_upto" expression="\r?\n[a-zA-Z]"/>
<extract type="re_over" expression="\r?\n"/>
<extract type="upto" expression=" " register5"ti_1"/>
<extract type="re_upto" expression="[0-9a-zA-Z]"/>
<extract type="upto" expression=" " register5"ti_2"/>
<extract type="re_upto" expression="[0-9a-zA-Z]"/>
<extract type="upto" expression=" " register5"ti_3"/>
<extract type="re_upto" expression="[0-9a-zA-Z]"/>
<extract type="upto" expression=" " register5"ti_4"/>
<extract type="re_upto" expression="[0-9a-zA-Z]"/>
<extract type="re_upto" expression="\r?\n"
register="ti_5"/>
```

Box 7.3 The output XML file

```
<product>
    <Type>CAPSULEline Pro</Type>
    <Wattage>10W</Wattage>
    <Cap_base>G4</Cap_base>
    <Voltage>12V</Voltage>
    <Ordering>409706 50</Ordering>
</product>
<product>
    <Type>CAPSULEline Pro</Type>
    <Wattage>20W</Wattage>
    <Cap_base>G4</Cap_base>
    <Voltage>12V</Voltage>
    <Ordering>402103 50</Ordering>
</product>
<product>
    <Type>CAPSULEline Pro</Type>
    <Wattage>20W</Wattage>
    <Cap_base>GY6.35</Cap_base>
    <Voltage>12V</Voltage>
    <Ordering>402196 50</Ordering>
</product>
```

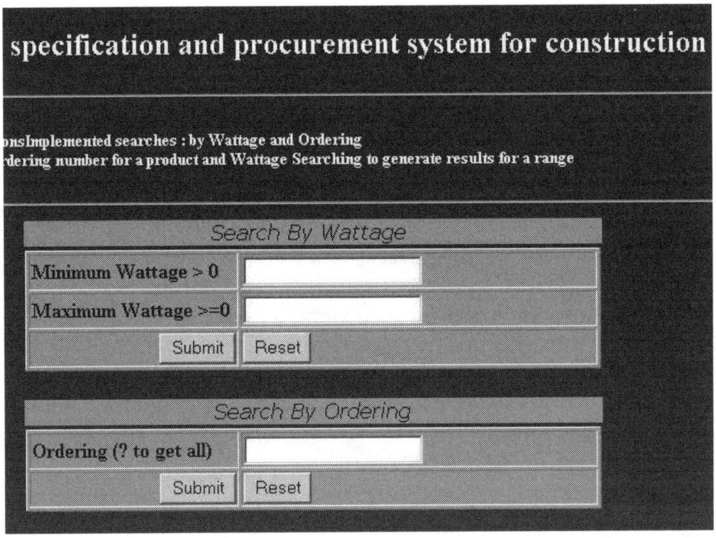

Figure 7.4 Web-based search form.

7.4.2.3 Accessing the structured data

An important goal of APRON is to simplify tasks for end-users. Such parties can interact with APRON in two levels: (1) using a Web-based interface (2) using a client application. The Web-based interface comprises a Search Module and a Procurement Module. The former presents them with a search form (depicted in Figure 7.4) that can be used to retrieve product specifications. This module maintains real-time access with the database holding structured product information.

Box 7.4 Agent dialogue in the procurement module

```
SELLER: Inform BUYER that I own: Philips Lighting
product-0:
bulbShopOntology.Product@1c6d11a product-1:
bulbShopOntology.Product@1ca209e product-2:
bulbShopOntology.Product@123a389

BUYER: Information received from SELLER. Message is

(INFORM
:sender (agent-identifier :name agent@pc2000-
cveao:1099/JADE)
:receiver (set (agent-identifier :name agent@pc2000-
cveao:1099/JADE))
:content "((OWNS (agent-identifier :name agent@pc2000-
cveao:1099/JADE)
(bulb :serialID 123456 :name \"Philips Lighting\"
:products
(sequence (PRODUCT :type \"CAPSULEline pro\"
                   :wattage \"10W\"
                   :cap_base G4
                   :ordering_no \"345689 90\")
          (PRODUCT :type \"CAPSULEline pro\"
                   :wattage \"20W\"
                   :cap_base G4
                   :ordering_no \"345689 91\")
          (PRODUCT :type \"CAPSULEline pro\"
                   :wattage \"30W\"
                   :cap_base G4
                   :ordering_no \"345689 92\")))))"
:language FIPA-SL
:ontology Bulb-shop-ontology)

Owner is: ( agent-identifier: name agent@pc2000-
cveao:1099/JADE)
```

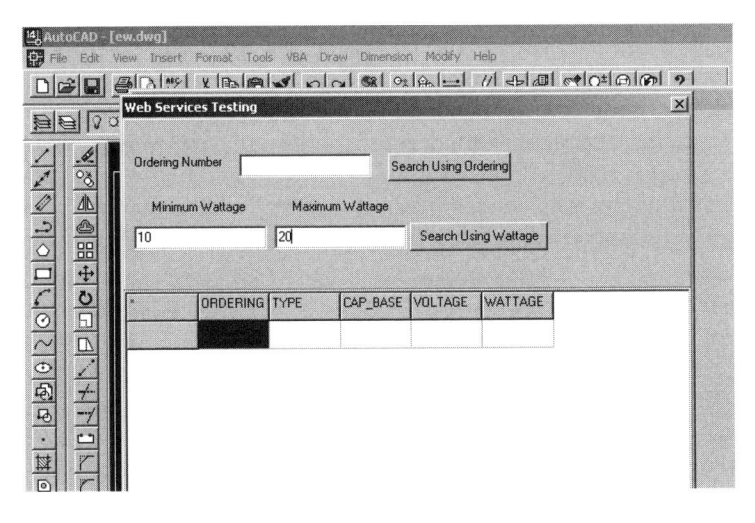

Figure 7.5 AutoCAD-based search form.

An additional assumption made in the APRON prototype is that once an end-user has specified product, he/she would like to transact with the seller of the product to finalise the deal as soon as possible. The requisite tasks are delegated to software agents. The interactions in the pilot scenario commence with the seller's agent informing the buyer's that he owns the specified product. A transaction then ensues and if the terms are agreeable to both parties, ownership of the products is passed from the seller to the buyer. An excerpt of the communications between these two agents is shown in Box 7.4.

As pointed out in Sections 7.1.2 and 7.1.3, the APRON project was an extension of work that had been done in creating 'intelligent' components in AutoCAD and also the GENIAL project. AutoCAD was therefore selected as the end-user application in the pilot scenario. All the components necessary to support end-user interactions run from the server. The client application has been packaged into a Visual Basic Application (VBA) that interfaces with the APRON Web Service using a dynamic link library. The VBA is only new component an end-user requires to exploit APRON. Loading and running the VBA from AutoCAD presents the end-user with a form shown in Figure 7.5 that can be used to search for detailed product specifications using known attribute definitions.

7.5 The evaluation of APRON

This section reports on the approach adopted in the evaluation of the APRON prototype. Two types of evaluation – formative and summative evaluation – were undertaken.

7.5.1 Formative evaluation

Throughout the product development, there were a number of 'in-house' evaluation exercises, which resulted in a number of modifications depicted in the spiral model shown in Figure 7.6. This strategy of reconfiguring and improving the prototype through several solution cycles was an adaptation of the method used by Böttcher and Suhl (1999). In the basic scenario, a general problem was defined: integrating the Philips Lighting Catalogues into a Web Portal. This was developed into a basic APRON model

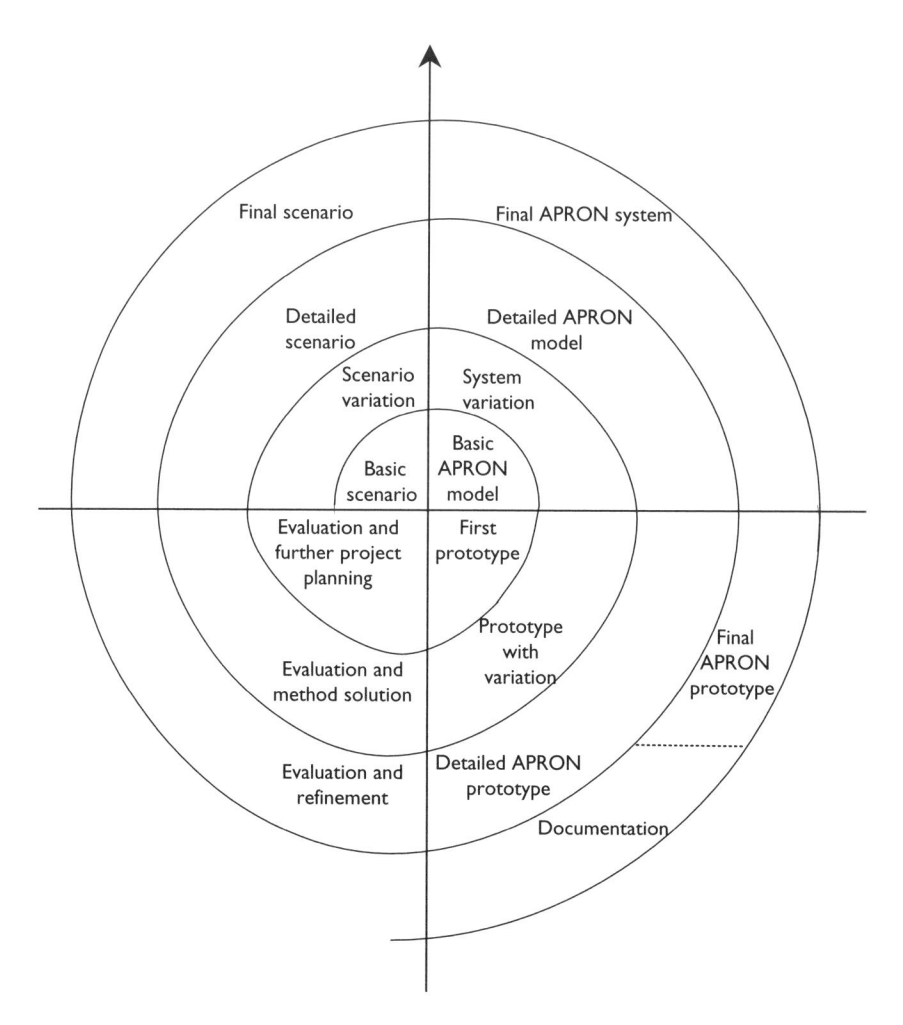

Figure 7.6 The spiral model.
Source: Adapted from Böttcher and Suhl, 1999.

comprising of a number of discrete tasks as given in the following list:

- accepting requests from the end-users;
- reasoning on the information provided by the manufacturers; and
- displaying responses to the end-users.

It was then established that construction product manufacturers have adopted a semi-structured way of presenting information in their Web site. It therefore became necessary to include a number of modules to download the files, extract the information and transform it into a structured format. These functionalities were packaged into an application that constituted the first prototype. Through subsequent evaluation and project planning, scenario variation and solution variation, some manual aspects of the first prototype were automated. This resulted in a prototype exhibiting the core functionalities. The next step involved detailing a scenario for hosting the prototype on the Web. A detailed APRON model, in which the prototype was packaged as a Web Service, was developed. This was extended through further evaluation and refinement to include a client application for the end-users.

7.5.2 Summative evaluation: focus groups

Two focus group sessions were held as part of a summative evaluation. The primary purpose of the group evaluations was to determine the adequacy of the APRON prototype through soliciting the advice of parties external to the project on

- what they liked about the system;
- what they thought would work; and
- what they thought would not work in a commercial scenario.

A total of 15 people participated in the small group evaluation sessions. The first group comprised 12 researchers affiliated to Loughborough University. The second group comprised experts in construction industry-specific e-commerce and knowledge management. Prior to each group session, the participants were given a brief brochure describing the APRON prototype. This was followed by a demonstration of the software. The participants were then given time to complete a brief questionnaire and the last phase of each session was a group discussion.

7.5.3 Group A

7.5.3.1 Results for analysis of questionnaire

The objective of the questionnaire completed by participants in Group A was to test the functional performance of the APRON prototype. Table 7.3

Table 7.3 Results from questionnaire

Construct	Mean	Max score	Count
Familiarity with Internet-based systems	3.36	5	12
Initialisation and navigation	3.77	5	12
Scope and functionality	3.78	5	12

represents the mean scores across the participants for the constructs of familiarity of the participants with Internet-based systems, and their views on initialisation and navigation within APRON and the scope and functionality of the implemented prototype. The various questions under these constructs were rated in questionnaires using a scale 1 to 5 for low to high rating respectively. On average, the participants were fairly knowledgeable on Internet-based systems, since most of them are researchers in construction information technology and related areas. The mean ratings for the two main constructs (1) initialisation and navigation, and (2) the scope and functionality of the implemented prototype were 3.77 and 3.78 respectively out of 5. The functional performance of the implemented APRON prototype in terms of the selected constructs was, therefore, very good.

7.5.3.2 Results from group discussion

The opinions documented in this section emerged from open-ended questions, comments made during a demonstration phase and remarks made during group discussion. These opinions can generally be broken into three categories:

- things the participants liked about the APRON prototype;
- things they did not like about the prototype; and
- suggestions for additions or extensions.

The results from the questionnaire are consistent with the views that were expressed during the discussion in the focus group. The main strengths of the system identified by participants in Group A are outlined here.

- the system is flexible – it provides the end-users with the option of either searching from the Web or to load a client application into their system;
- APRON filters through the vast amount of information provided and results in the retrieval of just relevant data;
- APRON can be used to retrieve cost information from various vendors for existing product schedules; and
- the APRON approach of having one data store on a server rather than having many individual designers creating local data stores.

The main suggestion for further work was the integration of the APRON prototype with the use of 'blocks' in AutoCAD.

7.5.4 Group B

7.5.4.1 Evaluation results

The second group review was with a team of three experts from Taylor Woodrow Plc. Taylor Woodrow was an appropriate choice for expert review of the system because of two primary reasons:

- the connection between APRON, GENIAL and eConstruct as cited in Section 7.3;
- Taylor Woodrow had developed in-house competency in various facets of construction-specific e-business.

7.5.4.2 Evaluation results

The industrial experts did a comparative evaluation between the APRON approach and the approach adopted in the GENIAL and the eConstruct projects. One participant pointed out that APRON utilised 'sophisticated' techniques to achieve discrete product search using attributes. The agent approach had resulted in some better solutions, for example, the extraction of data from legacy applications within APRON, was more effective and efficient than their approach in certain aspects. In their approach, they had written a sub-routine that read legacy data and populated an MS Excel worksheet. This was then used to create XML pages. The Taylor Woodrow approach was tedious, and took a very long time.

Being industry-based, the participants in the expert review were very keen to identify very specific commercial application scenarios for the APRON prototype. From their experiences in GENIAL and eConstruct, they had established that construction product manufacturers generally 'resisted' the deployment of tools that enabled end-users to compare their products with those offered by their competitors. Manufacturers of construction products had been known to deliberately refuse to adopt strategies that would support the widespread use of tools such as APRON. It was therefore appropriate to adopt an approach that did not depend on direct involvement of manufacturers.

The group further pointed out that successful implementation of a commercial application from the APRON project would be impeded by the trends in adoption of industry standards for the definition of product attributes. The group had established that:

- it would take significant amount of time to fully develop the standards;
- it would be very difficult to create a critical mass of standard users even when the standards eventually matured; and

- many organisations and many Architecture, Engineering and Construction (AEC)-specific information portals already had their product classification systems that they would not easily abandon in favour of emerging industrial standards such as the IFCs and the bcXML.

Such challenges do not necessarily mean that the industry is not ready for systems such as the APRON prototype system. Elements of the work done by Taylor Woodrow had, for example, been adopted and developed into an in-house solution by one European company. This suggests that there are Use Cases in which the APRON solutions can be applied in the industry despite problems related to the use of standards. One of the identified Use Cases was government contracts. Being a major player in the generation of building stock, the government is in an ideal position to enforce their business partners to define their products and services using some agreed standards. This would then form an ideal test-bed for implementing APRON.

The group also identified the potential usefulness of the APRON system in building maintenance and facilities management. In such scenarios, although the end-user have detailed product schedules, they could still use the APRON system to, for example, perform price checks and price comparisons across a number of product suppliers. It can also be used to place orders for specialities such as wallpaper. Other commercial scenarios identified by the group included adopting the APRON system in a company's procurement strategy:

- for ordering construction products by job;
- to transact with preferred product suppliers; and
- in the development of large housing schemes.

In summary, the group established that the approach used in the APRON prototype was efficient and effective in achieving its set goals. The APRON prototype would face the same challenge as GENIAL and eConstruct: a large-scale commercial application can only be developed when industry standards mature and gain popularity. However, there are specific scenarios in which the tool can be deployed as a commercial application.

7.6 Discussion and conclusion

The APRON prototype system is essentially a context-specific information retrieval framework. This chapter demonstrated its use to provide services that enhance the use of the Internet to support the specification and procurement of construction products. The use of the APRON system results in filtered responses to an end-user's queries. It increases efficiency through the automation of routine tasks such as product data and information search. From the evaluation that was carried out, it is clear that APRON is a useful and helpful tool. It finds information that conventional search engines would not. An important aspect of the system is its extensibility.

The APRON project resulted in a prototype that was rigorously tested and evaluated. It was, however, not implemented in a commercial scenario. The subsequent paragraphs discuss the connection between the research results in the APRON context and the intractable problem of returns on investments in potential efforts to develop APRON into a commercial tool.

Although findings of the APRON were neither arbitrary nor abstract, developing the prototype into an application in a commercial scenario will not be an easy task. Latest surveys indicate that the adoption of agent technology in e-business has been very slow (Shehory, 2003). One of the primary factors hindering rapid adoption of agent technology is the current global economic trends towards recession. Companies have generally postponed their major IT investments and become more risk averse and more savvy in their IT investments. Web-based technologies in particular have been adversely affected. Investors are presently disillusioned – the IT market has witnessed the end of the era of Internet euphoria, peaking between 1999 and 2001, and characterised by 'over-spending' (Benjamins et al., 2003). This problem will only be significantly reduced with the passage of time. In the long run, the economy cycle will tend towards prosperity and IT investment will increase once more.

The slow uptake of agent technology in commercial settings can also be attributed to 'failures' among researchers in the agent community. Leading researchers in agent technology have now acknowledged there was excessive and somewhat misleading enthusiasm among agent technology pundits in the late 1990s (Luck et al., 2003). This resulted in unrealistic expectations and disappointments when the promised benefits were only modest. Researchers in this realm must impoverish resolutely any notions in the commercial context that software agents hold the ultimate answer to the problem of 'digital anarchy'. Researchers had earlier cautioned that the construction IT research community should guard against the temptation to see agent paradigm as the panacea to all construction problems. For example, Ugwu et al. (2000b) pointed out that it is necessary to avoid the temptation to over-sell agent technology from commercial software vendors, to avoid the industry fallout from similar expectations on the potentials of expert systems technology in the 1980s. Thus although agent technology contributes to the solution, it must be applied in conjunction with other building blocks such as RDFs, ontologies and XML. Agents must also be integrated with technologies such as Web Services that would provide the necessary infrastructure for deploying large-scale applications.

The problem related to unrealistic expectations and resulting disappointments may have been compounded by the time lag between research activities and commercial implementation. In general, research prototypes in agent technology do not immediately result in commercial applications. It has in fact been established that current commercial agent implementations are generally based on publications and prototypes from 3 to 5 years back (Luck *et al.*, 2003). Prospective investors in agent technology must therefore be guided into developing sombre expectations. This can be achieved through educational workshops and courses in agent technology. Agentcities and AgentLink have championed awareness creation in the European context. The two groups have promoted awareness in agent technology in universities and companies across Europe, through, for example, awarding deployment grants for commercial agent-related projects.

Construction industry-specific projects are conspicuously missing from the ventures supported by the two organisations. Such initiatives provide an opportunity to form inter-disciplinary teams that would allow professionals in the construction industry to exploit the skills and competencies that have already been established by experts in agent technology. Top-level managers in construction sector should be encouraged to participate. When such key players appreciate what software agents are and what they are capable of, they would quickly identify the existing business opportunity for agents in their mainstream business processes.

Several important issues that would hamper commercial deployment of agent-based systems in an e-business setting emerged in the APRON project. Many of them are consistent with the general facts discussed earlier. These are outlined here:

- *Issue 1* – development platforms are not stable enough for operational environments;
  - agent technology is still maturing and there is not a single development environment that could have been used on its own to create the designed APRON prototype;
- *Issue 2* – the value in agents can be best described in terms of business processes: software agents are just a subset of the new evolving technologies that support e-business and other micro-level process automation;
  - software agents on their own would not make a good business case for a commercial application: the approach adopted in implementing the APRON system, therefore, integrated agent technology with the existing information systems and other micro-level process automation;
- *Issue 3* – existing industry standards such as IFCs and bcXML are yet to be sufficiently developed to support e-business;
  - the construction industry is, in general terms, not quite ready for large-scale adoption of software agents: however, specific commercial scenarios with a restricted end-user group can be targeted;

- *Issue 4* – the key players in the construction industry generally lack awareness on the potential of agent-based systems;
  - a significant amount of literature available focuses on technical issues such as agent architecture, learning and negotiation protocols – early adopters of agent technology should provide benchmarking and case studies of their experiences;
- *Issue 5* – as a result of Issue 4, 'cultural' resistance in terms of organisations protecting investment in hardware, software and skills can be expected;
  - the APRON prototype has defined a model with agents in the background with provision for gradual migration into a new technology.

Sections 7.1.1 and 7.1.2 established that given the Web's semi-structured nature and enormous size, a whole new industry has emerged based on the need for tools capable of searching the Web and executing business transactions in a more 'intelligent' way. However, despite the great potential, much work needs to be done to make agent technology more acceptable for end-users in a commercial context. A recent construction industry study in Hong Kong also identified related key enablers for IT adoption in the industry. These include: clear and unambiguous understanding of end-user requirements by the development. Thus there is need for end-user participation during the development process of any IT project including agent-based system (Ugwu *et al.*, 2003a,b). This section has presented a view on the future outlook for 'agents in business'. It has also identified the significant hurdles in the adoption of agent technology, particularly in the selected domain. But such hurdles only limit the scope for applying agents in a commercial scenario. As established in the evaluation there are specific scenarios that can benefit from the use of APRON.

In general, full automation through agents in construction industry-specific e-business cannot be attained at present. There is however a 'latent potential' (Radjou, 2003) for agent-based technology in the selected domain. It is particularly important to understand the contextual business process issues, collaborative working practices (workflows), knowledge and organisational models in deploying agent systems for decision-support in the construction sector (Ugwu *et al.*, 2003b).

References

Aylett, R., Brazier, F., Jennings, N., Luck, M., Nwana, H. and Preist, C. (1997) *Agents Systems and Applications*, Report from the panel discussion at the Second UK Workshop on Foundations of Multi-agent Systems (FoMAS' 97).

Baeza-Yates, R. A. (1998) 'Searching the World Wide Web: challenges and partial', in H. Coelho (ed.) *Progress in Artificial Intelligence – IBERAMIA 98*, in the Sixth Ibero-American Conference on AI, Lecture Notes in Computer Science, 1484 Springer, pp. 39–51.

Balasubramanian, S., Maturana, F. and Norrie, D. H. (1996) 'Multi-agent planning and coordination for concurrent engineering functionality', *International Journal of Cooperative Information Systems*, 5(2–3): 153–79 (Special Issue on Agent-based Information Management).

Bayardo, R., Bohrer, W., Brice, R., Cichocki, A., Fowler, G., Helal, A., Kashyap, V., Ksiezyk, T., Martin, G., Nodine, M., Rashid, M., Rusinkiewicz, M., Shea, R., Unnikrishnan, C., Unruh, A. and Woelk, D. (1996) *Semantic Integration of Information in Open and Dynamic Environments*, MCC Technical Report, MCC-INSL-088-96.

Benjamins, V. R., Contreras, J. and Prieto, J. A. (2003) 'Agents and the semantic web', *AgentLink Newsletter*, An AgentLink Publication.

Berners-Lee, T. (1989) *Information Management: A Proposal*, CERN, Geneva, Switzerland.

Berners-Lee, T., Hendler, J. and Lassila, O. (2001) 'The semantic web', *Scientific American*, May 2001 Issue, [Online] Available from <http://www.sciam.com> [Accessed on 15 January 2002].

Blake, M. B. (2002) 'AAAI-2002 workshops: agent-based technologies for B2B electronic commerce', *AI Magazine*, AAAI Press, 23(4): 113–14.

Blake, M. B. and Gini, M. (2002) 'Guest editorial: agent-based approaches to B2B electronic commerce', *International Journal on electronic commerce*, 7(1): 5–6.

Böttcher, S. and Suhl, L. (1999) 'Integrated modeling, development, and evaluation of an agent-based production planning system for a network of mechanical engineering companies', Research Summary, [Online] Available from <http://www.wirtschaft.tu-ilmenau.de/wi/wi2/SPP-Agenten/beitraege.html> [Accessed on 4 May 2003].

Brustoloni, J. C. (1991) *Autonomous Agents: Characterization and Requirements*, Carnegie Mellon Technical Report CMU-CS-91-204, Carnegie Mellon University, Pittsburgh.

Deloitte Research (2002) *Directions in Collaborative Commerce, Managing the Extended Enterprise, A Deloitte Research Collaborative Commerce Viewpoint*, Deloitte Research Report.

Ferber, J. (1999) *Multi-agent Systems: An Introduction to Distributed Artificial Intelligence*, Addison-Wesley, UK.

Fingar, P. (1998) *A CEO's Guide to E-commerce Using Object-oriented Intelligent Agent Technology*, CommerceNet Research Report, No. 98-19, USA.

FIPA Architecture Board (2001) *FIPA Agent Software Integration Specification*, 2001 [Online] Available from <http://www.fipa.org> [Accessed on 30 September 2002].

Franklin, S. and Graesser, A. (1996) 'Is it an agent, or just a program? A taxonomy for autonomous agents', in J. P. Müller, M. J. Wooldridge and N. R. Jennings (eds) *Intelligent Agents III, Agent Theories, Architectures and Languages (ATAL)*, Muller, Springer-Verlag, Berlin, Germany, pp. 21–35.

Froese, T. (2003) 'Future directions for IFC-based interoperability', *ITcon*, 8: 231–46 (Special Issue IFC on Product models for the AEC arena).

Hammer, J., Garcia-Molina, H., Cho, J., Aranha, R. and Crespo, A. (1997) *SIGMOD Record*, 26(4): 24–31.

Hendler, J., Berners-Lee, T. and Miller, E. (2002) 'Integrating applications on the semantic web', *Journal of the Institute of Electrical Engineers of Japan*, 122(10): 676–80.

Jennings, N. R. and Wooldridge, M. (1998) 'Applications of intelligent agents', in N. R. Jennings and M. Wooldridge (eds) *Agent Technology: Foundations, Applications, and Markets*, Springer-Verlag, Berlin, pp. 3–28.

Jennings, N. R., Sycara, K. and Wooldridge, M. (1998) 'A roadmap of agent research and development', *Autonomous Agents and Multi-agent Systems*, 1(1): 7–38.

Kashyap, N. (1997) *An Expert Systems Application in Electromagnetic Compatibility*, Master's Thesis, University of Missouri-Rolla, MO.

KPMG Consulting (2000) 'Knowledge management research report', Available from <http://www.kmadvantage.com/docs/km_articles/KPMG_KM_Research_Report_2000.pdf> [Accessed on 3 July 2003].

Lieberman, H. (1997) 'Autonomous interface agents', in S. Pemberton (ed.) *Proceedings of the ACM Conference on Computers and Human Interface*, ACM/Addison-Wesley, pp. 67–74.

Luck, M., McBurney, P. and Preist, C. (2003) *Agent Technology: Enabling Next Generation Computing*, A Roadmap for Agent-based Computing, Version 1.0, AgentLink Report.

Madhusudan, T. (2001) 'Enterprise application integration – an agent-based approach', *IJCAI Workshop on AI and Manufacturing*, Seattle, WA, August.

Maes, P. (1997) 'Agents that reduce work and information overload', in J. M. Bradshaw (ed.) *Software Agents*, AAAI Press/MIT Press, Menlo Park, CA.

Mahapatra, T. and Mishra, S. (2000) *Oracle Parallel Processing*, 1st edn, O'Reilly, UK, chapter 1.

Nwana, H. S. (1996) 'Software agents: an overview', *The Knowledge Engineering Review*, 11(3): 205–44.

O'Brien, P. (2003) 'Agent technology: experiences from the telecom industry', Presented at the Agent Technology Conference: Agents for Commercial Applications, Barcelona, Spain http://www.agentlink.org/agents-barcelona/presentations/index.html [Accessed on 23 December 2003].

Paranuk, H. V. D. (1996) 'Applications of distributed artificial intelligence in industry', in G. M. P. O'Hare and N. R. Jennings (eds) *Foundations of Distributed Artificial Intelligence*, Wiley/Inter-Science, New York.

Petrie, C. (1997) *ProcessLink Coordination of Distributed Engineering*, Technical Report, Centre for Design Research, Stanford University, Stanford, CA.

Radeke, E. (ed.) (1999) *Final Report, GENIAL Global Engineering Networking Intelligent Access Libraries*.

Radjou, N. (2003) Software Agents in Business, *AgentLink Newsletter*, Issue 13, An AgentLink Publication.

Samtani, G. (2003) *B2B Integration*, A Practical Guide to Collaborative E-Commerce, World Scientific Pub Co Inc, USA.

Shehory, O. (2003) 'Agent-based systems: do they provide a competitive advantage?', *AgentLink Newsletter*, An AgentLink Publication.

Stephens, J., Dimtas, N., Lima, C., Steinmann, R., Woestenenk, K., Bohms, M. and Tolman, F. (2002) *ECommerce and eBusiness in the Building and Construction Industry*, Preparing for the Next Generation Internet, Final Report.

Thiels, L., Bragoni, R. and Shamsi, T. A. (2002) *Efficient Bidding and Procurement in the Tile Industry – Practical Trading Tools and Broker Services for the Exchange of Product Characteristics*, e-bip Final Report.

Ugwu, O. O. and Anumba, C. J. (2000) 'An e-commerce framework for integrating product design, manufacture, and procurement in construction', *Second International Conference on Decision Making in Civil and Urban Engineering*, Lyon, France, 20–22 November 2000, pp. 933–44.

Ugwu, O. O., Kamara, J. M. and Anumba, C. J. (1999) *Prototype Application for Matching Customer Requirements to Available Products and Services*, GENIAL Project Report No: G_D16_V00_99_TW.DOC (for Taylor Woodrow Plc, UK), Loughborough University, November 1999.

Ugwu, O. O., Anumba, C. J., Newnham, L. and Thorpe, A. (2000a) 'Agent-oriented collaborative design of industrial buildings', in R. Fruchter, F. Pena-Mora and W. M. K. Roddis (eds) *Proceedings of the Eighth International Conference of Computing in Civil & Building Engineering*, ICCCBE-VIII, pp. 333–40.

Ugwu, O. O, Anumba, C. J., Newnham, L. and Thorpe, A. (2000b) 'The application of DAI in the construction industry', in G. Gudnason (ed.) *Proceedings of International Conference on Construction Information Technology – CIT2000*, Icelandic Building Research Institue, Reykjavik, Iceland, 28–30 June 2000, 2, pp. 950–70.

Ugwu, O. O., Kamara, J. M., Anumba, C. J. and Leonard, D. (2002) 'Electronic procurement of construction products', *International Journal of Services, Technology and Management*, 3(2): 222–37.

Ugwu, O. O., Ng, S. T. and Kumaraswamy, M. M. (2003a) 'Key enablers in IT implementation – a Hong Kong construction industry perspective', in I. Flood (ed.) *Proceedings of Fourth Joint International Symposium on Information Technology in Civil Engineering*, Nashville, Tennessee USA, 15–16 November 2003.

Ugwu, O. O., Kumaraswamy M. M., Ng, T. S. and Lee, P. K. K. (2003b) 'Agent–based collaborative working in construction: understanding and modelling design knowledge, construction management practice and activities for process automation', *Hong Kong Institution of Engineers (HKIE) Transactions, Tenth, Anniversary Edition* (Special Issue on Emerging Technologies).

URL1: http://www.agentcities.org/
URL2: http://www.agentcities.org/
URL3: http://www.agentlink.org/
URL4: http://www.x-fetch.fi/

Chapter 8

Agent-based virtual marketplace for AEC-bidding

*M. Schnellenbach-Held, H. Denk and
A. Albert*

8.0 Introduction

A first step towards automating Architecture Engineering and Construction (AEC)-bidding is to make use of the Internet respectively of technologies applied in e-commerce. Within the last years several virtual marketplaces for the building industry have been established. These virtual marketplaces support bid- and (online) project management. Time-consuming tasks like locating contractors and bidding out projects are facilitated by publishing bid documents on the marketplace, receiving responses electronically, etc. Here, the documents are no longer chapter based. Furthermore, building owners and construction professionals can find qualified and competitive trade partners.

These marketplaces also offer procurement solutions, allowing contractors to source and purchase building materials, equipment and supplies. It can be pointed out that these marketplaces act as document exchanges and provide little automation compared to what is conceivable.

Software-agents could be used to facilitate and to accelerate the bidding process. For example, an owner wants to bid out a project. He sends his agent to the virtual marketplace with specifications of the project in eXtensible Markup Language (XML). On the marketplace agents from contractors and/or suppliers read the specifications, look for subcontractors if necessary and negotiate with them, calculate a price and finally file a tender. The owner's agent receives all the bids. After the final date, the owner's agent looks for the best offer, which is not always the cheapest one. It creates a list of all the offers and sends the list to the owner, who decides which company will be awarded. Afterwards, the owner's agent enters a contract with the agent of the awarded company that again enters contracts with the agents of its subcontractors and so on.

This chapter deals with software models for an agent-based virtual marketplace for AEC-bidding which has been developed at the Institute of Concrete Structures and Materials (Darmstadt University of Technology). More precisely concepts for an adequate ontology, for providing security for an agent-based marketplace in terms of the (German) Digital Signature

Act, and for satisfying the official contracting terms for the award of construction performance contracts will be presented.[1]

8.1 Scenario

The following scenario shall provide a closer insight into the sequences of a potential agent-based virtual marketplace for AEC-bidding.

In order to ensure a constant workload for its employees building company A (BA) is always interested in public invitations to tender. Therefore BA generates a software-agent which looks for new calls for bids on the virtual marketplace for AEC-bidding.

Services and building materials which cannot be provided by BA itself are bought from subcontractors. In the past, BA co-operated very well with the steel structure company (SB), and therefore BA wants to continue this co-operation in the future. Nevertheless, BA solicits bids from other steel structure companies for price comparison.

On the agent-based virtual marketplace the agent of BA (BA-Agent) quickly finds interesting invitations to tender for public works and receives the specifications for tenders from the agent of the contracting authority. It immediately starts calculating a proposal and looks for subcontractors that can provide the services that BA is not able to provide.

The BA-Agent has been configured in such a way that it does not accept proposals from subcontractor-agents at once, but starts negotiating the price using a specific strategy which has been implemented while the agent has been generated. Again these agents can look for further subcontractors and so on.

After consultation with their users the subcontractor-agents file their tenders to BA. Yet, a contract is not placed, but the proposals of the subcontractor-agents have a temporally limited validity.

In consideration of these proposals and its own prices the BA-Agent calculates a bid price and informs its user BA about the invitation to tender and about its proposal. The responsible employees of BA decide that a bid shall be filed for the project which has been announced by the contracting administration (ADM).

Due to legal regulations, all submitted proposals have to be preserved in such a way that they cannot be examined by the ADM and its agent before the time-limit set for submitting has expired.

At the final date the ADM-Agent decrypts all the bids by using the symmetric session keys which it receives exactly at the final date. It lists and informs its user about all the received proposals.

After ADM has reviewed the proposals and has come to the decision that the BA-agent respectively BA shall be awarded it informs its ADM-Agent. Subsequently the ADM-Agent sends an *accept*-message to the awarded BA-Agent and *reject*-messages to the other agents which have submitted proposals.

Because BA has calculated its proposal with involvement of subcontractors the BA-Agent informs the subcontractors-agents about the award and enters into legally binding contracts with them.

8.2 General conditions

In order to develop a virtual marketplace for AEC-bidding on which public tendering procedures can be legally performed by software agents the general conditions have to be analysed at first.

The agent-based marketplace is mainly influenced by legal regulations. Within this chapter the authors refer to the 'directive of the European parliament and of the council on the coordination of procedures for the award of public supply contracts, public service contracts and public works contracts', the German 'official contracting terms for the award of construction performance contracts' (VOB, May 2000), and the German Digital Signature Act. Nevertheless, the developed software models should meet equivalent regulations of other countries, too.

8.2.1 Different types of auctions

Generally auctions can be divided into traditional and inverse auctions. In traditional auctions goods and/or performances are offered for sale. Potential buyers are in competition and the price is rising in the course of the auction. In inverse auctions goods and/or performances are demanded. Here, the sellers are in competition and the price is decreasing in the course of the auction.

For traditional as well as for inverse auctions different types of auctions are distinguished. In the following list the four primary types are introduced (KPBM01):

- *English auction* English auctions are open-bid auctions. In traditional English auctions the bids are incremented successively. The item is sold to the highest bidder. In inverse English auctions the bids are decremented and the seller with the lowest bid is awarded.
- *Dutch auction* The traditional Dutch auction is a descending price auction and uses an open format. Bidding starts at an extremely high price and is progressively lowered until a buyer claims an item. The inverse type starts at a low price which is rising successively. The seller, who has been the first to interrupt the price-rising process, is awarded.
- *First price sealed bid auction*

    Traditional – the potential buyers send a sealed bid to the auctioneer. All bids are opened at the same time by the auctioneer and the winner is determined. The name 'first-price' derives from the fact that the award is made at the highest offer.

Inverse – the potential sellers send a sealed bid for the demand to the auctioneer. All bids are opened at the same time by the auctioneer and the winner is determined. The award is made at the lowest offer.

- *Vickrey auction (second price sealed bid auction)* The difference between the traditional first and the traditional second price sealed bid auction is that the goods or performances are awarded to the highest bidder (buyer) at a price equal to the second highest bid (or highest unsuccessful bid). Respectively, in inverse auctions the lowest seller is awarded at a price equal to the second lowest bid.

The procedures for the award of public construction performance contracts correspond mainly to the inverse first (or second) price sealed bid auction.

8.2.2 Official contracting terms

In Germany public tendering procedures are regulated by the 'official contracting terms for the award of construction performance contracts' (VOB, May 2000) in which the usage of electronic means for AEC-bidding is permitted. Public contracting authorities are bound by law to apply the public tendering procedures which are regulated in the VOB.

The VOB, May 2000 was influenced by the 'directive of the European parliament and of the council on the coordination of procedures for the award of public supply contracts, public service contracts and public works contracts'. In case that electronic means are applied the VOB refers to the German Digital Signature Act.

Although the usage of electronic means is permitted, the regulations are not sufficient yet. Especially, software agents within the AEC-bidding process were not considered at all. Therefore several regulations within the VOB are not reasonable for an agent-based AEC-bidding procedure.

Consequentially the agent-based marketplace considers the primary ideas and regulations of the VOB: Contracting authorities shall take all necessary steps to ensure compliance with the principles of (Beck-Texte, 2001):

- equality of treatment
- transparency
- non-discrimination.

Furthermore, the following aspects have been considered: the public invitations for tenders have to be announced in newspapers, journals, bulletins or via the Internet. All interested contractors can ask for the delivery of the bidding documents, in which the construction project is described and specified in detail.

The content of the proposals shall remain confidential until the time-limit for reply has expired and the bidder must have the opportunity to withdraw his proposal before the final date is reached.

8.2.3 Digital signature act and ISIS-MTT

The (German) Digital Signature Act creates framework conditions for electronic signatures. In its newest version this Digital Signature Act meets the requirements of the 'Directive 1999/93EC of the European Parliament and the Council of 13 December on a Community Framework for Electronic Signatures' (European Parliament, 2000). It makes high demands on public key infrastructures, providing the means for the signatures, that is, on signature devices, on signature software as well as on services of the certificate authority (CA).

In order to execute legally effective actions on the agent-based marketplace the requirements of the Digital Signature Act have to be observed. In Germany, the newest technical standard which translates these requirements into technical specifications is the ISIS-MTT Standard (ISMT01) (Figure 8.1).

The ISIS-MTT standard builds on the most common form of certificate repository, the X.500 directory. As for the transport of protocol information between repository and clients, ISIS-MTT restricts itself to the TCP/IP-based

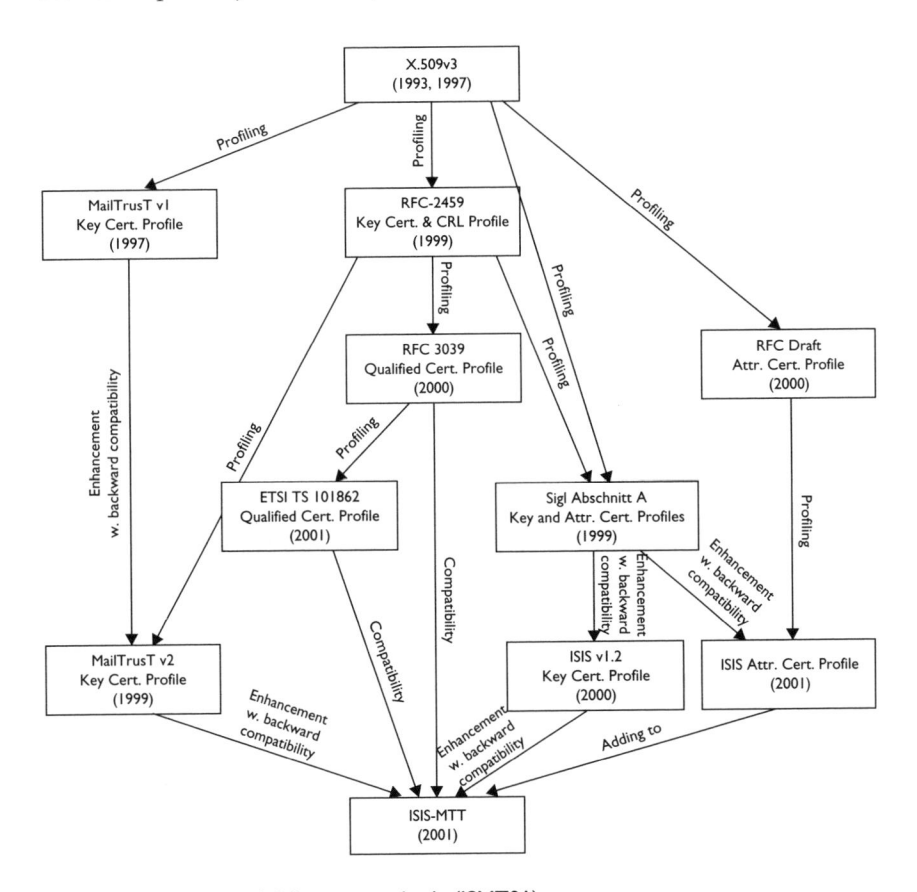

Figure 8.1 Overview of different standards (ISMT01).

protocols Lightweight Directory Access Protocol (LDAP) (for LDAP-access) and HTTP for Online Certification Status Protocol (OCSP).

Among other cryptographic algorithms ISIS-MTT requires and/or recommends the use of the following algorithms:

- SHA-1 (one-way hash function)
- RSA (public-key algorithm)
- SHA-1WithRSAEncryption (signature algorithm)
- DES and Triple-DES (content encryption algorithms).

8.3 Brief introduction to cryptography

Cryptography is the science of developing methods that allow information to be securely sent in such a way that the only person able to retrieve this information is the intended recipient (Schneier, 1996).

In traditional secret key cryptography one secret key which is used for the encryption and decryption of messages is created. The communicating parties have to possess this symmetric key to encrypt or decrypt the messages. This key system has the significant flaw that if the symmetric key is discovered or intercepted by a maleficent third party, the encrypted messages are not secret and secure anymore. Therefore, the symmetric key has to be stored secretly and it has to be transferred via a secure channel (Schneier, 1996).

In public key cryptography, a public and a private key are created simultaneously using the same algorithm by a CA. The private key is given only to the requesting party and the public key is made publicly available as part of a digital certificate in a directory that all parties can access. The private key is never shared with anyone.

The following scenario describes how encryption and message integrity is achieved in a public-key system: Alice uses her private key to decrypt text that has been encrypted with her public key by Bob (who can obtain her public key from the CA). In addition to encrypting messages, which ensures privacy, Alice can digitally sign a message by using her private key for the encryption. The receiving party (Bob) can use her public key to decrypt it, because Alice is the only person who has access to her private key, it is ensured that the message has been sent by her.

As asymmetric cryptography is much slower than symmetric cryptography, so-called hybrid systems are used to gain the advantages of both methods: Alice generates a symmetric key pair (session key) and encrypts the message by using this session key. Then the session key itself is encrypted by Bob's public key for whom the message is dedicated. Both, the encrypted message and the encrypted symmetric key are sent to Bob. Only Bob is able to decrypt the session key by using his private key. Possessing the session key Bob can decode Alice's message. Within one session Alice and Bob are using the same session key for encrypting their messages.

Modern cryptography provides confidentiality, integrity, non-repudiation and authentication. Meanwhile, transmitting information via computer networks is an integral part of several modern applications. In most of the cases, the TCP/IP protocol is used which enables bi-directional data transfer between two computers. While IP handles the actual delivery of the data, TCP keeps track of the individual units of data that a message is divided into for efficient routing through a network (e.g. Internet). In order to provide integrity, confidentiality and optional non-repudiation and authentication, the secure sockets layer (SSL) protocol can be located as an additional layer between the application and TCP (Lipp et al., 2000). For establishing an SSL connection between a client and a server the so-called 'handshake'-dialog has to be performed.

8.4 Development of an agent-based virtual marketplace

Within this research project the agent-based virtual marketplace SiReAM[2] has been developed. On SiReAM software-agents are divided into utility agents and task agents (Figure 8.2). For implementing SiReAM the Java classes of the ZEUS Agent building tool kit have been used (see Section 4.2). Therefore, the models and concepts of ZEUS have influenced the development of the agent-based virtual marketplace SiReAM.

SiReAM is operated by three different kinds of utility agents:

1. The Agent Name Server (ANS) manages the IP-addresses of the distributed software-agents. Depending on the number of agents which are registered on the marketplace one or more ANS can be created for fulfilling this task.
2. The Security Agent (SEC) provides the security relevant services for a public key infrastructure which are required by the Digital Signature Act.
3. The Facilitator (FAC) asks in regular intervals newly registered task agents for their abilities and stores these information in a database. Task agents which have to co-operate with other agents in order to achieve their goal can obtain the names of other task agents which have the required ability. The FAC represents the 'yellow pages'.

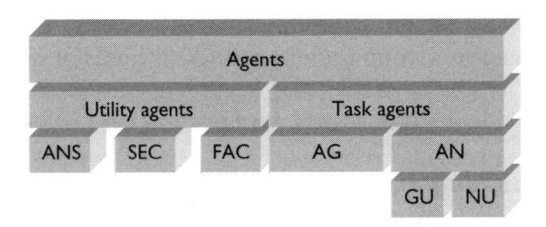

Figure 8.2 Classification of agents on SiReAM.

The task agents are subdivided into owner-agents (AG-Agents) and contractor-agents (AN-Agents).

AG-Agents publish specifications for tenders on the marketplace, receive sealed bids from AN-Agents, decrypt the bids at the final date, inform their users (owners) about the received proposals, inform the AN-Agent which is awarded and reject the proposals of AN-Agents which are not awarded, respectively.

AN-Agents again can be subdivided into agents for general contractors (GU-Agents) and agents for subcontractors (NU-Agents). After an AN-Agent has been registered on SiReAM it waits for calls for proposals (cfp) which are published by the AG-Agents. For example, if a GU-Agent receives a cfp-message it starts calculating a bid based on the specifications for tenders. Services which cannot be provided by the GU-Agent itself are purchased from NU-Agents. The GU-Agent can either negotiate the price or can directly accept the proposed price of the NU-Agent. Corresponding strategies can be implemented.

The development of the agent-based virtual marketplace for AEC-bidding (SiReAM), requires the design of new and the extension of existing software models. In the following subsection SiReAM's architecture and underlying concepts are described.

8.4.1 Architecture of SiReAM

SiReAM is an open marketplace for distributed software-agents. As already mentioned, the legal regulations affect the architecture of the marketplace to a great extent. Especially, the principle of non-discrimination demands that everybody can participate in the agent-based public tendering procedures. Therefore, SiReAM is divided into a user section and an agent section (Figure 8.3).

After an owner-agent has been successfully registered on the marketplace the digitally signed specifications in XML are uploaded to the user section.

The user section provides access to the currently published specifications. Potential contractors enter this section by using a common browser. The software for participating in the agent-based AEC-bidding on SiReAM can be downloaded there, too. The administrator of the marketplace can start, control and terminate the utility agents via a password protected secure connection.

On the server side of the agent section the three different kinds of utility agents are located which ensure a stable and secure operation of the marketplace. On the client side the task agents are operating.

8.4.2 The ZEUS agent building tool kit

ZEUS is an open source software for building multi-agent systems (MAS) (URL2). It has been developed at the British Telecommunication (BT)

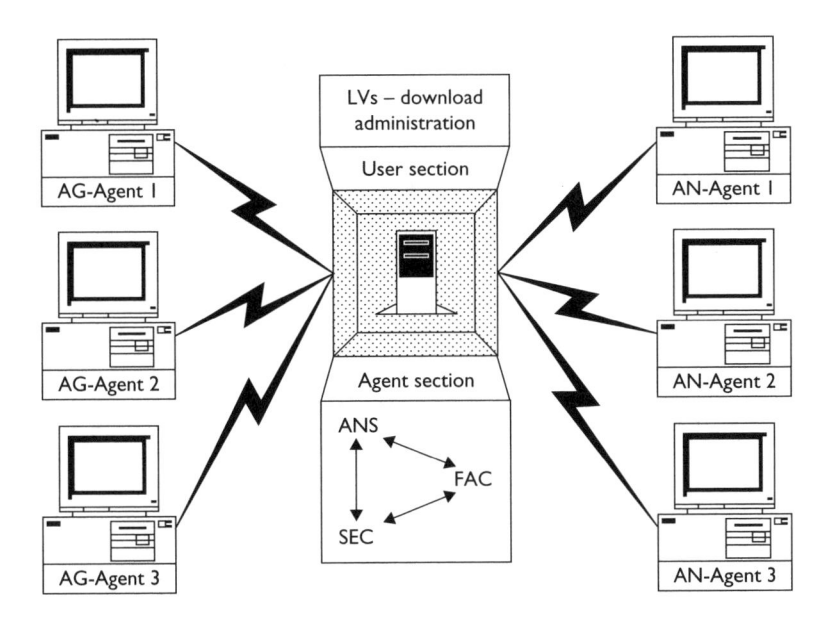

Figure 8.3 Architecture of SiReAM.

Laboratories. ZEUS is a synthesis of established agent technologies with novel solutions to provide an integrated collaborative agent building environment (Nwana *et al.*, 1999). ZEUS-agents act deliberatively, goal directed and rational. They are always truthful when dealing with other agents and they are versatile, that is, they can pursue many goals and can be engaged in a variety of tasks.

The tool kit consists of three main components:

- *Agent Component Library* – this library is a collection of Java classes that form the building blocks of individual (task-)agents.
- *Utility Agents* – ZEUS provides three different types of utility agents: an ANS, a FAC and a Visualiser. All three utility agents are constructed using basic components of the Agent Component Library (Nwana *et al.*, 1999).
- *Agent Building Software* – this component provides a visual environment for developing MAS.

ZEUS does not support any kind of security issues and its architecture is not sufficient for the requirements of an agent-based public tendering procedure for the award of construction performance contracts. But on the other hand, its open source structure enables developers to modify, evolve and extend the source code. This makes ZEUS very suitable for this research project.

8.4.3 Ontology

The word ontology describes a 'theory concerning the kinds of entities and specifically the kinds of abstract entities that are to be admitted to a language system (URL4)'. A software-agent using a particular model of these entities will only be able to perceive that part of the world that its ontology is able to represent. Only the things in its ontology exist for that agent, that is, an ontology becomes the basic level of a knowledge representation scheme for an agent (URL3).

8.4.3.1 Specifications for tenders in XML

On an agent-based virtual marketplace for AEC-bidding software agents have to deal with building materials and performances. Thus, the vocabulary of the agents must be composed by the contents of standard specifications for tenders. In order to develop a vocabulary in which the format as well as the data of the specifications are described the XML has been used to create the LVML.[3] The structure of the LVML is defined by a Document Type Definition (DTD). The LVML consists of six hierarchically ordered main elements (Bitzan, 2001):

1. LV – the name of the contracting authority, the name of the project and final date for the submission are declared by attributes of this element.
2. Title – building shell, in-house engineering...
3. Division – sitework, concrete, masonry...
4. Section – concrete formwork, concrete reinforcement...
5. Position – structural cast-in-place concrete, reinforcing steel...
6. LVText, amount, unit, unit cost, cost.

8.4.3.2 Conversion of the LVML

The ontology defines the facts that the agents have to deal with. Because the ontology used by generic ZEUS agents is not implemented in XML yet, the LVML had to be converted into the format a generic ZEUS agent can understand.

For the conversion it also had to be considered that the structure of the ontology is of crucial importance for the definition of the *tasks* which are specified by the *facts* and which the agents have to conduct in order to achieve their goal.

Three different fact types have been defined. Fact-Type 1 describes the whole project. It obtains its name from the attribute LV-Type of the LV-element. The attributes of Fact-Type 1 are derived from all *Positions* which are listed in the LVML with all their descriptions and costs. This Fact-Type is used to specify the effects of the owner-task (see Figure 8.4 AG-Task).

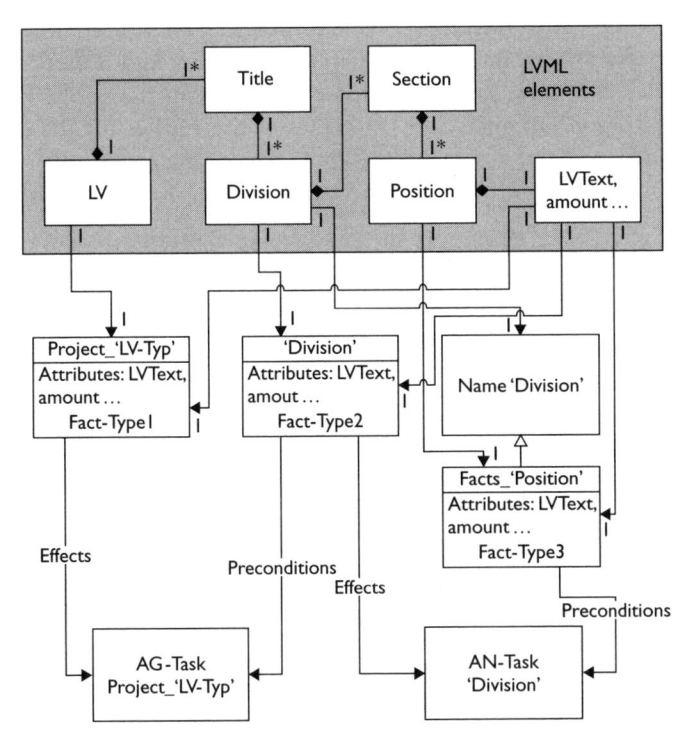

Figure 8.4 Using data from the LVML to create facts and tasks.

For each *Division* a fact of the Fact-Type 2 is created. Its attributes are composed by the elements LVText, amount, etc. which belong to the corresponding *Division*. The Fact-Types 2 define the preconditions of the owner-task as well as the task effects of the contractor-tasks (see Figure 8.4 AN-Task). These facts act as an interface for the negotiation between the owner-agent and the contractor-agents. The third Fact-Type specifies the preconditions of the contractor-tasks. Its structure is depicted in Figure 8.4.

8.4.4 Public-key infrastructure of SiReAM

In general, a public-key infrastructure (PKI) enables users of a basically unsecured public network such as the Internet to securely and privately exchange data. This exchange is conducted through the use of an asymmetric key pair (private and public key) that is obtained and shared through a trusted authority. The PKI provides for a digital certificate and directory services. The digital certificate is used to identify an individual or an organisation. Directory services are needed to store and, when necessary, revoke the certificates (Denk and Schnellenbach-Held, 2001).

Usually a PKI consists of:

- a CA that issues and verifies digital certificates and a certificate that includes the public key and/or information about the public key;
- a registration authority (RA) that acts as the verifier for the certificate authority before a digital certificate is issued to a requestor;
- one or more directories where the certificates with their public keys are held;
- a certificate management system.

In a PKI a so-called chain of trust has been built. At the beginning, an accredited CA is located which confirms the trustworthiness of the next entity within the chain. This entity again can certify another entity and so on. In order to keep the chain of trust relatively short, some entities receive end-user certificates, that is, they are not allowed to certify other entities.

Employees or associates of any company only receive end-user certificates. Hence, the task agents of SiReAM cannot obtain an own certificate according to the Digital Signature Act, that is, a different way has to be chosen for authenticating the agents.

Therefore, on SiReAM agents obtain a self-generated certificate which is digitally signed by the end user. The agent's self-generated certificate is sufficient for establishing SSL connections and the agent is authenticated due to the signature of the end-user. In order to distinguish the self-generated certificate from the accredited end-user certificate, in the following the first one is named agent-certificate and the second one user-certificate.

The user-certificate is not only used for authenticating the agent, but also to enter contracts: before any contract is made legal and binding by an agent, the user has to grant its permit. Then the contract is signed with the private key which belongs to the user-certificate (Denk and Schnellenbach-Held, 2002).

8.4.4.1 Task-agents – adjustment of the architecture

The architecture of a task-agent is based on the concepts of a generic ZEUS-agent. Extensions of the architecture had to be performed in order to integrate task-agents into the PKI. The components *Security Module*, *Keystore* and *Truststore* have been added (Figure 8.5).

The *Security Module* handles all security relevant tasks like establishing SSL-connections, encrypting and decrypting proposals by using a symmetric session key and transmitting this session key to the contracting authority at the final date. Also the agent-certificates of task-agents which have been classified as trustworthy by the SEC are managed by the *Security Module*. Locating the Security Module in front of the component *Mailbox*

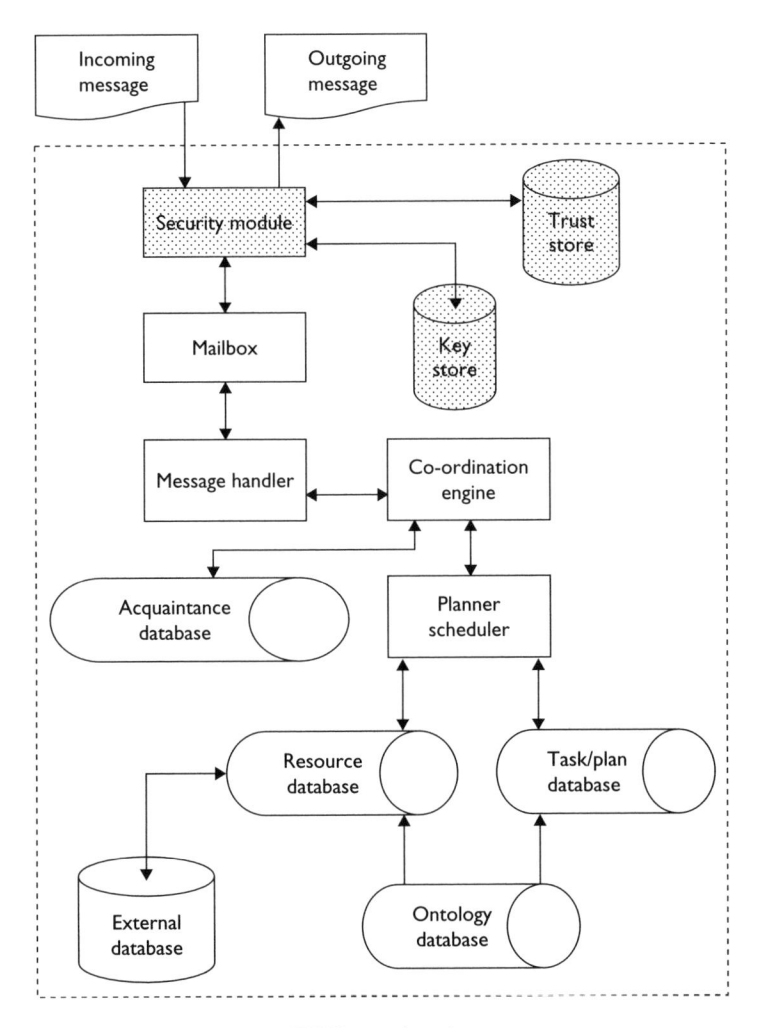

Figure 8.5 Extensions of a generic ZEUS agent's architecture.

has the additional effect that undesirable messages are filtered out before they are executed by the *Mailbox* and the *MessageHandler*. In a way this can be compared to a firewall for the task-agents.

The agent's asymmetric keys are stored in the component *Keystore* whereas the certificates of other agents which are successfully registered on the marketplace are stored in the *Truststore*. Messages which do not pertain to the PKI are forwarded to the responsible component of a generic ZEUS-agent.

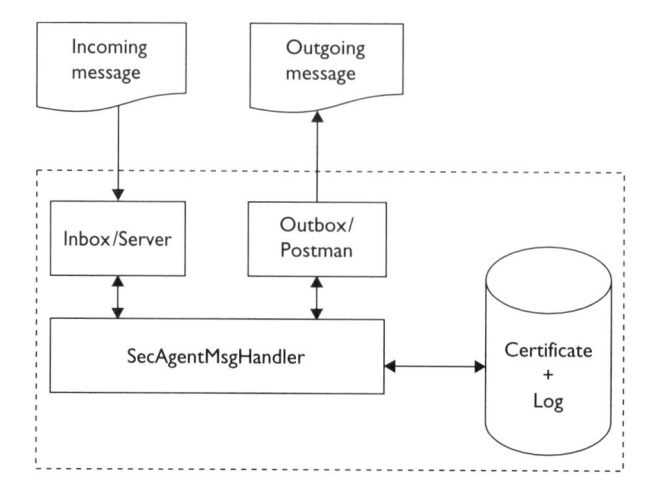

Figure 8.6 Architecture of a security agent.

8.4.4.2 Security agent

Due to the PKI of the marketplace, a new kind of utility agent had to be developed which must be able to perform certification services.

The architecture of the SEC consists of the following main components (Figure 8.6): Inbox/Server, Outbox/Postman, SecAgentMsgHandler and a database for storing agent-certificates, user-certificates and the user's signature of the associated agent-certificate. Furthermore, the whole correspondence of the SEC is logged in this database. In order to verify certificates and digital signatures the SEC has to contact the CA which issued the user-certificates.

In the current implementation, the LDAP is used. The SEC checks the Certificate Revocation List (CRL) in regular intervals. In case a user-certificate is classified as invalid the associated agent loses its authentication and the permission to participate in the marketplace. Therefore, the SEC sends a *revocation*-message to all agents which are in possession of the revoked agent-certificate. In future versions the OCSP will be supported, too.

8.4.4.3 Registration and establishing contact

As already mentioned, owners and contractors have to possess a certificate from an accredited CA in order to participate in SiReAM. Then the registration process can be described by the scenario as shown in Figure 8.7.

After Company A has received its certificate it digitally signs its Agent A or more precisely the agent's self-generated certificate. Now, Agent A sends this agent-certificate, the signature, and Company A's user-certificate to the SEC. The SEC verifies the signature by contacting the CA. If the signature

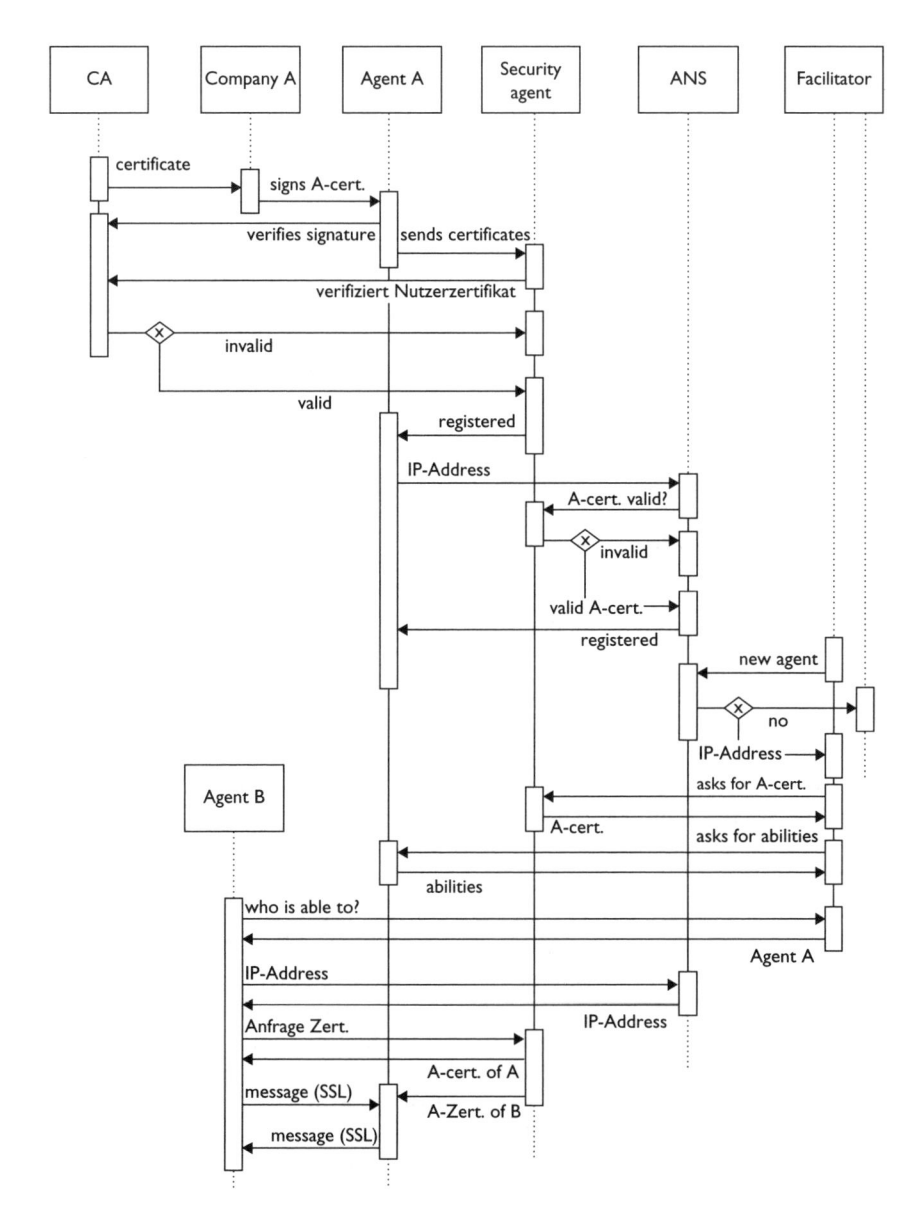

Figure 8.7 Sequence diagram for registration and establishing contact.
Source: URL1.

is valid the SEC can clearly assign Agent A to Company A and stores the agent-certificate in its trust store.

In order to register, Agent A has to contact the ANS which manages the IP-addresses of all agents. The ANS asks the SEC for the agent-certificate of

Agent A and confirms the registration if the SEC has confirmed its validity. The FAC frequently asks the ANS if new agents have been registered and receives the IP-addresses of the new agents. In this scenario, the FAC gets aware of Agent A and asks the SEC for the valid agent-certificate of Agent A. Without this valid agent-certificate the FAC cannot contact Agent A by using a SSL connection (Denk and Schnellenbach-Held, 2002).

Now an already registered Agent B is looking for an agent with a specific ability. It asks the FAC for an agent with the required ability. Here, the answer is Agent A. Agent B receives the IP-address of Agent A from the ANS and the valid agent-certificate from the SEC. Because client/server-authentication is used for establishing a SSL-connection also Agent A needs to know Agent B's agent-certificate. Therefore, the SEC sends Agent B's agent-certificate to Agent A which stores it in its trust store. Finally the SSL-connection between Agent B and Agent A can be established.

8.4.5 Public tendering procedure

In this sub-section co-ordination concepts for the following procedures are introduced (Denk and Schnellenbach-Held, 2002):

- Submitted tenders have to be preserved in such a way that they cannot be examined by the contracting authority before the time-limit set for submitting has expired.
- Contractors must be able to withdraw their tenders before the time-limit set for submitting has expired.

In Figure 8.8 a sequence diagram depicts these procedures: after a contractor's agent (AN-Agent) has calculated a proposal it either can refuse the cfp or can file a signed and encrypted tender. For the encryption, a symmetric session key is generated by the AN-Agent and the proposal is enciphered using this session key. The AN-Agent stores this session key and sends it to the owner's agent (AG-Agent) at the final date.

After the proposal has been received by the AG-Agent, an acknowledgement is sent to the AN-Agent. This confirmation of the receipt is of crucial importance because of the legal regulations the sender must always prove that the message has been received by the corresponding party. In order to confirm a message an agent digitally signs the received message and sends this signature back to the original sender of the message. The AG-Agent stores all submitted proposals, decrypts them after the session keys have been received at the final date, and informs the owner (AG) about all decrypted proposals.

The owner decides which bid will be awarded and notifies its AG-Agent which on its part sends an *accept*-message to the awarded AN-Agent

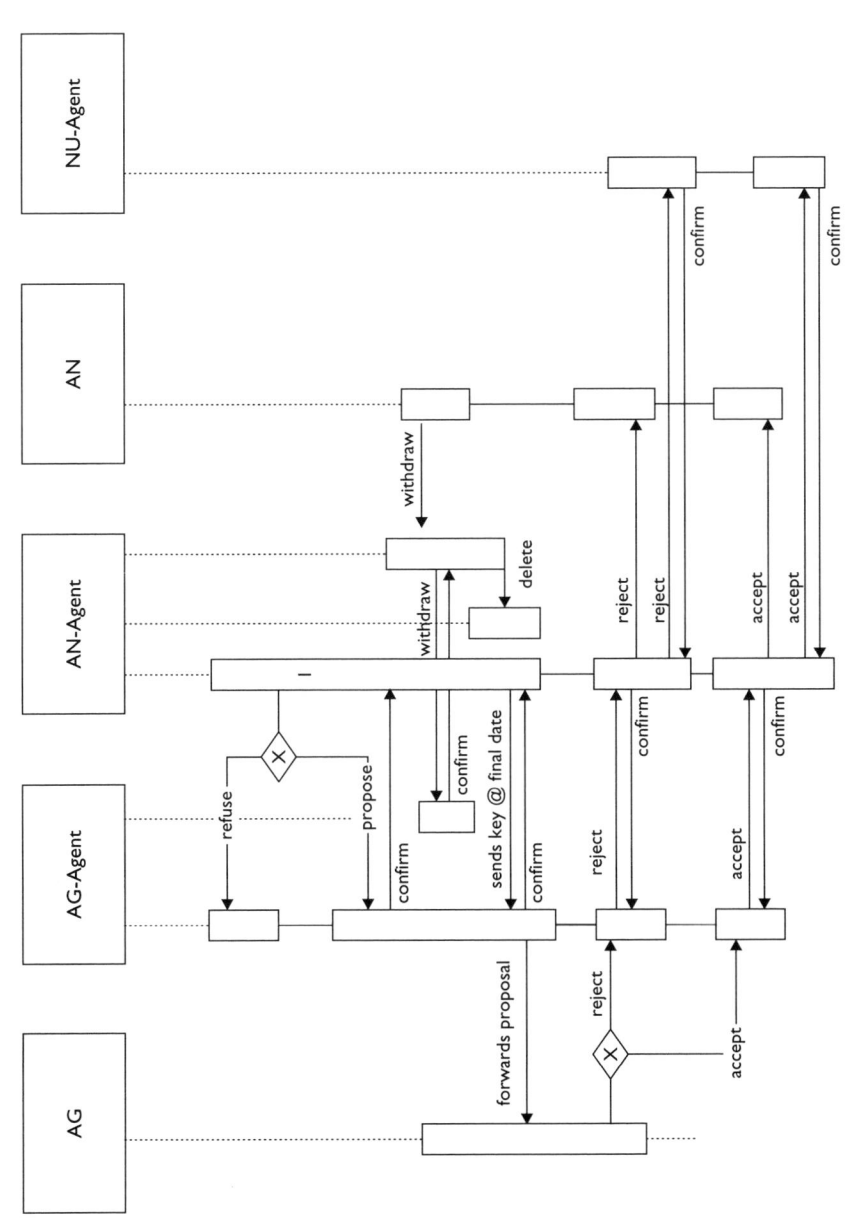

Figure 8.8 Co-ordination concepts.
Source: URLI.

and *reject*-messages to the other bidders. The receipt of the *accept*- and *reject*-messages has to be confirmed by the AN-Agents, too.

Before the final date is reached, the contractor (AN) can withdraw the submitted proposal by informing his AN-Agent. The AN-Agent sends a withdraw-message to the AG-Agent and deletes the session key with which the proposal has been encrypted. The AG-Agent acknowledges the receipt of the withdraw-message, identifies the withdrawn proposal and deletes it.

8.4.5.1 Submission of variants

In Figure 8.9 a sequence diagram for submitting variants is depicted. After the AN-Agent has calculated a proposal based on the published specifications for tenders, it asks its AN for his agreement. In case the AN wants to submit a variant he either sends a *delete-proposal*-message or a *store-proposal*-message to its AN-Agent. Depending on this message the AN-Agent either deletes its previously calculated proposal or stores it in order to submit the original proposal along with the variant later on.

After the AN has changed the specifications for tenders, he sends these information to its AN-Agent. The AN-Agent immediately starts trying to achieve its new goal. In case abilities are required that the AN-Agent cannot provide, it asks the FAC for other qualified agents. In Figure 8.9 the AN-Agent receives the name of a NU-Agent by the FAC, the IP-address by the ANS, and the agent-certificate by the SEC. The SEC also sends the agent-certificate of the AN-Agent to the NU-Agent.

Via a client/server-authenticated SSL-connection the AN-Agent contacts the NU-Agent and starts negotiating about the price (different strategies can be implemented). Here the AN-Agent accepts the second offer. Now the NU-Agent asks its user for his agreement. If the NU agrees the NU-Agent sends a digitally signed acknowledgement (using the private key of the NU) to the AN-Agent which again confirms the acknowledgement. Now the AN-Agent can achieve its goal and again asks the AN for the permission to file the proposal to the AG-Agent. If the AN agrees, the encrypted and digitally signed (private key of the AN) variant is submitted. At the final date, the session key is sent to the AG-Agent which decrypts the variant and informs the AG about all submitted proposals and variants.

Of course, the receipt of the variant and the receipt of the session key are confirmed, too. The next transactions correspond to the public tendering procedure which has been described in Section 8.4.5.1.

8.4.6 Agent communication language

In ZEUS the (FIPA) ACL is used. But herein, the defined performatives do not satisfy the needs of the PKI of SiReAM. Therefore, the class *pkiPerformative* extends the ZEUS class *performative*. The class diagram is depicted in Figure 8.10.

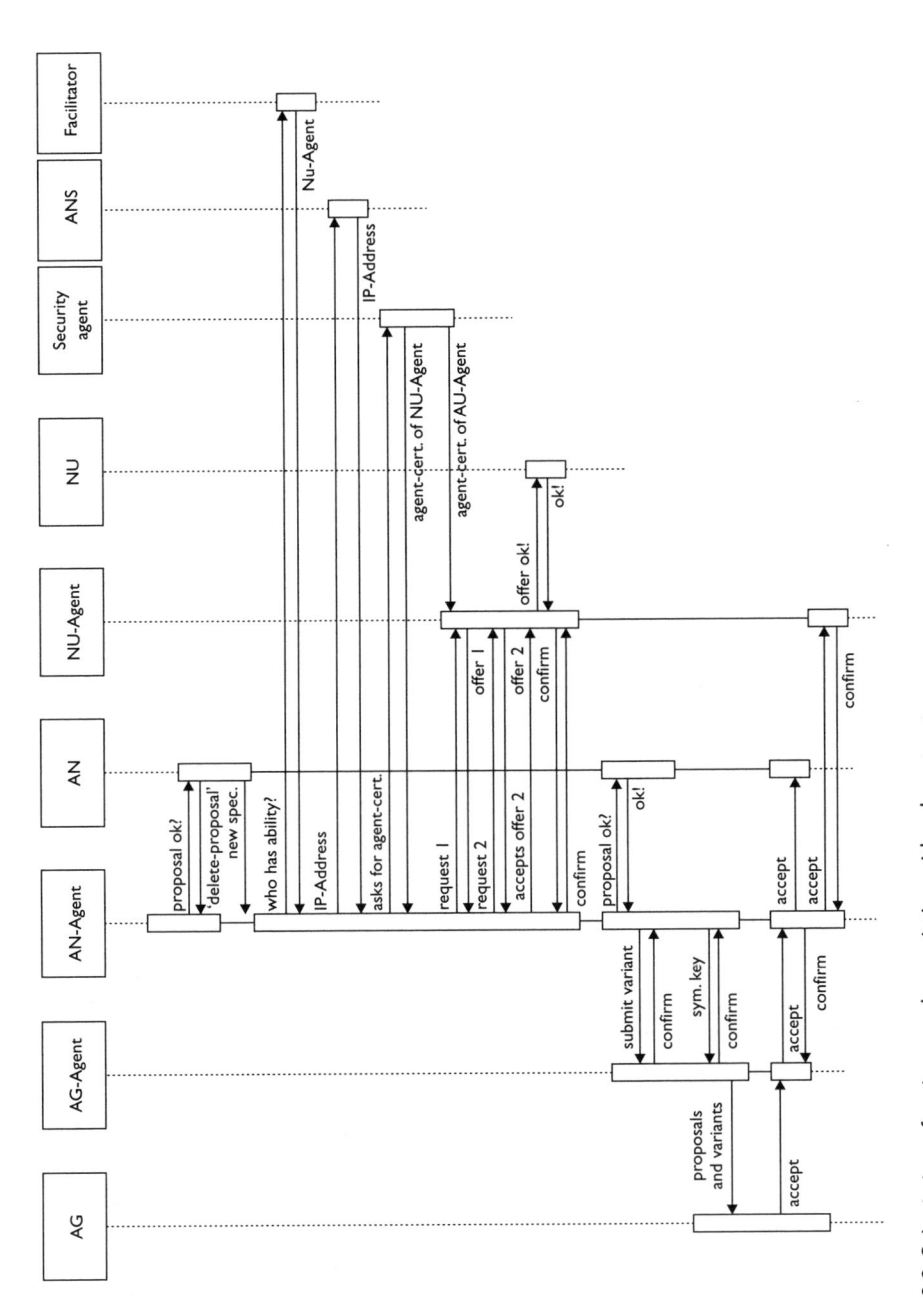

Figure 8.9 Submission of variants and negotiating with subcontractors.

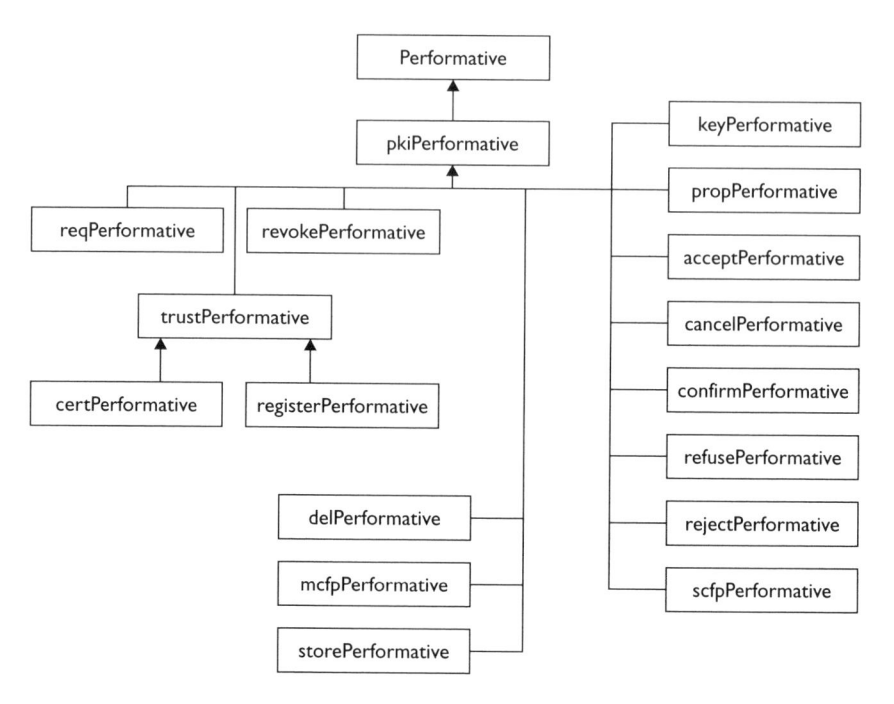

Figure 8.10 Class diagram pkiPerformative.

Depending on the pki-messages which are supposed to be generated different message-types are instanced. All new pki-message-types are listed in Table 8.1.

8.5 Summary

The acceptance of an agent-based virtual marketplace for AEC-bidding mainly depends on a reliable and trustworthy PKI. Only if confidentiality, integrity, non-repudiation and authentication are provided reliably, users will assign sensitive business processes like public tendering procedures to software agents.

The developed PKI for the distributed agent-based marketplace SiReAM provides these essential security issues. Here, secure transactions between software-agents are made legal and binding according to the Digital Signature Act.

Furthermore, models for an agent-based public tendering procedure according to the official contracting terms for the award of construction performance contracts have been presented. The current implementation of SiReAM proves that AEC-bidding can be conducted by multi-agent societies.

Table 8.1 PKI-message-types

Message-type	Class	Specification
CERTREQUEST	reqPerformative	Message for requesting an agent-certificate from the SEC
REVOKE	revokePerformative	The SEC informs other agents that a specific certificate has been revoked
REGISTERPERFORMATIVE	registerPerformative	Message for registering at the SEC
CERTIFICATE	certPerformative	Certificate and signature are sent to the requesting party
NOTSPECIFIED	pkiPerformative	In case a message cannot be allocated it is caught by this type
PROPPERFORMATIVE	propPerformative	Message-type for sending an encrypted proposal
KEYPERFORMATIVE	keyPerformative	Message-type for sending the session key at the final date which is required by the AG-Agent for the decryption of the corresponding proposal
SIGNEDACCEPT	acceptPerformative	Signed message-type for accepting a proposal
SIGNEDCANCEL	cancelPerformative	Signed message-type for excluding a proposal, i.e. an AN-Agent, from the award procedure
SIGNEDCONFIRM	confirmPerformative	Signed message-type for sending an acknowledgement
SIGNEDREFUSE	refusePerformative	Signed message-type for refusing a calls for proposals (cfp) and for withdrawing an already submitted proposal before the time-limit set for submitting has expired
SIGNEDREJECT	rejectPerformative	Signed message-type for rejecting a proposal (AG-Agent rejects)
SIGNEDCFP	scfpPerformative	Signed message-type for sending a cfp
DELETEPROPOSAL	delPerformative	Message-type for deleting the proposal which has been calculated by an AN-Agent based on the original specifications for tenders
STOREPROPOSAL	storePerformative	Message-type for storing the proposal which has been calculated by an AN-Agent based on the original specifications for tenders
MODIFIEDCFP	mcfpPerformative	Message-type for informing the AN-Agent about the cfp which has been modified by the AN

Comprising the advantages of an agent-based AEC-bidding procedure it can be pointed out that the bidding process is facilitated and accelerated. Due to the regular attendance of a company's agent on the marketplace companies can react faster and more often to invitations to tender. Vice versa, contracting authorities can reach more bidders in a relatively short period of time.

Notes

1 This research project which is sponsored by the German Science Foundation (DFG) is conducted in co-operation with Prof. Dr jur. Rossnagel (Kassel University, Germany).
2 SiReAM: abbreviation for 'Sicherer Rechtsgemäßer Agentenbasierter Marktplatz'. Translated into English this means 'secure agent-based marketplace according to legal regulations'.
3 LV is the German abbreviation for standard specifications for tenders.

References

Beck-Texte (2001) Vergaberecht, 3. Auflage.
Bitzan, S. (2001) *Konvertierung eines XML-Leistungsverzeichnisses in das Format der ZEUS-Ontologie mittels Document Object Model (DOM)*, TU Darmstadt (unpublished).
Denk, H. and Schnellenbach-Held, M. (2001) *Ontology and Security for Software-Agents in AEC-bidding*, Darmstadt Concrete, Darmstadt.
Denk, H. and Schnellenbach-Held, M. (2002) 'Models for an agent-based public tendering procedure according to German regulations', in M. Schnellenbach-Held and H. Denk (eds) *Advances in Intelligent Computing in Engineering,* in *Proceedings of the Ninth International EG-ICE Workshop*, Darmstadt.
European Parliament (2000) *Official Journal of the European Communities*, Series D, Debates of the European Parliament, Office for Official Publications of the European Communities, Luxembourg, 19(1).
KPMG, BMWI: *Chancen und Risiken inverser Aukitonen im Internet fuer Auftraege der oeffentlichen Hand*, Abschlussbericht, 2001.
Lipp, P., Farmer, J., Bratko, D., Platzer, W. and Sterbenz, A. (2000) *Sicherheit und Kryptographie in Java*, Addison-Wesley, 2000.
Nwana, H. S., Ndumu, D. T., Lee, L. C. and Collis, J. C. (1999) 'ZEUS: A Toolkit for Building Distributed Multi-Agent Systems', British Telecommunication Laboratories, in *Applied Artificial Intelligence Journal*, 13(1), 129–86.
Schneier, B. (1996) *Angewandte Kryptographie*, Addison-Wesley.
URL1: http://www.auml.org
URL2: http://more.btexact.com/projects/agents.htm
URL3: Gruber, T. *What is an Ontology*, http://www-ksl.stanford.edu/kst/what-is-an-ontology.html
URL4: The ontology Page, http://www.kr.org/top/
VOB: *Verdingungsordnung fuer Bauleistungen*, Ausgabe (2000) Deutsches Institut fuer Normung e.V.

Chapter 9

Agent-based information search and retrieval

M. Sun

9.1 Introduction

Construction is an information intensive process. Many activities involve the creation and consumption of large amount of information ranging from product data to technical publications, from building regulations to best practice guides. With the increased use of computers in the industry, much of the information is available in digital forms. There is a growing demand for exchanging construction information electronically over computer networks. In recent years, the World Wide Web (WWW) has emerged as the most popular vehicle for information dissemination on the Internet. It is estimated that the volume of information on the WWW is doubling every 50 days. The amazing growth of the WWW is an indication of its strength. However, as more and more information becomes available, the task of finding the right information becomes more and more difficult. This chapter discusses the existing and emerging solutions to the on-line information search and retrieval problem.

9.2 Development of the World Wide Web

Although the Internet has been in existence for more than three decades, it is the emergence of the WWW in the last 10 years that made Internet accessible for ordinary users. Schatz (1997) estimated that the number of Internet users increased from 1 million to 25 million in the five year leading up to January 1997. The trend of increase has been confirmed by a number of other estimates and surveys since. For example, Matrix Information and Directory Services, an Internet measurement organisation, put an estimate of the Internet users worldwide as 57 million in April 1998. The figure was estimated to research 377 million by year 2000. Nua reported a survey conducted in September 1999, which put the total Internet users worldwide as 201 million (URL1). Despite some differences between different surveys and estimates, all evidences show a continued and rapid growth in the number of people using the Internet.

The main reason for the user increase is the tremendous growth of the information available on the Web. Both Lynch (1997) and Zakon (1996) have reported that the number of Internet hosts (servers) has been doubling every year in the 1990s. The number of public accessible Web pages has grown at an even faster pace. Today, the WWW can be seen as a vast hyper database with the following characteristics:

Vast size – The amount of information on the Web is increasing all the time. There are no accurate statistics for its actual size. Any such statistics would become outdated as soon as they are published. Some estimates put the total available documents at 2 billion.
Dynamic – Constant change is another feature of the Web. New WWW servers are created every day. New documents are added; existing documents are modified and removed all the time.
Disorganised – There are a large number of information providers on the WWW. There is no unified quality control of the contents.
Heterogeneity – The documents on the WWW vary in formats, including text, graphics, audio and video, databases, etc.

Given the enormous volume of the Web documents and the dynamic nature of the Web, it comes as no surprise that most Internet users are increasingly relying on information search tools to find specific information.

9.3 Current Internet information searching solutions

When a collection of information reaches a certain size, there is a need to find ways to organise the information and retrieve it when required. In the paper-based information age, card catalogues and classification systems like the Dewey Decimal System (Fowler, 1997), are effective information retrieval tools. Today, the amount of digital information is growing exponentially and much of it is available through the Internet. The WWW has become an expanding hypermedia database where information in various formats can be found on many related and unrelated topics. The traditional methods alone are no longer able to manage this information growth. A large number of new tools have emerged aimed at helping users to locate information on the WWW. These tools can be broadly divided into three types:

1. Yellow Page type of *hierarchical directories*
2. Robot-based Internet *search engines*
3. *Meta-search engines*.

The first is a 'yellow-pages-like' hierarchical directory in which the on-line information is organised in logical categories. The user can locate the right information by browsing through the listings. The second type of solutions is the use of Internet search engines. A search engine is a special WWW server that gathers information about documents located on other WWW servers. A user can locate the required information by supplying keywords to the search engine that in turn finds the documents by matching keywords. Meta-search engines are special type engines that do not have databases of their own. They pass user's query to several other search engines in order to perform the search in a wider scope.

9.3.1 Hierarchical directories

The 'Yellow Pages' is a well-established business directory publication. Using the Yellow Pages, a user can locate address and telephone information of a firm by looking through a classification hierarchy. At the initial stage of the WWW development, surfing and browsing are the main navigation methods. Types of information gateways such as Yellow Pages were popular, ranging from systematically grouped hotlists to corresponding help on home pages, from starting points or lists of interesting resources to international services covering multiple publication-types and subjects (Brümmer *et al.*, 1997). Good examples of this type of information gateways include the WWW Virtual Library (Manning, 1999) and Yahoo (URL2). There are also subject-specific gateways for example the Construction Industry Gateway in the UK, an initiative to provide a gateway service for construction-specific information on the WWW (Lockley and Amor, 1998).

There is no doubt about the usefulness of these systems. However, they also have some major weaknesses. Each of these services usually only covers a very small fraction of the Internet resources. There are no widely accepted standards for classification. Each information provider decides the structure and content of its materials governed by chance, occasional decisions and staff responsible for the implementation. The consequence is a lack of consistency and reliability and a lack of independence from the individuals performing the task. Another weakness of information gateway solutions is the difficulty of keeping information updated. Due to the rapid changing nature of the WWW, many resources are quickly becoming non-usable. The lack of automatic updating mechanisms makes it impossible to follow the constant changes of web document contents, addresses, appearance and disappearance.

9.3.2 Internet search engines

Internet search engines are programs that roam the Internet (with flashy names like *spider, worm* or *searchbot*) to build up an index of meta-information about everything available on the net (Gilster, 1996). The gathered information, characterised by a number of keywords (references) and perhaps some supplementary information, is then put into a large database. The search solution is based on the principle of keyword matching. A user describes his/her information needs using a number of words. When a user specifies a query, the system returns the documents containing all or some of the words in the query. An Internet search engine has three components, a crawler, an indexer and a query server. The crawler collects pages from the Web. The indexer processes the retrieved documents and represents them in an efficient searchable data structure. The query server accepts queries from the user and returns the result pages by consulting with the search data structure.

While search engines are best available information search solutions, several weaknesses have become more and more evident (Hermans, 1996). All these search engines rely on the user being able to formulate queries effectively. To use advanced queries, a user needs to apply sophisticate Boolean functions, for example, 'window AND (NOT (comput* OR Microsoft)) AND building'. This is often beyond the capability of ordinary construction professionals. Search engines are domain-independent in the way they treat gathered information and in the way they enable users to search for it. Terms in gathered documents are lifted out of their context, and are stored as a mere list of individual keywords. A term like 'information broker' is most likely stored as two separate terms 'information' and 'broker' in the meta-information of the document that contains them. Someone searching for documents about an 'information broker' will therefore also get documents where the words 'information' and 'broker' are used, but only as separate terms, for example, as in 'an introductory information text about stock brokers'. There is no good rating mechanism to rank the many hundreds or even tens of thousands of hits in response to each user's query.

9.3.3 Meta-search engines

Most of the current popular Internet search engines only cover a small proportion of the Internet, less than 20% according to the best estimate. There are now meta-search engines available, which pass user's query to multiple search engines in order to perform searching in a wider scope. Examples of meta-search engines include MetaCrawler (URL3), Dogpile (URL4), Copernic (URL5), etc. Using these meta-search engines, a user does not have to visit each Internet search engine separately. The user can also specify the number of hints returned from each engine. The advantage is obviously saving searching time and widening search scope. However, there are also some disadvantages. Since a meta-search engine needs to integrate several Internet search engines, its user interface is usually very simple, using 'Least-Common-Denominator.' Some of the rich functionalities of specific search engines are often discarded. It is difficult for users to input complicated queries and retrieve specific information. In addition, because a meta-search engine needs to send queries to multiple engines, sometimes the response time is slow.

9.4 Advanced information searching techniques

In recent years, a number of advanced techniques have emerged which are aimed at improving information search and retrieval on the Internet. They are introduced in the following sections.

9.4.1 Information filtering

The goal of an Information Filtering (IF) system is to sort through large volumes of dynamically generated information and present to the user those that are likely to satisfy his or her information requirement (Newell, 1997). IF has been originated from the manual alerting services that brought new information to the attention of users of special interests. With the growth of the Internet and other networked information, research in automatic filtering of networked information has been on the increase in recent years. Using a filtering system, a user does not need to search for information using queries. Instead, the user specifies his or her interests and the system monitors the information sources to inform the user when information of the required nature becomes available.

One of the early applications of IF is the Content-based IF systems for the Usenet News (Konstan *et al.*, 1997). The Usenet News system provides the capability to a user to post an article in a common bulletin board that can be accessed by any other users. Despite the fact that each bulletin board is related to a specific area of interest, the number of articles posted each day can reach several hundreds making the task of finding and reading the interesting ones quite difficult. A Content-based IF system tries to solve this problem by automatically selecting those articles that suit a user's long-term interests. This selection is based on a user profile, which is a representation of the user's long-term interests, and on methods that match this profile with the incoming articles. A user profile consists of keywords and Information Retrieval methods can be used to match articles and profiles.

The simplest systems require the user to create this profile manually or with limited assistance. Examples of these systems include: 'kill files' that are used to filter out advertising, e-mail filtering software that sorts e-mail into categories based on the sender, and new-product notification services that request notification when a new book or album by a favourite author or artist is released. More advanced IF systems may build a profile by learning the user's preferences. A wide range of agents including Maes' agents for e-mail and Usenet news filtering (Maes, 1995) and Lieberman's Letizia (Lieberman, 1997) employ learning techniques to classify, dispose of or recommend documents based on the user's prior actions. Similarly, Cohen's Ripper system has been used to classify e-mail (Cohen, 1996); alternative approaches use other learning techniques and term frequency (Boone, 1998). IF techniques have a central role in recommending systems and to build a profile of user preferences that is particularly valuable when a user encounters new content that has not been rated before.

9.4.2 User profile

User profile technology is originally developed for the purpose of IF, which now has a wider use in a number of information search and retrieval areas. A user profile is usually represented by a set of keyword vectors or keywords. These keywords can be specified by the users manually. Alternatively, some more advanced systems can extract them automatically based on the user's search history using software agent and machine-learning technologies. For example, the WWW browsing assistant Syskill & Webert constantly observes the user and suggests which WWW pages are of interest (Pazzam and Billsus, 1997). The system learns the user's profile, and uses this to guide its suggestions for interesting WWW pages. Initially a user supplies the name of the topic, and the URL of an index of WWW pages for that topic. The browsing assistant represents interests as keyword vectors. As users browse the WWW they can manually rate pages as good or bad to give positive (interesting) and negative (uninteresting) training examples. The updated user profile is then used to compare documents linked from the current page to suggest which are of interest to the user and should be followed.

Autonomy's AgentWare (URL6) agent can accept textual descriptions of interests (keywords, phrases or a set of documents) and then applies neuro-fuzzy pattern matching to identify the key concepts within the document in order to build up a description for the user's interests. A single 'agent' represents user interests as some form of encoded concept representation, which can be shared between agents within the Autonomy AgentWare suite. Autonomy's agents can also determine user profiles by observing the users' actions (assuming the users' read items in which they are interested).

9.4.3 Collaborative filtering browsing

Various studies show that in real life when people search for information, the first thing they usually do is to ask other colleagues who have done similar searching before. The existing browsing and searching systems focus on technical information discovery methods while neglecting perhaps, the single most important method of discovery that people rely on – other people. Collaborative browsing, sometimes also known as social resource discovery, assumes the existence of others, who have located and evaluated relevant resources. The goal of collaborative browsing systems is to aggregate and share the fruits of individual activity and knowledge of Internet information retrieval (Twidale and Nichols, 1998).

Searching for information would be much more efficient if people were able to share their knowledge about information resources. The problem is to bring together people who can benefit from each other. If someone browses for information, there is a high probability that someone else is interested in the same subject at the same time, but people browsing the WWW are unaware of the presence of any fellow-browsers. Even if they are submitting the same keywords to a search engine or are searching at the same starting point, there is no possibility of getting in touch with each other. The goal of CF/browsing systems (also known as recommendation systems), such as CoBrow (Sidler *et al.*, 1997) is to bring these people together.

A Collaborative IF system, instead of comparing the user profile with the information items, compares a user profile with other profiles to identify groups of users having similar interests. Having identified such groups, it predicts the importance of an information item for a particular user on the basis of the degree to which other users of the group have found this document interesting. The user profiles in CF systems can consist of keywords describing the users' information needs or pairs of item-rating that the user gave in the past. These ratings can be sometimes implicitly calculated by the system.

The CF systems build a database of user opinions of available items. They use the database to find users whose opinions are similar (i.e. those that are highly correlated) and make predictions of user opinion on an item by combining the opinions of other like-minded individuals. While Tapestry (Goldberg *et al.*, 1992), the earliest CF system, required explicit user action to retrieve and evaluate ratings, automatic CF systems such as GroupLens (Konstan *et al.*, 1997) provide predictions with little or no user effort. Later, systems such as Ringo (Shardanand and Maes, 1995) and Bellcore's Video Recommender (Hill *et al.*, 1995) became widely used sources of advice on music and movies, respectively. More recently, a number of systems have begun to use observational ratings; the system infers user preferences from actions rather than requiring the user to explicitly rate an item (Terveen *et al.*, 1997). In recent years, a wide range of Web sites, such as amazon (www.amazon.com), have begun to use CF recommendations in a diverse set of domains including books, grocery products, art, entertainment and information.

Collaborative IF systems have several advantages over the content-based ones. Because they are based on human's evaluation of the information items they can be used for items that are not in a machine readable form, for example, sound and photographs, and can select items based on some assessment of quality, style or semantics. In addition, they can select items the knowledge of which the user did not have before (serendipitous finds).

However, the disadvantage of these systems is that the number of item ratings given by the users must reach a certain level before the system starts making sensible selections.

9.5 Intelligent agent-based information search

Agent technology originated from the branch of Artificial Intelligence (AI) known as Distributed Artificial Intelligence (DAI). In recent years, it has grown into a fast expanding research area. As any other new research field, there are diverse definitions of the term 'agent'. Here we want to quote a very generic definition given by Janca (1995), 'an agent is a software that knows how to do things that you could probably do yourself if you had time'. There is a growing consensus in the Internet community that one of the most promising solutions to the problem of Internet information retrieval is the use of software-agent technology.

The main feature of an agent-based WWW is that information agents perform the role of managing, manipulating or collating information from many distributed sources (Nwana, 1996; Davies *et al.*, 1997). Some information agents have intelligent features, for example, they can be mobile, can learn and can co-operate with the other agents. Mobile agents are able to roam the WWW, interact with WWW servers, gather information on behalf of their owner and return home after performing duties set by their users. The learning ability of information agents refers to the fact that they can react and interact with external environment and improve their efficiency over time. While individual agent is able to work on its own, a number of agents can co-operate with each of them performing a role to achieve a collaborative goal.

Although there is still scepticism considering software agent as just another buzzword, the initial development in this area offers promising potentials. The benefits of software agents' approach include (Hermans, 1996):

- Agents are capable of searching information more intelligently, for instance, because tools (such as a thesaurus) enable them to search on related terms as well or even on concepts. Agents will also use these tools to fine-tune or even correct user queries (on the basis of a user model or other user information).
- Individual user agents can create their own knowledge base about available information sources on the Internet, which is updated and expanded after every search. When information (i.e. documents) has moved to another location, agents will be able to find them, and update their knowledge base (k-b) accordingly. Furthermore, in the future agents will be able to communicate and co-operate with other agents (see previous applications areas discussed in Chapters 5–8). This will

enable them to perform tasks, such as information searches, quicker and more efficient, reducing network traffic. They will also be able to perform tasks (e.g. searches) directly at the source/service, leading to a further decrease of network traffic.
- Agents can relieve their human user of the need to worry about 'clerical details', such as the way the various Internet service have to be operated. Instead, he or she will only have to worry about the question what exactly is being sought (instead of worrying about where certain information may be found or how it should be obtained). The user's agent will worry about the rest.
- As a user agent resides on a user's computer, it is always available to the user. An agent can perform one or more tasks day and night, sometimes even in parallel. As looking for information on the Internet is such a time-consuming activity, having an agent do this job has many advantages, one of them being that an agent does not mind doing it continuously. A further advantage of agents is that they can detect and avoid peak-hours on the Internet.
- Software agents will be able to search information based on contexts. They will deduce this context from user information (i.e. a built-up user model) or by using other services, such as a thesaurus service. See Chapter 4 and 6 for more detailed information about this.
- User agents can adjust themselves to the preferences and wishes of individual users. Ideally, this will lead to agents that will more and more adjust themselves to what a user wants and wishes, and what he or she is (usually) looking for, by learning from performed tasks (i.e. searches) and the way users react to the results of them. Furthermore, agents are able to continuously scan the Internet for (newly available) information about topics a user is interested in.

Figure 9.1 shows an agent-based information search and retrieval system for the construction domain proposed by Sun *et al.* (2000), called a Collaborative Construction Information Network (CCIN). The purpose of the system is to provide a value-added gateway for construction related information on the Internet. It is acknowledged that the WWW-based information network will remain a distributed resource with information hosted on many servers. The gateway seeks to provide a recognised starting point for the user to search for information and a repository for shared knowledge base that helps the information search. The figure shows a typical three-layer architecture with the CCIN servers playing the intermediary role between the information users (clients) and information suppliers (WWW servers).

The described information network consists of CCIN servers and client-agents. The CCIN client-agent is a plug-in package to standard WWW browsers, such as Microsoft Internet Explorer and Netscape, to provide

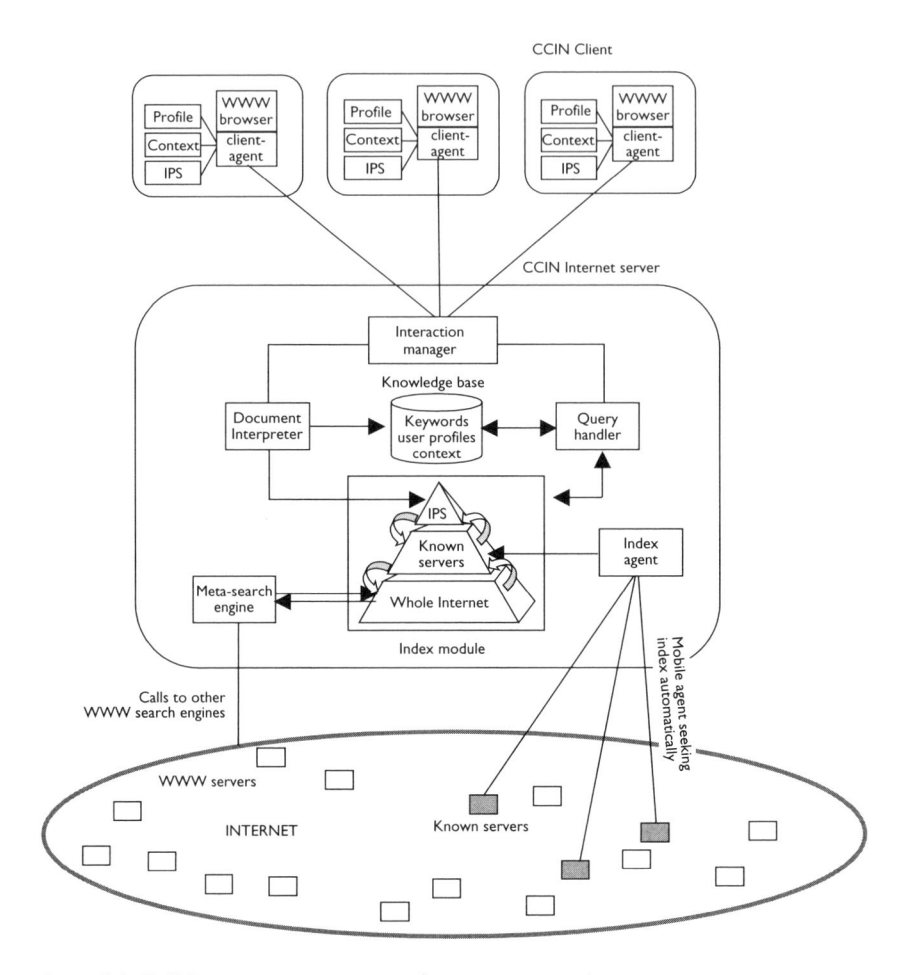

Figure 9.1 Collaborative construction information network.

additional functions. Once it is installed on a user's computer, the agent will facilitate the communication with the CCIN server and provide intelligent information search and retrieval services. To achieve this, the agent needs to gather information of user profile, information query context and information search history. Part of this information is communicated to the CCIN servers to improve the information search task of the user community as a whole. The CCIN server is in essence a construction-specific WWW search engine. Its main function is to match information sources to users search queries.

The key feature of the server is the framework that enables construction domain knowledge and user knowledge to be accumulated and used in intelligent information searching. The server consists of the following components.

Layered index module. At the centre of the CCIN server is an index module that consists of three hierarchical layers. The top layer stores information of documents that have been accessed by users and recommended to the server through the CCIN client agents. The data are stored in a meta-information format similar to Jasper's Intelligent Page Store (IPS) (Davies, 1997). For each WWW page, the index module stores at least the following information gathered automatically by the client-agents with a degree of user manual input:

- the document title
- a summary of the content
- a set of keywords
- a set of user profiles who recommended the page
- information type (product data, technical publication, news, project data, etc.)
- users' annotations
- universal resource locator (URL) and
- date and time of storage or update.

The middle layer contains an index of documents located on known WWW servers which are more likely relevant to the construction users' interests. These servers become known to the gateway server through two ways: (1) The information providers register their servers explicitly. (2) When the client agent submits a document to the top layer of the index module, the host server becomes known to the gateway implicitly. The CCIN server agent, a mobile information agent, visits all the known servers periodically and gathers information about documents hosted on these servers.

The bottom layer of the index module covers most of the accessible Web hosts on the Internet. Given the constant growth of the Internet, the gateway server does not store any documents. Instead, it will rely on a Meta-search engine to call upon other Web search engines, for example, Alta vista, Infoseek, etc.

The hierarchical layering implies that the amount of information accessible increases as one moves down from the top to the middle and the bottom layers, but the average potential relevance to the users' interests decreases. The advantage of the layering approach is that it allows a user to specify the scope and manner of a query. One can raise a query just for the IPS layer where more criteria can be applied apart from keywords. Alternatively, the user can make a general query using enhanced keywords supported by the server's knowledge base to a wider index in the middle layer or even the whole Internet.

A document's position in the index module is not fixed. There are migration paths through which a document can be moved from one layer to another as a result of users' search and retrieval actions. For example, when a document in the middle layer is retrieved by a user, the client-agent will communicate that fact to the server. The server will upgrade the document to the top layer in the index module. On the other hand, if a document in the top layer has not been used by any user for a period of time (pre-defined threshold), the server will degrade it to the middle or even bottom layer. The purpose is to ensure the efficiency of the index system.

Query handler. The task of the query handler is to handle more complex queries other than merely keyword search. It aims at acquiring more knowledge about the user and the user's information needs so that more accurate queries can be formulated. The system provides several searching methods of different levels of complexity catering for users with different levels of computer competence. It enables the end-users to perform information search in a variety of ways such as using information concept, information types, user profiles, as well as the conventional keyword methods. To achieve this, the CCIN server needs to possess and accumulate knowledge about its users and construction domain-specific information.

Knowledge base. If the server knows more about a user's information needs, it can provide more accurate results for the user's search queries. The (k-b) of the CCIN server keeps information of three aspects that are construction domain specific.

- *Keywords* – Keyword-based searching remains an important method of information retrieval on the WWW. Instead of considering it as a simple syntax comparison, the CCIN server identifies a set of keywords and associates them with conceptual meanings. Relationships between these keywords are also analysed. As a result, the server is able to support concept-based searching. For example, when a user elects to search for information about 'enclosing building element', he/she can choose to have extended the search to include sub-types 'wall', 'window', 'door', 'roof', or association-types 'building material', 'component product', etc. These keywords are identified based on recent data and processing modelling works of other research projects and the common classification standards used in the industry.
- *User profiles* – A user profile articulates the features and the information needs of a distinctive type of users. The CCIN server has general profiles for the common types of construction information users, such as architects, engineers, project managers, researchers, academics, etc. The CCIN users can choose one or a combination of these profiles for their searching query. They can also derive personalised profiles based on these general ones.

- *Context* – Context is a description of the purpose behind a user's search for information. It may be about stages of the construction process, design tasks or a research topic area.

The (k-b) is populated at the server setup stage. The server also has the ability to learn from its interactions with the client-agents. The (k-b) is updated constantly.

Document Interpreter. When a user retrieves a document of interest, the client-agent installed on the user's computer extracts some information from the document, such as the title and a set of keywords. This information is stored locally in an IPS format. At the same time the client-agent sends it to the CCIN server together with the user profile information and its context. The Document Interpreter on the server is responsible for processing this information and storing the document summary in the index module and updating the (k-b) if necessary.

Index agent. This is a mobile information agent. Its function is to traverse the list of known WWW servers and build up an index for all the documents on these servers. It has the ability to handle different data formats, HTML documents, postscript files, compressed files, databases, etc. In the case of interaction with databases on a third-party server, minimum human involvement may be required for the interpretation of the data structures until industry-wide data standards are universally adopted.

Meta-search engine interface. If a user wants to expand the scope of search to the whole Internet, the meta-search engine routes the query to other general purpose WWW search engines, such as Alta Vista, Infoseek, etc. In this event, the CCIN server still uses the local (k-b) in processing the search results so that the returned documents can be appropriately ranked according to their significance.

The CCIN client-agent is implemented as a plug-in package to standard WWW browsers, such as Microsoft Internet Explorer and Netscape, to provide additional functions. It is available for free to the users. Its main purpose is to monitor the user's information retrieval activities and communicate with the CCIN servers with the explicit permission of the user.

User profiler. When a user is connected to the CCIN server, he or she is presented with a collection of built-in user profiles characterising the main type of users related to the construction industry. Each profile is associated with a set of keywords and information context, and can be used as independent key for formulating queries. The user can select one or more standard profiles according to his/her interests. The user's profile is further personalised during the process of querying. The local computer stores the personal profile.

Context. There is always a reason behind a user's search for information. The reason could be performing a design task, writing building specifications, doing research on a particular topic, and so on. The context seeks to use a structured framework to capture knowledge of this aspect. It is another channel for collaborative-information search. For example, a user can request for documents associated with a stage or related stages of the context. Similarly, the CCIN server has an initial built-in context model that will be modified with constant input from the client-agents.

Intelligent page store. This is a local repository facility for the end-user providing enhanced storage which is more comprehensive than the existing WWW browser's bookmarks function. It helps the user to organise downloaded Internet documents for quick recall in the future. It adopts the same structure as the IPS layer of the CCIN server Index Module.

At present, the CCIN system is at 'proof of concept' stage. Much work is still needed to develop it into a practical tool for construction users.

9.6 Conclusion

The emergence of the WWW as a global information network brings with it a number of challenges for helping non-technical users to retrieve information effectively. A number of characteristics of the WWW make the information-retrieval task difficult. The WWW is a very fragmented network with millions of information providers and even more users. It is very difficult to enforce any unified classification schemes over the WWW. The contents of the WWW are constantly changing, new materials are being added and existing materials are being removed or modified.

This chapter reviewed the state-of-the-art of the Internet search and retrieval technologies. The more established information search and retrieval solutions, such as search engines and hierarchical directories, continue to improve. Several advanced techniques, especially agent-based systems, have shown particular promise. It is safe to predict that more intelligent and user-friendly information search and retrieval systems will soon emerge utilising these techniques.

References

Boone, G. (1998) 'Concept features in re-agent, an intelligent email agent', in *The Second International Conference on Autonomous Agents*, Minneapolis/St. Paul, MN, ACM, pp. 141–8.

Brümmer, A., Day, M., Hiom, D., Koch, T., Peereboom, M., Puulter, A. and Worsfold, E. (1997) 'The role of classification schemes in Internet resource description and discovery', *Research Report*, NetLab, Lund University Library, Sweden, http://www.ub2.lu.se/desire/radar/reports/D3.2.3

Cohen, W. W. (1996) 'Learning rules that classify e-mail', in *Proceedings of the AAAI Spring Symposium on Machine Learning in Information Access*, AAAI Press.

Davies, J., Weeks, R. and Revett, M. (1997) *Jasper: Communicating information agents for WWW*, technical report, BT laboratories.

Fowler, A. (1997) *Dewey Decimal System*, Children's Press Inc., USA.

Gilster, P. (1996) *Finding it on the Internet; The Internet Navigators Guide to Search Tools and Techniques*, John Wiley & Sons, UK.

Goldberg, D., Nichols, D., Oki, B. M. and Terry, D. (1992) 'Using collaborative filtering to weave an information tapestry', *Communications of the ACM*, 35(12): 61–70.

Hermans, B. (1996) 'Intelligent software agents on the Internet', Tilburg University, Tilberg, The Netherlands, http://www.hermans.org/agents

Hill, W., Stead, L., Rosenstein, M. and Furnas, G. (1995) 'Recommending and evaluating choices in a virtual community of use', in *Proceedings of ACM CHI'95*, ACM, Denver, CO, pp. 194–201.

Janca, P. (1995) *Pragmatic Application of Information Agents*, BIS Strategic Decisions, Nowell, United States.

Konstan, J. A., Miller, B. N., Maltz, D., Herlocker, J. L., Gordon, L. R. and Riedl, J. (1997) 'Applying collaborative filtering to usenet news', *Communications of the ACM*, 40(3): 77–87.

Lieberman, H. (1997) 'Autonomous interface agents', in *Proceedings of ACM CHI'97*, ACM, pp. 67–74.

Lockley, S. R. and Amor, R. (1998) 'The construction information gateway', in R. Amor (ed.) *Proceeding of ECPPM'98 Product and Process Modelling in the Building Industry*, BRE, UK, pp. 337–48.

Lynch, C. (1997) 'Searching the Internet', *Scientific American*, 276(3): 52–6.

Maes, P. (1995) 'Agents that reduce work and information overload', in *Readings in Human-Computer Interaction, Toward the Year 2000*, Morgan Kauffman.

Manning, G. (1999) *About the Virtual Library*, On-line WWW page: http://vlib.org/AboutVL.html

Newell, S. C. (1997) 'User models and filtering agents for improved Internet information retrieval', *User Modeling and User-Adapted Interaction*, 7(4): 239–56.

Nwana, H. S. (1996) 'Software agents: an overview', *The Knowledge Engineering Review*, 11(3): 205–44. October/November http://www.labs.bt.com/project/agents/publish/papers/review1.htm

Pazzam, M. J. and Billsus, D. (1997) 'Learning and revising user profiles: the identification of interesting Web Sites', *Machine Learning*, 27(3): 313–31.

Schatz, B. (1997) 'Information retrieval in digital libraries: bring search to the Net', *Science*, 275, 17 January, 327–34.

Shardanand, U. and Maes, P. (1995) 'Social information filtering: algorithms for automating "word of mouth"', in *Proceedings of ACM CHI'95*, ACM, Denver, CO.

Sidler, Gabriel, Scott, Andrew and Wolf, Physiker Heiner (1997) 'Collaborative browsing in the WWW', in *Proceedings of the Eighth Joint European Networking Conference*, Edinburgh, 12–15 May, http://www.tik.ee.ethz.ch/~cobrow/papers/jenc8/jenc8.html

Sun, M., Bakis, N. and Watson, I. (2000) 'Intelligent agent based collaborative construction information network', *International Journal of Construction Information Technology*, 7(2): 35–46.

Terveen, L., Hill, W., Amento, B., McDonald, D. and Creter, J. (1997) 'PHOAKS: a system for sharing recommendations', *Communications of the ACM*, 40(3): 59–62.

Twidale, M. B. and Nichols, D. M. (1998) 'Computer supported cooperative work in information search and retrieval', *Annual Review of Information Science and Technology*, 33: 259–319.

Zakon, R. H. (1996) 'Hobbes' Internet Timeline v2.3a., http://info.isoc.org/guest/zakon/Internet/

URL1: Nua's Surveys: can be accessed at http://www.nua.ie/surveys

URL2: Yahoo: 1999, On-line WWW Page: http://www.yahoo.com

URL3: http://www.metacrawler.com

URL4: http://www.dogpile.com

URL5: http://www.copernic.com

URL6: http://www.agentware.com/main/agent/index.html

Chapter 10
Agents for standards processing
H. Kiliccote and J. H. Garrett, Jr

10.1 Introduction

Representing and reasoning with engineering design standards, regulations and codes is a highly active research area. Various methods and environments have been proposed and developed in the last 30 years. The existing proposals for standards processing include: the SASE Model (Fenves *et al.*, 1987); knowledge-based (k-b) models (Lopez and Elam, 1989; Topping and Kumar, 1989; Garrett *et al.*, 1995); logic-based models for standard representation (Jain *et al.*, 1989; Rasdorf and Lakmazaheri, 1990); object-oriented models (Garrett and Hakim, 1992; Badrah *et al.*, 1998); description logic models (Hakim and Garrett, 1993); hybrid models (Yabuki and Law, 1993; Neilson and Kumar, 1998); context-oriented models (Kiliccote *et al.*, 1994); and fuzzy logic-based models (Schnellenbach-Held and Albert, 2000). A more detailed description and history of most of the earlier methods and environments can be found in (Fenves *et al.*, 1995). Most of these methods and environments employ a single representation and reasoning method with a single locus of control, focus of attention and knowledge base. Each of these representation and reasoning methods has advantages and disadvantages compared to the other proposed methods. There are certainly other representation and reasoning methods that have not yet been used in standards processing. However, as identified by Kiliccote (1997), there is no single representation and reasoning method that is general, powerful and complete enough to represent and reason with complex design standards, such as the BOCA National Building Code (BOCA, 1993). Even if a unifying representation and reasoning method were developed, it would be difficult to use this single representation to represent and reason with all parts of a design standard. Also, as identified by various researchers (Guha and Lenat, 1990; Johnson and Mead, 1991; Sowa, 1991), the development of general representation and reasoning methods is not simply a matter of scaling up from simple ones.

To identify the limitations of the existing methods for representing design standards, we analysed a large number of a standards. Our analysis showed that a major weakness of existing proposals is that they cannot accommodate indeterminate provisions. The most frequent cause of indeterminacy of standards is caused by open-textured concepts used in expressing the provisions. An open-textured concept cannot be applied automatically. It requires judgement and is context-dependent. Another weakness of the existing proposals is that they ignore higher-order provisions of the design standards. Even the ratio of higher-order provisions vary from standards to standards, we found that higher-order provisions constitute 20–30% of the statements found in a design standard. For example, chapter 10 of the 1993 BOCA National Building Code (BOCA, 1993) contains 429 provisions, 167 of which are not first order. Another weakness of existing proposals is that they only support one method representation and reasoning strategy. The lack of support for heterogeneous representation and reasoning strategies severely limit the developers from the best representation and reasoning methodology for any given problem. For example, some strategies suit indeterminate reasoning better than some other methodologies (e.g. fuzzy reasoning, neural networks, statistical methods, machine learning). Another major limitation of the existing proposals is that they do not provide access to multiple design standards and supplementary programs modelled or developed by different organisations. For example, designers use a variety of supplementary programs, such as analysis packages (e.g. structural-analysis packages that use matrix and finite element methods), simulation programs (e.g. emergency escape simulations) or other standards-processing tools while applying the knowledge.

We propose to solve the scalability and other identified problems by using a distributed framework to represent and reason with design standards. The main theme of this approach is to develop models of design standards from a collection of agents, each using possibly a different representation and reasoning method rather than using a single, very general representation and reasoning method.

In the construction industry (actually in most industries), design evaluation has become an increasingly co-operative endeavour carried out by multiple agents with diverse kinds of expertise (Klein, 1991). A single person, team or even a large company does not have sufficient knowledge or resources to perform the design evaluation for all the aspects of a large construction project. The traditional approach to dealing with this problem is to apply specialised knowledge workers with specialised knowledge bases to different aspects of the problem. Different phases in the life cycle of design evaluation are handled with different experts possibly located at different places. Most of these experts use various software tools to perform their tasks. While most of these software tools provide their users with adequate support when used in isolation, there is increasing demand for tools that can operate together, that is, to exchange information and services with other programs and thereby solve problems that cannot be solved alone (Wilson, 1993; Genesereth *et al.*, 1994).

As various companies are involved in the development of these programs, they are written by different people at different times in different languages and run on different machines. Getting programs to work together often necessitates extensive work on the part of the users of those programs or their programmers – to learn the characteristics of completed programs and to negotiate communication protocols for programs under development (Genesereth *et al.*, 1994).

To facilitate the task of getting existing and future programs related to standards representation and processing to work together, a standards processing framework (SPF) has been developed. A prototype of this framework has been implemented by the authors in C++ and consists of more that 50K lines of source code. SPF is a distributed network of SPF Agents. Examples of SPF Agents are design systems, standards processor servers, supplementary programs, standards facilitators, and guides and heuristics supplied by designers, companies or standard organisations. The SPF can be viewed in two ways. When viewed as a single entity, the framework is a system that has a certain number of capabilities, which is the union of the capabilities of all the agents. Alternatively, SPF can be viewed as a collection of independent agents that communicate using a specified communication protocol.

One immediate benefit of distributing the representation and reasoning to multiple agents is that the modellers can selectively deploy other proposed methods for representing design standards as agents in the same system. In this case, SPF functions as a glue between these methods. Such a framework can also accommodate various types of knowledge bases and provide access to multiple-design standards. Guides and heuristics supplied by the designers, companies or standard organisations may be agents in this distributed framework. These agents may be very broadly distributed and running on multiple computers all over the world, or they may be very simple and be running in the same address space on a single computer.

In this chapter, this agent-based approach is presented. Section 10.2 presents the architecture of the framework. Section 10.3 presents the communication protocol among the agents of this framework. Section 10.4 presents the architecture of the agents of this framework. Section 10.5 presents the concept of external functions and their use in SPF. Section 10.6 presents a common processing engine for the agents in this framework. Section 10.7 presents an example. The last section summarises this chapter.

10.2 Architecture of SPF

The overall architecture of SPF is shown in Figure 10.1. SPF Agents may be located in the same process, in different processes (communicating using interprocess communication), on different machines in the same organisation (communicating using a local network) or on different machines in different organisations distributed all over the world (communicating using the Internet).

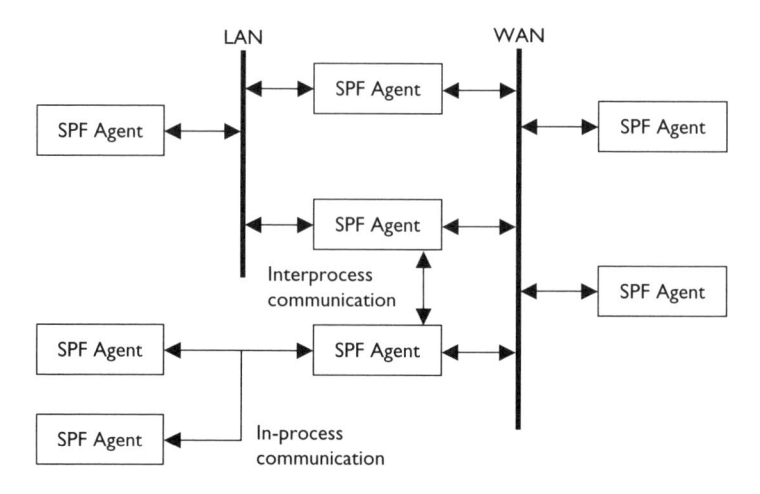

Figure 10.1 The overall architecture of SPF.

SPF does not impose an overall hierarchy or organisation. The SPF agents are free to co-ordinate their activities by communicating with each other directly. In order to save resources when they act collectively, agents may organise their activities. Organising conserves resources by providing a basis for expectations and commitments about the behaviour of other agents (Gasser *et al.*, 1987). In SPF, the term organisation refers to the abstraction that allows agents to treat a collection of other agents as being part of a known concerted effort (definition adapted from Gasser *et al.*, 1987).

Since the SPF does not impose an overall hierarchy or organisation, it is able to accommodate multiple types of organisations. A possible architecture is the *federation architecture* to support anonymous interaction between agents. (The federation architecture in this chapter refers to the architecture explained in Genesereth *et al.*, 1994.) In this architecture, agents relinquish some of their autonomy to another SPF Agent and agree to abide by additional constraints. To each agent, it appears that there is a single SPF Agent that handles all requests directly. This SPF Agent, called a 'facilitator', acts as a virtual agent with the capabilities of all other agents. In this architecture, the interaction of the SPF Agents (one being the facilitator) can be as shown in Figure 10.2.

In the federation architecture, messages from the SPF Agents to the facilitator are undirected, that is, they have content, but they do not specify addresses. It is the responsibility of the facilitator to route these messages to agents able to handle them. We discovered that when there are many agents and heavy interaction between agents, two major problems degrade

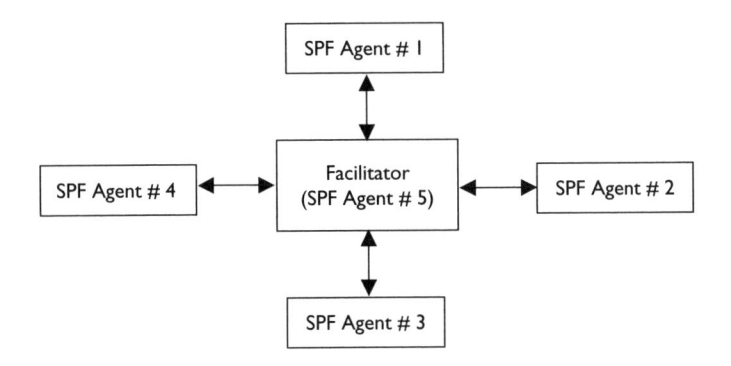

Figure 10.2 Federation architecture.

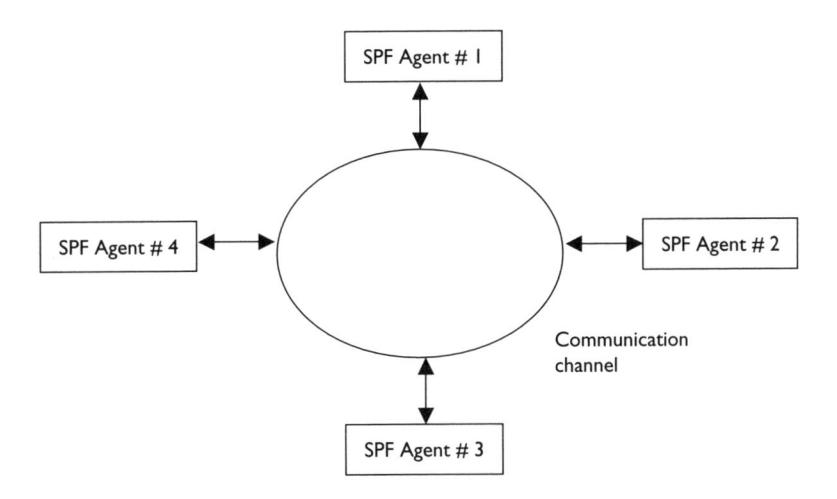

Figure 10.3 Agent community.

the performance of this architecture:

- the facilitator can be easily overloaded because it has to handle all the messages sent from all other agents in the organisation and
- all the messages are unnecessarily sent and parsed twice (first in the facilitator and then in the agents able to handle them).

To solve these problems, SPF is designed to also support another type of architecture: an *agent community*. This architecture is shown in Figure 10.3. In this architecture, agents communicate with each other directly. A single agent acts like a 'librarian' to store the addresses of other agents and the basic information shared by all other agents. In this architecture, a new agent registers itself to the librarian by sending its network address. The librarian returns the addresses of all other agents in the network to the agent and informs other agents of this new agent.

SPF is not a blackboard architecture. Blackboard architectures have three defining features: 'a global database called the blackboard, independent knowledge sources that generate solution elements on the blackboard, and a scheduler to control knowledge source activity' (Hayes-Roth, 1987). In SPF, none of the above assumptions are made. Although SPF is not a blackboard architecture, some of the agents can be specialised to implement a blackboard and a scheduler to mimic a blackboard architecture.

10.3 Communication protocol between the SPF Agents

Similar to other distributed systems, SPF mandates a single universal communication language, one in which inconsistencies and arbitrary notational variations are eliminated. This universal language is based on the idea that communication can be best modelled as the exchange of declarative statements. To allow the exchange of declarative statements, the communication language is based on semantic networks, which are networks of nodes and links used for representing knowledge for use by various types of reasoning methods (Woods, 1990). Semantic networks are labelled, directed acyclic graphs in which nodes represent entities and labelled arcs represent binary relations between the entities (Shapiro, 1991). Higher-order relations (example ternary relations) can be reduced to binary relations by decomposing the relation into a node and deploying a link between the node and each argument of the relation. Semantic networks provide modularity that facilitates entering and checking knowledge, and supports algorithms that seem in practice to provide reasonable performance (Thomason and Touretzky, 1991).

The communication language between the SPF Agents is called the SPF Communication Language (SPF-CL). The messages expressed in SPF-CL are encapsulated in expressions in the Knowledge Query and Manipulation Language (KQML) (Finin and Wiederhold, 1991; Finin et al., 1995) which is a language and protocol for exchanging information and knowledge in multi-agent systems (MAS). KQML is part of a larger effort, ARPA Knowledge Sharing Effort (Genesereth and Singh, 1993), which is aimed at developing techniques and methodologies for building large-scale knowledge bases. KQML is both a message format and a message-handling protocol to support knowledge sharing among agents. KQML uses a set of operators (called performatives) to define permissible operations that agents use to access each other's knowledge. KQML is indifferent to the language used to express the message content. In KQML, such languages are called 'content languages' (Finin *et al.*, 1995). In this sense, SPF-CL is a content language in KQML.

SPF-CL allows information processing in two ways. In the first case, an agent contacts another agent and requests a piece of information or service that it requires to complete its function. The contacted agent collects all the information that it requires to complete this service, performs this service and returns the result to the agent. We call this type of information processing *information shifting*. In the other case, the contacted agent sends all the information required to perform the service to the agent that needs the service. The actual service is performed by the requesting agent. From Guha (1991), we call this type of information-processing *information lifting*.

SPF-CL is used to describe and ask for

- information about agents and organisations (e.g. to define that there is an agent that can compute the maximum allowable height of a building);
- design standards and individual provisions (e.g. to define requirements, including their applicability conditions, and data needs); and
- design information (e.g. to specify the total gross area of a building).

These three descriptive functions of SPF-CL are provided by three separate languages that together comprise SPF-CL:

- the Agent Description Language (ADL);
- the Standards Modeling Language (SML); and
- the Standards Usage Language (SUL).

ADL is used to describe knowledge about agents and organisations. SML is used to describe the logical content of a design standard to an agent. SUL is used to describe the design to an agent and request information about the result of the evaluation of a standards provision against that design.

The syntax and semantics of these three languages are described in Kiliccote (1997).

10.4 SPF Agents

In the SPF, an agent is a self-contained, active entity that communicates with other agents through messages. All SPF Agents must be able to act in four ways:

- they interact with other agents through messages expressed in SPF-CL;
- they contain knowledge;
- they manipulate knowledge; and
- they provide gateways to local reasoning methods available on the machines on which each agent resides.

As long as an SPF Agent can communicate by SPF-CL, they can be implemented using any representation and reasoning method and can be implemented in any language deemed fit by the implementor of that agent.

Since implementing an agent from scratch is a time-consuming and a tedious task, we developed a generic skeleton by which SPF Agents can be easily built and integrated with other agents. The overall architecture of this skeleton is shown in Figure 10.4. The main component of this generic skeleton is *A Standards Processing Engine* (ASPEN). ASPEN supports the four ways in which an SPF Agent must act through

- an SPF-CL parser;
- a knowledge base;
- a planning system; and
- external functions linked with the agents.

The parser transforms the SPF-CL commands into nodes and links in a semantic network in the knowledge base of ASPEN. The manipulation of the knowledge expressed using the semantic network is performed by a planning system, which generates and executes plans to satisfy the requests sent by other SPF Agents using SPF-CL. The external functions, which are perceived as black-box abstractions outside the agent, provide a way to access local resources and reasoning methods available to that agent. The modeller is free to use any available reasoning method to implement these external functions. This will allow for the coexistence of various reasoning methods. One of the most important features of ASPEN is the ability to integrate ASPEN with external functions written in C, C++, or any language or application that provides C bindings, such as ADA, FORTRAN and CLIPS (Johnson Space Center Software Technology Branch, 1993). In the following two sections, we present the properties of the external functions linked with the agents and the planning system used by ASPEN.

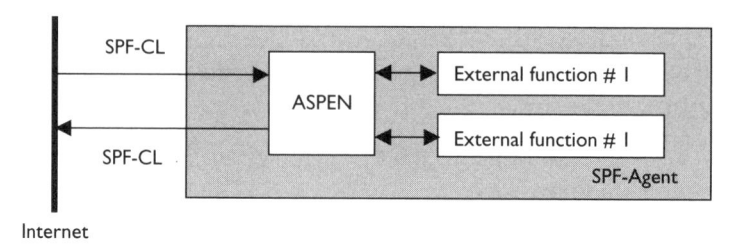

Figure 10.4 A generic skeleton for SPF Agents.

10.5 External functions

In this section, an overview of how external functions can be integrated with SPF is presented. External functions in ASPEN are described in detail in (Kiliccote, 1997).

One of the most important features of ASPEN is its ability to be integrated with external functions or applications. The external functions can be used in various contexts.

- *To support reasoning with open-textured concepts* – open-textured concepts are the ones that cannot be automatically applied to factual situations, but that requires judgement and that are context-dependent (Berman and Hafner, 1988). There are various methods that can be used to deal with the open-textured concepts in design standards. Most of these methods require different computational methods. For example, methods that support case-based reasoning are different than the methods that support fuzzy reasoning. In SPF, such tools are attached as external functions.
- *To access supplementary programs* – a designer uses a variety of supplementary programs, such as analysis packages, simulation programs or other data processing tools. Such supplementary programs can be accessed through the use of external functions.
- *To define provisions that can be more easily represented procedurally* – some algorithms (such as the calculation of the shortest path between two points) can be very difficult to represent declaratively using SML. In such cases a procedural language such as C or FORTRAN may be better suited for the task. Such an algorithm can be implemented using a procedural language and attached as an external function to an agent.
- *To access the missing design information* – a design system does not need to supply all design information at once. It may provide a general description of the design. For example, it may supply the general properties of a building to an agent. In such cases, external functions can be used to access the missing design information whenever the agent needs to access it. Similarly, an external function may be used to ask questions to the designer about missing design information.
- *To extend the capabilities of ASPEN* – an external function can be used to provide a service that the current implementation of ASPEN does not automatically provide. For example, an external function may be used to add support for hypothesis generation.

A modeller can define external functions for use by ASPEN by first declaring it using SPF-CL and then by linking the compiled form of the external function with ASPEN. The external functions declared this way become equivalent to operators obtained from provisions of standards and can be used by other operators in the same manner. In fact, the vast majority of system defined operators provided by ASPEN are integrated with ASPEN in the exact same manner described in this section. The declaration of an external functions within ASPEN is done by describing the applicability conditions and consequents of the external function.

The applicability conditions, declared using SPF-CL, only specify the *minimal* conditions that must be true for this external function to be applicable, that is, the external function will never be deemed applicable if the conditions specified in the applicability conditions are not true. However, there may be other conditions not specified in the applicability conditions which may be checked by the external function to confirm the applicability of this function, that is the external function, by using these unspecified conditions, may deem itself not applicable. If the applicability conditions are more general than they should be and if the actual conditions are checked internally by the external function, the external function may be unnecessarily called. Since calls to external functions are slower than calls to internal functions, this may considerably slow down the execution of a request. If the conditions are more specific than they should be, then ASPEN may incorrectly deem the external function not applicable.

The consequents are used to describe what will happen if the external function is executed. Similar to the treatment of applicability conditions, the consequents declared using SPF-CL specify an approximation of what will happen if the external function is applied. The external function may add more or less information to the knowledge base of ASPEN than is specified in the consequent part of the generic operator. If the approximation of the consequents are more general than they should be (i.e. if they specify more information than it usually adds), the external function may be unnecessarily called. If the approximation of the consequents are more specific than they should be (i.e. if they specify less information than the external function usually adds), then ASPEN may incorrectly deem the external function not applicable.

The applicability conditions and consequents of external functions only describe approximations of the actual conditions and consequents of the external functions. Thus, an external function may need to access or change the nodes and links stored in the knowledge base. ASPEN provides various methods to access the nodes and links stored in its knowledge base.

10.6 Planning system

As described in Section 10.4, agents can send requests to other agents by using SPF-CL. To respond to a request sent to an SPF Agent by other SPF Agents, the agent may need to perform some calculations based on the request and the provisions it represents. Sometimes the request can be answered by searching the knowledge base for the correct answer. However, sometimes a set of provisions must be evaluated. Since the provisions that must be applied are not known in advance, this process must include a search process, which is called the planning phase of the reasoning process. The planning system in ASPEN generates sequences of provisions that must be applied to satisfy one or more explicit requests sent by other SPF Agents. In a classical planning system, such as STRIPS (Fikes and Nilsson, 1971) or SOAR (Laird *et al.*, 1987), an application domain is encoded in terms of a set of actions, their preconditions and effects. In ASPEN, actions are applications of provisions, preconditions are the condition parts of the provisions and effects are the consequent parts of the provisions. The form of the plan generated by ASPEN is a linear sequence of provisions that must be applied to respond to the request sent by another SPF Agent. In the planning phase, after a request is received from an SPF Agent, the search process generates a plan. The execution phase is entered if a plan can be generated in the planning phase. In the execution phase, the provisions selected in the planning phase are applied sequentially.

10.6.1 Planning in ASPEN

In order to solve most non-trivial requests sent by other agents, it is necessary to combine a knowledge representation mechanism with a basic problem-solving strategy. Most problem-solving processes can be described as a search through a state space, in which each state corresponds to a situation that might arise. The search usually starts with an initial situation and a sequence of allowable operations that can be performed until a situation that corresponds to a goal state has been reached (Rich and Knight, 1992).

It has been observed that if a problem is decomposable (i.e. if it is possible to divide the problem that must be solved into smaller pieces and solve these pieces separately), then *planning* can be used in solving the problem. Planning refers to the process of determining several steps of a problem-solving procedure before executing any of them. The main focus of planning is on ways of decomposing the original problem into appropriate subparts and on ways of recording and handling interactions among the subparts as they are detected during the problem-solving process.

In processing the provisions of standards, the solution steps (i.e. the provisions that will be applied) can be ignored or undone if they are deemed unnecessary, incorrect or not applicable. By using this property of standards processing, previous approaches for standards processing adopted a 'blind' search method by actually trying particular provisions. If a dead-end path was reached, then a new one was explored by backtracking to the last choice point. Blind search was adopted in the implementation of the Context-Oriented Method (Kiliccote *et al.*, 1994) and in SPEX (Garrett and Fenves, 1987). In these implementations, all applicable provisions were applied in no particular order until a state in which the original goal is satisfied was reached. In such a system, blind search was a marginally acceptable solution because it did not incorporate external functions as problem-solving methods.

In SPF, the representation of some provisions of a design standard is augmented by external functions that complete the representation of these provisions. Some of the problem-solving is performed by these external functions. These external functions may be located in different agents and may take a very large amount of time to be evaluated. Although the solution steps still can be recoverable (i.e. ignored or undone), since the execution of an external function may take a large amount of time, it is wise to first make sure that the application of an external function actually makes sense. For example, someone should not blindly apply an external function which calculates the stress at a point in a beam using a finite element analysis unless the value of the stress at that point is necessary to solve the problem at hand.

Since the execution of external functions may take a large amount of time, the planning system in ASPEN generates a plan that represents the sequences of provisions that must be applied to satisfy one or more explicit requests sent by other SPF Agents without calling the external functions, thus eliminating most of the unnecessary calls to external functions.

10.6.2 Representation of the operators

ASPEN, by parsing the provisions expressed in SML, generates multiple operators from a single provision. ASPEN describes for each operator, each of the changes it makes to the state description. The most common method to describe each change in state is by using a set of predicates representing the facts that will be added to or deleted from the new state. In this approach, the changes in the state are described by a list of new predicates that the operator causes to become true and a list of old predicates that it causes to become false. In STRIPS (Fikes and Nilsson, 1971), these lists are called the ADD and DELETE lists. A third list, the PRECONDITION list, contains those predicates that must be true for the operator to be applied. Any predicate not included on either the ADD or DELETE list of an operator is assumed to be unaffected by it. This approach is used in ASPEN, except for the following differences:

- Operators use semantic networks to describe PRECONDITION and ADD lists and not predicates as in STRIPS.
- Operators do not have DELETE lists, that is, they cannot be used to delete nodes or links already present in the knowledge base of ASPEN.
- The ADD list is called the *consequent part* of the operator.

- The PRECONDITION list is called the *condition part* of the operator.
- The consequent part can also be used to create new nodes. In all other planning systems, the ADD list can only be used to describe properties of existing instances, that is, they cannot be used to create new instances.

10.6.3 Choosing provisions to apply

ASPEN uses the most widely used technique for selecting appropriate rules to apply: means–ends analysis (Newell *et al.*, 1960). Means–ends analysis methods first isolate the set of differences between the desired goal state described by the initial request and the current state and then to identify those operators that are relevant to reducing those differences. For example, if the goal is to find the area limitation of a building, we would select provisions that calculate the area limitation of the building and not other irrelevant provisions. ASPEN uses the consequent parts of provisions to select the operators that can reduce the differences between the desired goal state and the current state.

10.6.4 Elaborating provisions

Once the operators that can reduce the differences between the desired goal state defined by the initial request and the current state are found, ASPEN checks the applicability of these operators. The application of an operator may depend on the existence of some nodes and links. The condition part of an operator describes the required nodes and links for an operator to be applicable, that is, the condition part of an operator describes a state in which this operator is applicable. The properties of nodes irrelevant to the application of an operator are not described in the condition part of an operator.

Since the state required by an operator may not be the initial state, ASPEN may need to find other operators that reduce the differences between the initial state and the state defined in the condition part of the operator. To be able to do that, ASPEN generates a request for the nodes and links described in the condition part of the operator.

Obviously, planning is a recursive process. ASPEN first finds the operators that can satisfy a request. Then it uses the operators to generate additional requests which can be satisfied by other operators. This process continues until the differences between the initial state and the desired goal state are completely eliminated.

In ASPEN, a request can only be considered as satisfied if it is satisfied by *at least one* operator that applies. An operator may be applied if *all* of the requests it generated are satisfied. This kind of reasoning is best represented as an AND-OR graph (or tree) (Rich and Knight, 1992). In an AND-OR graph, there are nodes and arcs, which are emerging from the nodes and pointing to other nodes. In an AND-OR graph, arcs can be AND arcs or

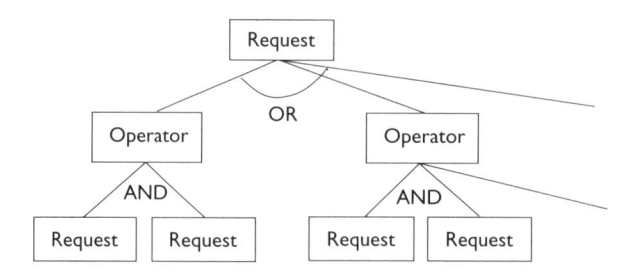

Figure 10.5 AND-OR graphs in ASPEN.

OR arcs. AND arcs may point to any number of successor nodes, all of which must be solved in order for the arc to point to a solution. OR arcs point to a variety of ways in which the problem might be solved.

An AND-OR graph in ASPEN is shown in Figure 10.5. As shown in this figure, nodes in ASPEN are requests or operators. From each request, multiple OR arcs may be emerging and pointing to operators at a lower level. From each operator, multiple AND arcs may be emerging pointing to other requests. If all requests emerging from an operator are satisfied, the operator can be applied. If an operator is applied, the parent request to which this operator is connected is considered satisfied.

The goal of planning in ASPEN is to find a plan that is composed of a set operators that, if applied, would satisfy the initial request. However, since the operators that represent external functions, which are simplified versions of the external functions, are used to generate the plan, a plan may fail in the execution phase.

10.6.5 Executing the plan

After a plan is generated, the execution phase begins. In this phase, the operators selected in the planning phase are applied sequentially. If an operator has an associated external function, the external function is used in the execution phase.

Since the effect of an external function may differ from the simplified version used in the planning phase, a plan generated by ASPEN may fail during the execution phase. There are two more reasons why a plan may fail during the execution phase:

- side effects and
- communication problems.

Side effects. The application of an external function may have an unspecified side effect that invalidates the plan. For example, the actual provision may specify a relation between two instances. If the lack of this relation is used to generate the plan, then a new plan must be generated.

Communication problems. Due to various reasons, an SPF Agent or links between two SPF Agents may be down. In this case, the provisions owned by the SPF Agents cannot be used in the plan. A new plan, which does not use this agent should be generated.

If a plan that does not use this provision cannot be generated, ASPEN deems that its knowledge base does not contain a set of provisions that can satisfy the last request. Depending on the request, this may indicate a success (e.g. the last request could have been issued to prove the non-existence of an instance of a given concept), or a failure (e.g. the last request could have been issued to find an instance of a given concept) or the existence of an insufficient data condition (e.g. the last request could have been issued to find an attribute of an instance).

10.6.6 Remote execution of an external function

When an SPF Agent is created, it downloads all of the information that can be lifted from other agents and it may need to generate a plan for a given request. However, since external functions cannot be lifted (i.e. downloaded to other agents), because they can use a software tool which is not available at the machine in which the agent runs, an agent may need to contact the agent that contains the external function and ask it to execute the external function. This is called remote execution of an external function.

In this case, the execution request, which is an SPF-CL request, is encapsulated in a KQML message and sent automatically to the agent that contains the external function. ASPEN uses only a subset of the performatives that KQML provides. The performatives that ASPEN currently uses are *tell* and *sorry*. The *tell* performative is used to encapsulate queries and their results. The *sorry* performative is used to indicate that a response to the query cannot be generated by the agent that received the query. The *tell* performative has three fields that ASPEN automatically fills: *sender* (i.e. the agent that sends the execution request), the *receiver* (i.e. the agent that will execute the external function) and *content* (i.e. the SPF-CL query).

Since the contacted agent may need a piece of information not present in its knowledge base to execute the request (e.g. an external function that calculates the exit access path from a point to an exit needs the location of the rooms, circulation areas and doors), it may contact the agent that has sent the original execution request and ask for additional information. These queries and their responses are also encapsulated in KQML messages that use the *tell* performatives.

After the remote agent gathers all the information it needs to execute the external function, it executes the external function and returns the result by using an SPF-CL command that is again encapsulated in a KQML message.

10.6.7 Advantages of planning

Generating a plan before executing provisions reactively, has two major advantages:

- planning eliminates many unnecessary computations and
- planning provides a way to use provisions defined in other SPF Agents.

Unnecessary computations. Planning provides a way to find the sequences of operators that will be applied to satisfy a request without executing any external function. The sequences of operators that would lead to a dead-end can be detected at the planning phase and not at the execution phase, without executing any external functions. This eliminates most of the unnecessary calls to external functions. However, notice that planning cannot eliminate all unnecessary calls to external functions because of the reasons described in Section 6.7.1, an external function may be deemed inapplicable at the execution phase.

Provisions defined in other SPF Agents. SPF-CL provides methods to access and use provisions defined in other SPF Agents. These agents can supply a simplified version of the actual provision to other agents. These other agents can then use this simplified version to generate a plan. At the execution phase, the actual agent can be accessed and the original provision can be used. Using the simplified version of a provision to generate a plan eliminates most of the unnecessary interactions between the agents that may occur during the planning phase.

Also, generating a plan in this way has other advantages that are not currently implemented:

- planning provides a way to find the cheapest plan;
- planning provides a way to select among multiple versions of the same provision depending to the problem at hand;
- planning would allow for the integration of a learning system; and
- planning provides a plan with which additional reasoning can be conducted.

Cheapest plan. Planning provides a way to investigate alternative ways of achieving the same result without executing them. Among these alternative ways, the cheapest way (in terms of duration and actual cost) to achieve a result can be chosen. Currently SPF-CL and ASPEN do not provide constructs to describe the cost of applying an operator.

Multiple versions. ASPEN currently maintains two versions of external functions: the simplified versions of external functions, which are used in the planning phase and the actual external functions, which are used in the execution phase. In early stages of design, access to actual external functions may be deemed unnecessary by the designer. In this case, access to the actual external functions may be delayed until the final evaluation phase, which would increase evaluation speed.

Learning system. Currently there are various methods (such as rote learning, learning with macro operators, learning by chunking and example-based learning) that allow a planning system to be integrated with a learning system. The idea behind these systems is to avoid expensive recomputation by recalling and reasoning with the successful plans. If integrated with a planning system, the speed of the planner used in ASPEN can be greatly improved.

Reason with the plan. Since the plan generated by an agent is considered as data, it is possible to reason with it. The plan can be compiled or translated into another language for faster execution. The plan can also be sent to the design system for *compliant design*. In compliant design, the design system uses the information present in the provisions during the design phase to generate designs compliant with the applicable provisions of a design standard.

10.7 Example

The example presented in this section will use a simple provision from BOCA93 that contains an indeterminate term. This example will be used to illustrate how SPF Agents can be set up and how planning can be used to access an external function that deals with the indeterminacy of design standards. In this example, SPF will be used to find the Use Group classification of a building occupied for transaction of business. The main problem is that such buildings can be classified as Use Group B or Use Group M depending on the quantities of the merchandise stored in the building. In this example, four SPF Agents, each running on different machines, will be used:

- Agent Librarian;
- BOCA Agent;
- BOCA Expert; and
- Design System.

Agent librarian. It contains the definitions of the building model used by the BOCA Agent including necessary world knowledge to correctly represent the model. Agent Librarian also maintains the list of other agents and their capabilities.

In this example, the *BOCA Agent* contains, among other concepts pertinent to BOCA such as the Use Group and the Type of Construction classifications, a simple provision from BOCA93. This provision is that of 304.1

from BOCA93, which contains an indeterminate concept, 'merchandise in limited quantities':

> **304.1 General:** All buildings [...] which are occupied for transaction of business [...] that involve stocks of [...] merchandise in limited quantities [...] shall be classified as Use Group B.

This provision can be modelled using SML by stating that if all transactions of business for which the building are used involve stocks of merchandise in limited quantities, then the building can be classified as Use Group B.

The *BOCA Expert* contains the definitions for the information used but not defined in BOCA. In this example, the BOCA Expert contains an external function that can be used to check if stacks of merchandise in a building can be classified 'in limited quantities'. This external function uses a neural network to classify the stocks of merchandise. This neural network has the following three input items:

- type of the merchandise being stocked;
- size of the stock; and
- size of the building.

The neural network has a single binary output node that returns whether the stock can be classified as being in limited quantities. The BOCA Expert contains the simplified version of the external function that is modelled using a generic operator that states that there is an external function that can assess the quantity type of stocks of merchandise in a building used for transaction of business, that is, whether the stocks of merchandise is in limited quantities or not.

The *Design System* will provide the design information. For this example, the design information will be a building that contains 3,000 pages of chapter.

When the design system is created, it downloads all of the information that can be lifted from other agents (i.e. all the list of agents in SPF from Agent Librarian, the definition of the provision 304.1 and concepts used in this provision from BOCA Agent and the generic operator which describes the functionality of the neural network from BOCA Expert). However, the neural network implemented in BOCA Expert cannot be downloaded (i.e. lifted) to design system because it uses a software tool (i.e. the neural network) which is not available at the machine in which the design system agent runs. When the user issues a command to find the use group of the building, the design system generates a plan that uses the provision from BOCA and the external function from BOCA Expert. During the execution phase, the design system contacts BOCA Expert and ask it to execute the external function. This function, by using previous cases used to train a neural network, deems that 3,000 pages of chapter in a building of this size is incidental to its usage and can be considered as being in limited quantities. Thus, the end result of this example would state that the use group of the building is *Use GroupB*.

The various details pertaining to this example can be found in Kiliccote (1997).

10.8 Summary and conclusion

Most of the previous research in standards representation and processing has been concentrated on arriving at a single representation and reasoning method. We realised that by taking an approach whereby the standards processing server is treated as an abstraction and its ability to respond to the communication protocol is its interface, a wide variety of representational techniques can be incorporated into the same model of a standard. As a demonstration of this approach we developed a multi-agent distributed Standards Processing Framework (SPF). SPF facilitates the task of getting existing and future programs related to standards representation and processing to work together. Examples of SPF Agents are design systems, standards processor servers, supplementary programs, standards facilitators and guides and heuristics supplied by designers, companies or standard organisations.

In SPF, an agent is a self-contained and active entity. It communicates with another through messages expressed in a common language, called the SPF Communication Language (SPF-CL). To facilitate the implementation of SPF Agents, we developed a generic skeleton by which SPF Agents can be easily built and integrated with the SPF. The main component of this generic skeleton is A Standards Processing Engine (ASPEN). ASPEN contains a parser which transforms the SPF-CL commands into nodes and links in a semantic network in the knowledge base of ASPEN. A planning system which generates and executes plans to satisfy the requests sent by other SPF Agents, manipulates the knowledge stored in the knowledge base of ASPEN. ASPEN can be integrated with external functions. External functions provide a way to access local resources and reasoning methods available to that agent and are perceived as black-box abstractions outside the agent. The modeller is free to use any available reasoning method to implement these external functions. This allows for the coexistence of various reasoning methods.

The major benefit of distributing the representation and reasoning to multiple agents over existing methods is that the modellers can deploy various representation and reasoning strategies to solve the scalability problem of existing methods. For example, it has been pointed out that existing proposals do not allow reasoning with indeterminate provisions. There are various methods that can be used to deal with the indeterminacy of design standards. SPF allows the co-existence of these methods. This is accomplished through the use of external functions attached to SPF Agents. Each of these external functions may use a different computational method. For example, one external agent may use Bayes's theorem (i.e. the external agent uses prior and posterior probabilities to represent indeterminate data and relations and then computes new probabilities in accordance with Bayes's theorem); one external agent may use fuzzy-logic (i.e. the external function represents the uncertainty of propositions such as beam 1 'is sensibly rectangular' with a distribution of values and then reasons about combinations of distributions); one external agent may use criterion tables (i.e. the external function assigns weights to clauses in rules based on their importance and then draws a conclusion if sufficient numbers of clauses in each rule are true); one external agent may use certainty factors (i.e. the external function assigns certainty factors to propositions and to associations among propositions, then uses formulas to determine certainty factors for inferred propositions); and one external agent may use a neural network (i.e. the external function generates an implicit function based on previous cases and then uses this function to classify new cases). SPF, as opposed to proposing a single approach to deal with indeterminacy, allows the modeller to select the best representation and reasoning method appropriate for each indeterminate provision.

Acknowledgements

This research is based on work supported by the National Institute of Standards and Technology (Computer Integrated Construction Program of Building Fire Research Laboratory) Project No. 08775, National Science Foundation under Grant No. DDM-8957493, US Army Corps of Engineers Research Laboratory under Grant No. DACA 88-93-D-0004 and the Carnegie Mellon University NSF-Sponsored Engineering Design Research Center.

This chapter was first reprinted in *International Journal of IT in Architecture, Engineering and Construction* (*IT-AEC*), Introduction Issue – January 2003, Special Issue on Agents and Multi-Agent Systems in construction as 'Agents for Standards Processing', pp. 9–26 and is reprinted here with permission by *International Journal of IT in Architecture, Engineering and Construction (IT-AEC)*.

References

Badrah, K., MacLeod, I. A. and Kumar, B. (1998) 'Using object-communication for design standards modelling', *Journal of Computing in Civil Engineering, ASCE*, 12(3): 153–61.

Berman, D. and Hafner, C. (1988) 'Obstacles to the development of logic-based models of legal reasoning', in C. Walter (ed.) *Computing Power and Legal Language*, Greenwood Press, Westport, CT.

BOCA (Building Officials and Code Administrators) (1993) BOCA National Building Code/1993, 12th edn, Country Club Hills, IL.

Fenves, S. J., Wright, R. N., Stahl, F. I. and Reed, K. A. (1987) Introduction to SASE: Standards Analysis, Synthesis, and Expression, NBSIR 873513, NBS, Washington, DC.

Fenves, S. J., Garrett, J. H. Jr, Kiliccote, H., Law, K. and Reed, K. (1995) 'Computer representations of design standards and building codes: a U.S. perspective', *International Journal of Construction Information Technology, Special Edition on the National Status of Computer Representations of Design Standards and Building Codes*, 3(1): 13–34.

Fikes, R. E. and Nilsson, N. J. (1971) 'STRIPS: A new approach to the application of theorem proving to problem-solving', *Artificial Intelligence,* 2(3–4): 189–208.

Finin, T. and Wiederhold, G. (1991) An overview of KQML: A Knowledge Query Manipulation Language. Research Report. Department of Computer Science, Stanford University, Stanford, CA.

Finin, T., Weber, J., Wiederhold, G., Genesereth, M., Fritzson, R., McGuire, J., Shapiro, S. and Beck, C. (1995) Specification of the KQML Agent-Communication Language. Research Report. Department of Computer Science, Stanford University, Stanford, CA.

Garrett, J. H. Jr and Fenves, S. J. (1987) 'A knowledge-based standards processor for structure component design', *Engineering with Computers*, 2(4): 219–38.

Garrett, J. H. Jr and Hakim, M. M. (1992) 'An object-oriented model of engineering design standard', *Journal of computing in Civil Engineering*, July, 6(3): 323–47.

Garrett, J. H. Jr, Mehrafza, M. J., Meinecke, C. and Scherer, R. J. (1995) 'Towards a standard-independent design process', in *Proceedings of the IABSE Colloquium on Knowledge Support Systems in Civil Engineering, Bergamo*, IABSE Publications, Zürich, Switzerland.

Gasser, L., Braganza, C. and Herman, N. (1987) 'MACE: a flexible testbed for distributed AI research', in M. N. Huhns (ed.) *Distributed Artificial Intelligence*, Pitman Publishing, London, UK.

Genesereth, M. R. and Singh, N. A. (1993) Knowledge Sharing Approach to Software Interoperation. Logic Group, Computer Science Department, Stanford University, Stanford, CA.

Genesereth, M. R., Singh, N. A. and Syed, M. A. (1994) A Distributed and Anonymous Knowledge Sharing Approach to Software Interoperation. Logic Group, Computer Science Department, Stanford University, Stanford, CA.

Guha, R. V. (1991) Contexts: A Formalization and Some Applications. Technical Report. Stanford University. Department of Computer Science. STAN-CS-91-1399. Stanford, CA.

Guha, R. V. and Lenat, D. B. (1990) 'Cyc: a mid-term report', *AI magazine*, 11(3): 31–59.

Hakim, M. M. and Garrett, J. H. Jr (1993) 'A Description Logic approach for representing Engineering Design Standards', *Engineering with Computers*, 9: 108–24.

Hayes-Roth, B. (1987) 'Blackboard systems', in S. C. Shapiro (ed.) *Encyclopedia of Artificial Intelligence*, Wiley, New York, 73–9.

Jain, D., Law, K. H. and Krawinkler, H. (1989) 'On processing standards with predicate calculus', in T. Barnwell (ed.) *Proceedings of the Sixth Conference on Computing in Civil Engineering*, ASCE, Atlanta, GA, pp. 259–66.

Johnson, P. and Mead, D. (1991) 'Legislative knowledge base systems for public administration – some practical issues', in *Proceedings of the Third International Conference on Artificial Intelligence and Law*, 25–28 June 1991, Oxford, UK, pp. 108–17.

Johnson Space Center Software Technology Branch. (1993) CLIPS Reference Manual. Houston, TX.

Kiliccote, H. (1997) A Standards Processing Framework. PhD thesis, Department of Civil Engineering, Carnegie Mellon University, Pittsburgh, PA.

Kiliccote, H., Garrett, J. H. Jr, Chmielenski, T. and Reed, K. A. (1994) 'The context-oriented model: an improved modeling approach for representing and processing design standards', in K. Khozeimeh (ed.) *Proceeedings of the First Congress in Computing in Civil Engineering*, June 1994, Washington, DC, American Society of Civil Engineers, Washington, DC, pp. 145–52.

Kiliccote, H., Garrett, J. H. Jr, Choi, B. and Reed, K. A. (1995) 'A distributed architecture for standards processing', in *Proceedings of the Sixth International Conference on Computing in Civil and Building Engineering*, 12–15 July 1995, Berlin, Germany.

Klein, M. (1991) 'Supporting conflict resolution in cooperative design systems', *IEEE Transactions on Systems, Man and Cybernetics*, 21(6): 1379–90.

Laird, J. E., Newwell, A. and Rosenbloom, P. S. (1987) 'SOAR: An architecture for general intelligence', *Artificial Intelligence*, 33(3): 1–64.

Lopez, L. A., Elam, S. and Reed, K. A. (1989) 'Software concept for checking engineering designs for conformance with codes and standards', *Engineering with Computers*, 5: 63–78.

McCarthy, J. (1980) 'Circumscription a form of non-monotonic reasoning', *Artificial Intelligence*, 13: 27–39.

Neilson, A., Kumar, B. and MacLeod, I. A. (1998) 'An environment for standards processing', *Journal of Computing in Civil Engineering, ASCE*, 12(4): 195–207.

Newell A., Shaw, J. C. and Simon, H. A. (1960) 'Report on a general problem-solving program for a computer', in *Proceedings of the International Conference on Information Processing*, UNESCO, Paris, France.

Rasdorf, W. J. and Lakmazaheri, S. (1990) 'Logic-based approach for processing design standards', *International Journal of Artificial Intelligence for Engineering Design, Analysis, and Manufacturing*, 4(3): 179–92.

Rich, E. and Knight, K. (1992) 'Artificial intelligence', (2nd edn) McGraw Hill, New York.

Schnellenbach-Held, M. and Albert, A. (2000) 'Representation of crisp and fuzzy knowledge for preliminary structural design', in *Proceedings of the Eighth International Conference on Computing in Civil and Building Engineering (ICCCBE-VIII)*, August, Stanford, CA.

Shapiro, S. C. (1991) 'Cables, paths, and "subconscious" reasoning in propositional semantic networks', in J. F. Sowa (ed.) *Principles of Semantic Networks*, Morgan Kauffmann, San Mateo, CA, pp. 137–56.

Sowa, J. F. (1991) 'Current issues in semantic networks', in J. F. Sowa (ed.) *Principles of Semantic Networks*, Morgan Kauffmann, San Mateo, CA, pp. 13–43.

Thomason, R. H. and Touretzky, D. S. (1991) 'Inheritance theory and networks with roles', in J. F. Sowa (ed.) *Principles of Semantic Networks*, Morgan Kauffmann, San Mateo, CA, pp. 231–66.

Topping, B. H. V. and Kumar, B. (1989) 'Knowledge representation and processing for structural design codes', *Engineering Applications of AI*, 2(9): 214–27.

Wilson, P. R. (1993) 'A view of STEP' (standard for the exchange of product model data), in *Proceeding of Geometric Modeling for Product Realization, September–October (1993)*, Rensselaerville, NY, pp. 267–96.

Woods, W. A. (1990) Understanding Subsumption and Taxonomy: A Framework for Progress. Technical report. Harvard University. Center for Research in Computing Technology. TR-19-90. Harvard University, Cambridge, MA.

Yabuki, N. and Law, K. (1993) 'An object-logic model for the representation and processing of design standards', *Engineering with Computers*, 9: 133–59.

Chapter 11

Multi-agent-based procurement in the construction material supply chain

C. E. Udeaja and J. H. M. Tah

11.1 Setting the scene

Arguably, construction is a key activity in any economy. Its practices involve various interrelated activities and cover a wide range of aspects that have been described in construction research literature. The industry is characterised as having many players from diverse disciplines who are brought together at different stages during the life cycle of unique projects (Austin *et al.*, 2001). The supporting activities result in a reliance on an enormous amount of collaboration between the many levels of abstraction and details. In general, existing software applications are not used to their current potential. This may be partially due to the organisational changes that must occur in order to maximise the benefits of their application (Bresnen, 1996). It is certainly due, in part, to the unavailability of suitable systems that can integrate and co-ordinate the activities of varieties of project participants (Udeaja and Tah, 2001).

There are moves towards integration and more collaborative working in the construction industry, with a view to reducing fragmentation, reducing cost and lead times, improving quality and achieving greater client satisfaction (Latham, 1994; Egan, 1998). To aid these changes in the industry, AI has been utilised; this has yielded some success but has been hampered by problems posed by the complex and dynamic nature of the construction supply chain (SC) environment (processes and entities) (Udeaja *et al.*, 2003). Recently, research in multi-agent systems (MAS) has brought an added dimension to distributed decision-making capabilities. The MAS approach provides a useful metaphor for reasoning about collaborative systems (Ndumu and Tah, 1998). This is evident from the fact that the approach typically involves tasks decomposition, negotiation and delegation of tasks to appropriate agents for execution, whilst proactively co-ordinating communications and exchanges between individual agents, monitoring their performances and re-scheduling any malfunctions in service delivery to more capable and willing agents. This is in essence virtual project procurement, which can be exploited in simulating the material SC prior to the actual start of construction.

In this chapter, the key concepts for developing MAS that support the argument for the use of this technology in material SC procurement are examined. The key issues identified from these concepts are used to develop a framework within which the complexity and organisational dynamics that characterise construction material procurement are captured. A hypothetical case study is developed to test and demonstrate the procurement of a material SC using MAS. In short, it is argued that the research is motivated by the need for a methodology and information technology tools that support and automate the task-oriented view of the material SC, which involves sourcing and other activities characterised by negotiation and collaboration.

11.2 Supply chain concept

The concept of SC has evolved enormously in the past few years (Handfield and Nichols, 1999). New (1997) argued that the idea of the SC owes much to the emergence from the 1950s onwards of system theory, and the associated notion of holism. This may be summarised by the observation that the behaviour of a complex system cannot be understood completely by the segregated analysis of its constituent parts. The use of this idea in this regard is neither consistent nor straightforward. Despite these different aspects of SC concepts that have been considered in the literature to date, there is a common ground on the principle tenets of SC. In the following section, the authors explore the development of the SC theme and explain why definitional problems reflect a deeper dilemma.

11.2.1 Supply chain definition

SC has received attention ever since, yet conceptually the SC is not particularly well understood. In providing a definition of the SC, this work supports Harland (1996) contention that within the SC literature there is a confusing profusion of overlapping terminology and meanings. As a consequence, in the literature many related terms can be found such as Value chain (Cox and Thompson, 1998), Supply Network (Harland, 1996), Supply Pipeline (Farmer and Ploos Von Amstel, 1991), Demand Chain (Blackwell and Blackwell, 1999), Logistics (Christopher, 1998) and more recently design chains (Austin *et al.*, 2001) etc. However, how this literature has conceptualised these terms, SC metaphor still remains the underlying theme for research in this area.

In an increasingly complex and dynamic world, organisations are changing their ways of exchanging goods; they co-operate with other organisations rather than internalise and control their activities and resources through vertical integration, or buy and sell remotely through arms-length procurement (Steven, 1991). Therefore, some organisations have started to look at the flow of goods from raw material to the end customer as an integrated process rather than a large number of independent transactions. To capture these full complexities of the flow of goods and to obtain a more holistic view of the process, therefore, organisations are increasingly starting to consider the big picture of SC. The situation has been illustrated in Harland (1996), who identified four specific description of the SC. These are illustrated in Figure 11.1. This illustration of all types of SC is, of course, a simplification of the actual flow of goods and services, as it does not capture the true complexity of the process. In reality, the process is often characterised by horizontal relationships, therefore, links between organisations at the same level or tier. There may also be other relationships, such as customers and suppliers vice versa, and customers with direct access to second tier suppliers. Grounded in the illustration by Harland (1996), this research defines SC as: 'A number of entities, interconnected for the primary purpose of supply of goods and services required by end customer.' This definition implies that the entities are somehow interconnected; consequently, a SC is not merely a group or a cluster of entities. What characterises a SC is that the entities are connected through transactions of goods and services.

11.2.2 Structure of the construction supply chain

The traditional SC structure has shifted, transformed and extended itself into dynamic and ever-changing processes. The transformation transcends the physical boundaries of the whole enterprise and reaches into the global and rapidly evolving series of network (Harland, 1996). Apart from broadening the perspective from single SC to networks, which results in a more holistic and strategic view of the process of supply, the use of the term 'supply chain' can also be linked to the growing complexity of the process. There are a number of ways of structuring SC, one of the most useful is node and link model, with plots usually representing movement over distance and nodes representing places/organisation where goods are stored or processed (DETR, 1998). It is an easy criticism to argue that the idea of SC structure is simplistic. This is because the process by which raw materials are turned into end products and services is rarely a simple linear process chain, and much more like a spaghetti or spider web of complex interconnecting relationships. To argue in this way is, however, to miss the point, the SC structure is a powerful metaphor. It simplifies a complex and dynamic reality. Furthermore, it provides an understanding that there must be a complex interplay of business to business relationships within the process that links raw material manufacture with the end products and services that are created to energise business relationships.

In the construction sector, the SCs are extremely complex, particularly on a large project where the number of separate supplying organisations will run into hundreds, if not thousands (Elliman and Orange, 2000). Most

LEVEL 1 - Internal Chain	The internal supply chain that integrates business functions involved in the flow of materials and information from inbound to outbound end of business.
LEVEL 2 - Dyadic Relationship	The management of dyadic or two party relationships with immediate suppliers.
LEVEL 3 - External Chain	The management of chain of businesses including a supplier, a supplier's suppliers, a customer and a customer's customer and so on.
LEVEL 4 - Network	The management of a network of interconnected businesses involved in the ultimate provision of product and service packages required by end users.

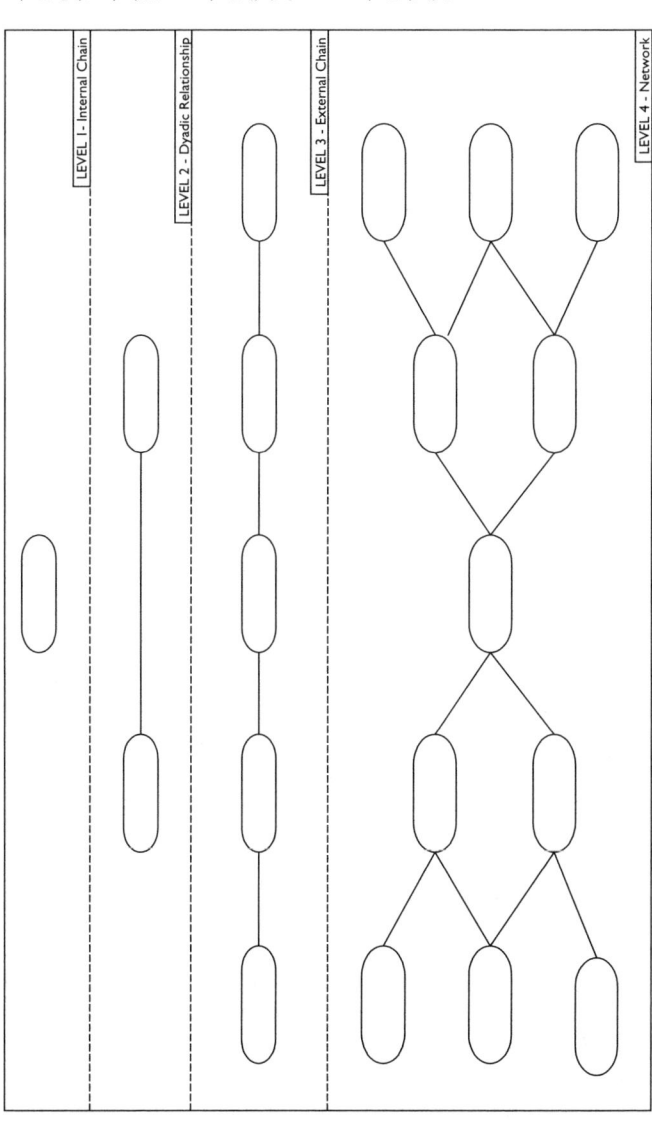

Figure 11.1 Four-main use of the term supply chain.

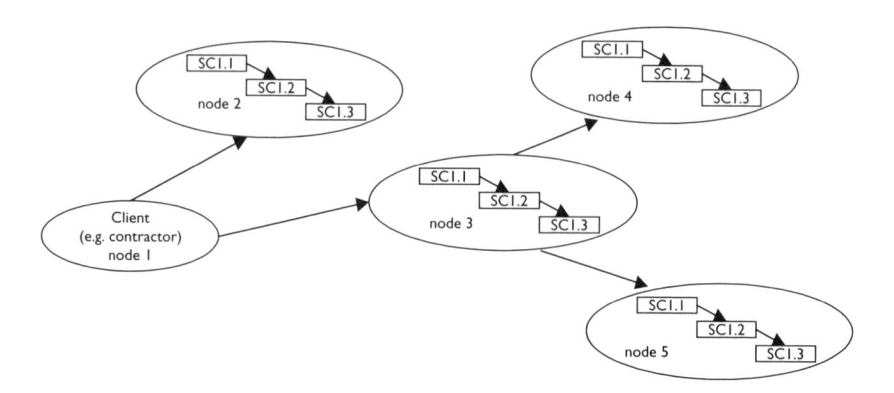

Figure 11.2 The supply chain structure.

organisations are simultaneously members of multiple SCs. An organisation in each chain typically offers a number of products and services, purchasing materials from a wide range of suppliers and selling to multiple customers. From the perspective of a typical organisation, each of its SCs will have both internal and external linkages. Figure 11.2 displays the main elements of a typical construction supply network, with the main contractor at the centre of the hub. There are links to the client, main supply agencies and to both design and any specialist management services, which are provided externally (Vrijhoef and Koskela, 2000). The illustration is a simplification of a real-world network. Clearly, the principle material supply organisations will also be dependent on many other organisations that provide raw material and component inputs to their production. Similarly, the main trade contractors will have their own SCs and many of these will further sub-contract out smaller work packages. The specialist construction sub-contractors will usually be much smaller firms, small to medium size enterprises and several of these may be providing labour only services. The composition of the network will tend to be unique to a specific contract, although some favoured suppliers would be used repeatedly by any given main contractor (Briscoe *et al.*, 2001). The figure illustrates a SC structure via its production graph. Node 1 is the client enterprise, whereas nodes 2–5 are its suppliers.

11.2.3 Typical material procurement process

The process of procuring materials is an area that witnesses a large number of changes throughout a project life cycle. The number and details of the steps involved in the material life cycle, indicate the complex nature of the process and the potential for problems in the process for even a moderate construction project. The series of approvals that must be obtained on a set time schedule, the lead time necessary to meet these set dates and the range of items that run through the life cycle process are all factors which must be incorporated during procurement (Tavakoli and Kakalia, 1993).

The material procurement process is best viewed within the general context of construction procurement. Thus, it commences with the receipt of a tender, running through the award/contract stage, onto completion and finally reconciliation and review of achievement (Meraghni *et al.*, 1996). The contractor serves as the key pointer for the procurement of the entire project for the client. They are responsible for acquiring the materials

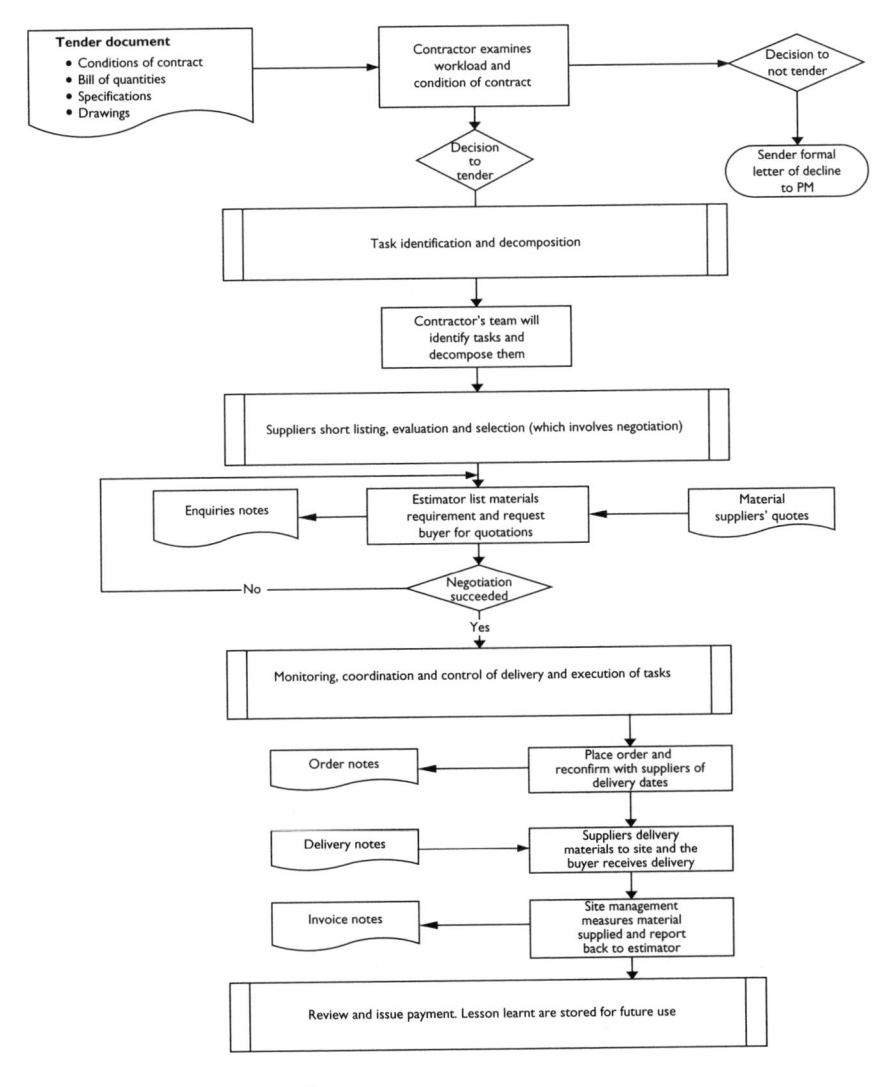

Figure 11.3 A typical material procurement process.

needed to execute specific parts of the project. Much of their time is devoted in translating materials request for the project into procurement order for sub-contractors/suppliers. Figure 11.3 depicts the activities and their sequence in a typical material procurement process. The description of the process presented here is divided into four main parts: task identification and decomposition; suppliers short listing, evaluation and selection; monitoring, co-ordination, and control of delivery and execution of tasks; and review and issue payment. The activities discussed in this section, are not the only activities identified by the industry, as an approach of representing the activities of a material procurement. However, it is the only approach in line with the SC objectives discussed in the previous sections. The approach describes the relationships between inter and intra-organisations and presents the dynamism and complexity in the material procurement process.

11.2.4 Characteristics of construction material supply chain

Having examined the nature of SC concepts and the material procurement processes, it is perhaps appropriate to draw together some of the characteristics that the research to date has reported. Analysis of a number of business processes from material SC domain resulted in several common characteristics being identified:

- Multiple organisations are often involved in the SC process. Each organisation attempts to maximise its own profit within the overall activity.
- Organisations are physically distributed, which may be across one site, across a country or even across continents. This situation is even more apparent for virtual organisations that form alliances for short periods of time and then disband when it is no longer profitable to stay together.
- In a project or inside organisations, there is a decentralised ownership of the tasks, information and resources involved in the SC.
- Different groups within the project or inside organisations are relatively autonomous. They control who, at what cost and in what time frame, consumes resources. They also have their own information systems, with their own idiosyncratic representations, for managing their resources.
- There is a high degree of natural concurrency. That is, many inter-related tasks are running at any given point of the SC process.
- There is a requirement to monitor and manage the overall business process (project). Although the control and resources of the constituent sub-parts are decentralised, there is often a need to place constraints on the entire process (e.g. total time, total budget, etc.).

- Material SC is highly dynamic and unpredictable. It is difficult to give a complete a priori specification of all the activities that need to be performed and how they should be ordered. Any detailed time plans which are produced are often disrupted by unavoidable delays or unanticipated events (e.g. people are ill or tasks take longer than expected).

11.2.5 Understanding the challenges

Despite the extensive studies on material procurement theories and practices, the increasing incidence of poor performance and lack of productivity identified by Latham (1994) and Egan (1998) imply that the current practices are yet to be tapped to their full potentials. Previous research work in this area, have addressed different aspects of traditional construction material SC activities, but there are problems and limitations of these practices, because of the changing nature of modern business processes. The modern business process faces ever-increasing problems in tracking and monitoring; responding to changes in the environment and the needs of clients. This work argues that the major reasons for these problems are as follows: the current tools fail to handle the distributed nature of the business process and lack facilities to manage the resources; the existing systems fail to handle process co-ordination, which is characterised by activities such as competitive negotiation; the current management system tools do not have the ability to cope with dynamic changes in resource levels and task availability and the complex nature of the business process limits the existing tool's ability to dynamically predict changes, due to external events, in both the volume and composition of work entering the business process.

However, research in intelligent systems has opened up new avenues by using software-agents for solving collaborative problems. Barbuceanu (1997) identified that heterogeneity in such a system cannot be tackled by a single centralised system. What is required are several agents (Nwana and Ndumu, 1996) co-ordinating their activities, in order to enable timely dissemination of information, accurate co-ordination of decisions and management of actions among people and systems. This will ultimately determine the efficiency and the viability of the whole material SC. This work addresses the construction material SC procurement in the same way, by organising the SC as a network of collaborating agents, each performing one or more tasks, and each co-ordinating their actions with other agents in the environment. The subsequent sections briefly discuss the principal issues of developing MAS technology, and then proceed to apply this technology in a material SC context.

11.3 Multi-agent systems

The limitations of current applications necessitate a paradigm shift and move from monolithic techniques such as the classic expert system decision-support system. The key concepts of MAS that are required in a distributed environment with particular relevance to the construction material SC procurement problem domain are presented in this section, together with their application to facilitate decision-making. A detailed discussion of MAS was given in Chapter 3.

11.3.1 Key concepts in developing multi-agent systems for a collaborative environment

The field of MAS is a collection of interacting autonomous agents, each having their own capabilities and goals that are situated in a common environment (Wooldridge and Jennings, 1995). This interaction may involve communication involving the passing of information, from one agent to another in an agent-based environment. Research here is concerned with co-ordinating intelligent behaviour among a collection of autonomous agents; how these agents co-ordinate their knowledge, goals, skills and plans to take action and to solve problems (Dijkstra and Timmermans, 2002). All agents' actions are derived from rules embodied into the agents, which depend on local information accessible to the agent. An agent possesses some sensors to perceive the environment within which it moves, and some effectors to act in this environment (Nwana et al., 1997).

In this field, agents in a MAS perform distributed problem-solving by applying concepts such as co-ordination, communication, ontology, reasoning techniques and information discovery. These concepts are well covered in research work by (Wooldridge and Jennings, 1995; Finn and Labrou, 1997; Green et al., 1997; Nwana and Ndumu, 1999; Dijkstra and Timmermans, 2002). However, the notion of autonomous agents collaborating to solve problems is currently popular in the AI research community and in distributed artificial intelligence (DAI) systems (Nwana and Ndumu, 1999). They are meant to help reduce the problem of information overload (Green et al., 1997), and importantly to facilitate interoperability of distributed heterogeneous systems (people and systems from different geographic location) (Ndumu and Tah, 1998). They emphasise certain attributes such as autonomy, flexibility (sociability, responsiveness and proactiveness), co-ordinating and high-level communication (Wooldridge and Jennings, 1995), which are relevant in a material SC procurement context (e.g. co-operation and/or negotiation), which in turn, bring in new tools and techniques that can augment traditional SC strategies.

11.3.2 Brief review of collaborative agent research

In general, business processes, Fox et al. (1991) integrated the SC in a manufacturing enterprise, using an agent-building shell to develop a co-operative SC environment. Their work identified an appropriate decomposition of SC functions and it did provide real-time performance for the SC. Jennings *et al.* (1996) developed an agent-based infrastructure for managing business processes called ADEPT. The project described the key technology of negotiation, service provision and autonomy that are relevant in business process management. The work has since been applied to British Telecommunications (BT) business process and issues raised were the need for richer and more flexible negotiation models, more scalable techniques and more flexible service scheduling algorithms. O'Brien and Wiegand (1996) developed an intelligent agent workflow management system. They adopted a 'service-oriented' view to meet the requirements of open distributed enterprises. The approach embraced and fully supported the decentralisation of responsibility and local autonomy. They concluded that the system automated the provisioning of business process as well as supported business process enactment. At BT Laboratory, the ZEUS building tool kit was used to build a simple SC environment for a PC manufacture (URL1). The test showed how a multi-agent framework lends itself to the organisational structure of a SC.

In construction, Cutkosky *et al.* (1993) described the Palo Alto Collaborative Testbed (PACT), which integrated four legacy concurrent engineering systems into a common framework. The experiment clearly demonstrated the potential of the agent-based approach in facilitating knowledge sharing between heterogeneous systems, but failed to provide support for negotiation. Case and Lu (1995) modelled a discourse model for collaborative engineering design. It contributes a new model for conflict-aware agents, dynamic identification, dissemination of agent interest sets, virtual workspace language, automatic detection of conflict and a unique protocol for negotiation. They identified that the model contributes new procedures for identifying agent interest sets, applying state transformations to the design model, identifying conflicts between designers and tracking resolved conflicts. Halfawy *et al.* (1998) developed MAS architecture that tackled foundation design problems. Their view is that MAS environment organises the design process as a co-operative MAS reasoning process, and employs concepts from the Co-operative Distributed Problem Solving to organise and co-ordinate the design activities. They are hoping to gain more insight into the multi-agent foundation design process and to understand the co-ordination of design activities among a set of co-operating agents. In the UK, Ndumu and Tah (1998) experimented using the ZEUS agent building tool kit (URL1) to build collaborative design systems for a building design. The research argued that the agent-based approach provides a useful metaphor for reasoning about design systems, as well as contributing new tools and techniques for facilitating the collaborative design process. Petrie *et al.* (1999) developed an approach for managing complex distributed projects using agents and a particular representation based upon a general model of design called 'Redux'. They showed how design, planning, scheduling and construction can be interleaved and distributed, but still co-ordinated centrally. The project enabled distributed projects to be completed more quickly, and also start early with less complete information because the system support change notification. Ren *et al.* (2002) developed a MAS to facilitate construction claim negotiation. It shows how a comprehensive agent negotiation mechanism could be developed to deal with complex and dynamic environment.

In the construction SC, there is very little or non-existent research activity into the application of collaborative agents. However, Ndumu and Tah (1998) further experimented with the ZEUS agent building tool kit (URL1) to cover construction SC provisioning. The problem was modelled with over thirty agents, distributed across a network of computers, and representing different disciplines. Their key argument was that agent-based approach effectively supports the procurement process and provided added value over and above Just-In-Time (JIT) techniques. Ugwu *et al.* (2003) discussed and emphasised the need to understand organisational structures, business process issues and workflow models as these are pre-requisites for successful application of agent systems for decision-support and collaborative working in construction. This observation also holds true for SC organisational structures and business processes. The next section presents the case for MAS in Construction Material SC.

11.3.3 A case for MAS-based procurement for the construction material supply chain

The brief description in the previous section has shown that there is a need to move away from treating modern business processes as homogeneous entity and to recognise the importance of distributed and dynamic nature of the environment. Clearly, there are a number of compelling reasons to use MAS to develop and solve problems in distributed and dynamic environment. Thus, the motivation for the increasing interest in MAS research includes the following:

- Process collaboration, which is lacking in most computer support approaches in business processes (such as competitive negotiation).
- Allocation of scarce resources among competing participants, coupled with the distributed nature of modern business processes.
- The modern business process is characterised by dynamic and complex activities, and existing tools fail to incorporate these functions, which may require plans and schedules to be repaired, even when execution is taking place.
- Existing systems are based on single-user model that views the decentralised business processes as one unit, with a view of one system accomplishing the whole task.

- Interoperability of legacy systems that abound the industry, such tools are generally incompatible with one another, and have failed to address the key need of the industry – that of support for collaboration.
- Multi-agent architecture matches the vision many have for the future of Internet computing. That is, the idea of intelligent entities communicating and co-ordinating with each other over wide area networks, which is popular within the Internet community.

It is evident from the discussion that MAS technology has the potential to provide practical solutions to problems in the construction domain. The major focus has been on design, with emphasis on collaborative work. However, in the construction material procurement process there are still significant gaps to be bridged to reach the best practice in its application of MAS. Fundamental changes are required in the existing approaches to incorporate complex internal processes of participants and the organisational dynamics that have often been neglected. The next section, presents the framework covering key issues that are required to model and design an effective and efficient decision support-system for material SC procurement.

11.4 Simulating the material supply chain

As stated earlier, material SC procurement consists of a number of related activities, which collectively realise some project objective. The procurement activities of materials consist of a number of related activities, which are performed in a logical step that could be a manual or automated activity. This in itself is a major problem associated with distributed and complex system. However, to tackle these problems, the author models the SC procurement system using MAS. The area addressed by the framework consists of a decomposition model for participants involved in the material SC environment, negotiation model that deals with selection of suppliers and negotiation of material. The penultimate issue deals with an execution model that incorporates the issue of ordering and payment/exchange of goods. The issue on common understanding and shared knowledge was also added to facilitate sharing and re-use of data and knowledge between complex and dynamic collaborating systems. Thus, the approach emphasises the procurement of a construction material SC that captures the existing structure and the autonomy of its sub-parts, thereby representing the complex and dynamic internal and external processes of organisations involved in the projects. Hence, it provides a more transparent, unsimplified and essentially indeterminate view of the structure and dynamics of SC relationships. In the remainder of this section, the research considers each of these domains in detail.

11.4.1 Task decomposition issue

When faced with a complex procurement problem, a contractor usually solves it by reducing it into a set of smaller more manageable sub-problems. These sub-problems are, in turn, such that a solution can be easily determined. This approach encourages developers to think of the problems in terms of roles that need to be played, and the responsibilities associated with the roles (Kendall, 1998). The problem of task decomposition in the material SC can be seen from several perspectives. In a typical decomposition process, a single super task is decomposed into smaller subtasks (take for instance an alumaco-door assembly as shown in Figure 11.4) each of which requires less knowledge or fewer resources. In general, there should be a choice among alternative task decompositions, depending on the ability of participants to perform such task.

The decomposition problem for multiple agents is more complex because of the need to match resources and capabilities of different agents with appropriate tasks; there must be sufficiency of knowledge, resources and control in the overall system to bring about an effective solution (Turner and Turner, 2000). This means that any decomposition regime will need to account for the capabilities and resources of agents, in making decisions about alternatives types and the granularity of decompositions. Sub-task aggregation, coalescing several tasks into one, large-grained unit, is one method of varying the granularity of tasks distributed to agents. It also becomes important in MAS, where multiple agents collaborate without centralised control and must aggregate their efforts for greater capability (Brazier et al., 1996a), rather than decomposing a global problem.

Brazier et al. (1996b) identify that difficult problems of decomposition arise because of dependencies among sub-problems and among the decisions and actions of separate agents. Conflicts over incompatible actions and shared resources may place ordering constraints on activities, restricting

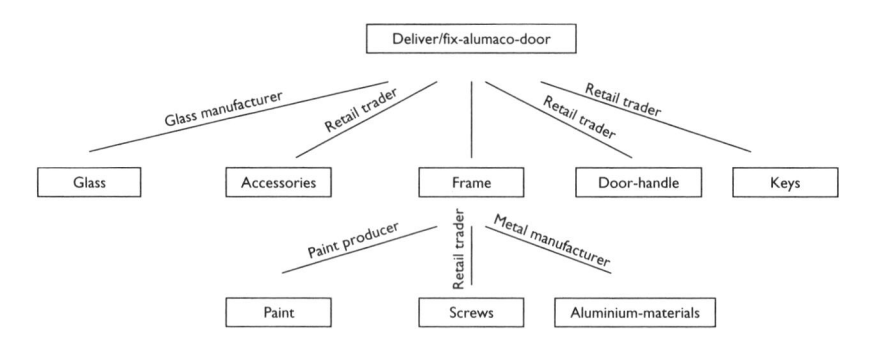

Figure 11.4 Pictorial specification of alumaco-door task decomposition.

decomposition choices and forcing a need to consider decomposition in a temporal (or other resource) dimension as well as the more usual dimensions of knowledge, location or abstraction dimensions. When several agents generate plans independently, there maybe problems in reconciling the sub-plans produced by each agent to reduce dependencies and conflicts (Brazier et al., 1996c). In such circumstances, the issues of redundancy are related to the trade offs between efficiency and reliability; redundancy should be eliminated to improve efficiency, but may be necessary for reliability (Bond and Gasser, 1988). In summation, this research acknowledges that there are various task decomposition approaches such as constrained heuristic search presented by Turner and Turner (2000). However, the work carried out here will use a formal hierarchical approach to model and specify complex dynamic tasks. The basic assumption is that complex (reasoning) tasks can be modelled by a compositional architecture, wherein (sub-)components correspond to (sub-)tasks. This technique assumes some intelligent approach to task decomposition, which need to consider the representation of tasks, several dimensions of decomposition, available operators that can be applied to perform sub-tasks, available resources and dependencies among tasks (Bond and Gasser, 1988).

However, the model of a material SC views the activities as a contractual obligation between two or more participants that can be decomposed into various tasks until the client's needs are satisfied. The model views this material SC environment as a contractual obligation between two or more SC participants described as SC predecessor and SC successor as shown in Figure 11.5.

In both form of SC offering, three types of roles exists: the SC head, the SC participant and the SC tail. In this scenario, the SC predecessor will

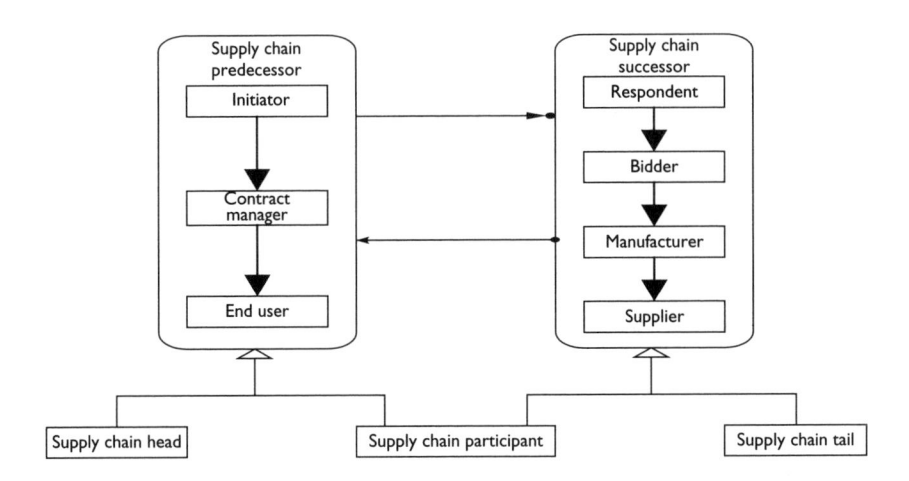

Figure 11.5 SC role model.

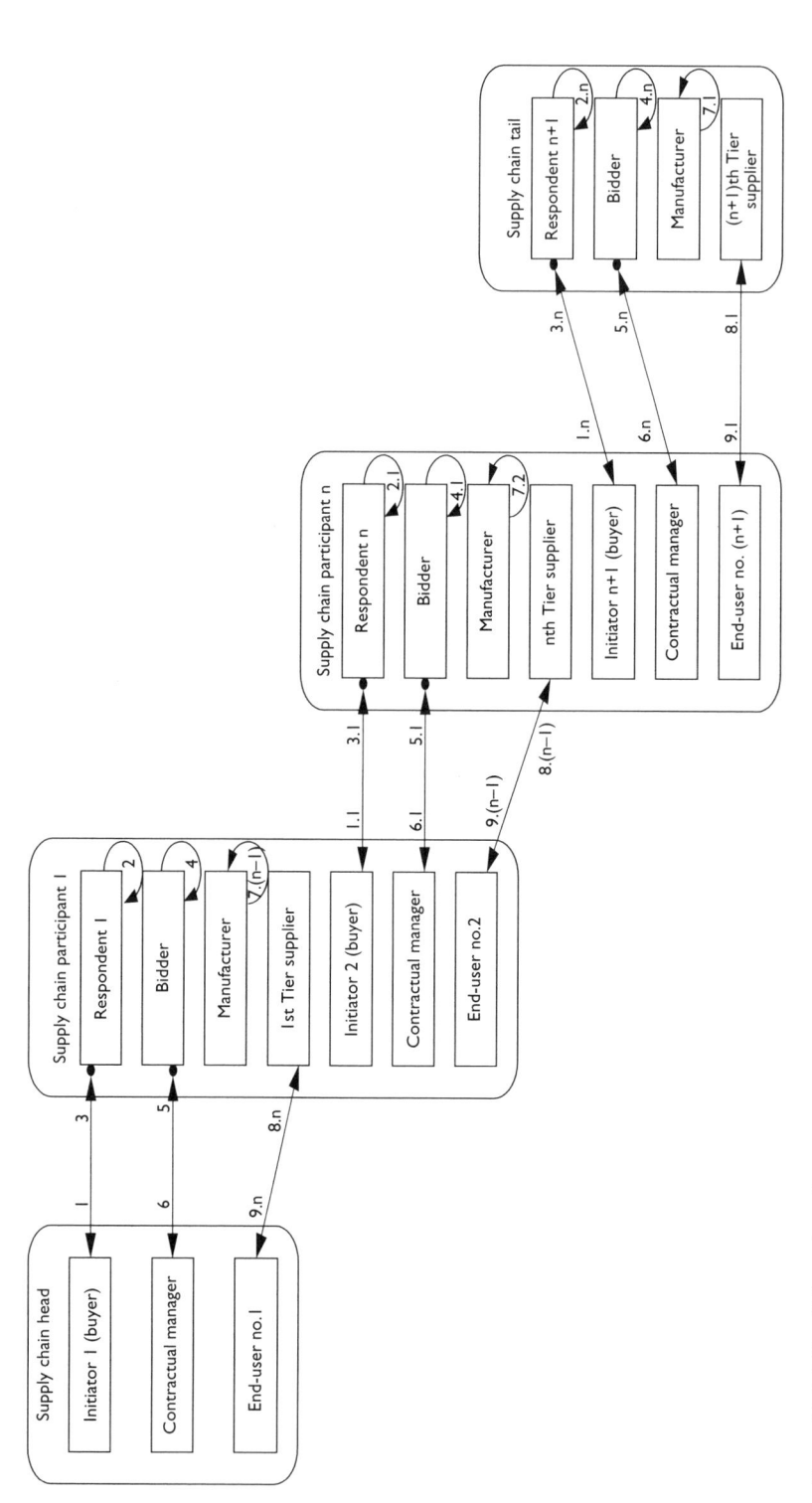

Figure 11.6 SC collaboration diagram.

assume the role of the SC head requesting resources (materials) from the successor SC. A typical example is Project Management Team requesting the services of a Prime Contractor or the Prime Contractor requesting a resource from a number of suppliers to satisfy the client's need (see Figure 11.6). Lastly, the SC tail exists, when a resource is assumed to be the last port of call on the chain that can be used to satisfy the client's need.

The figures show that there are three phases to a SC, one high-level activity involving task decomposition, and the others, which are not obvious from the figure is the low-level activity, involving negotiation and delivery. Under the task decomposition phase, the roles are decomposed into atomic roles depending on how the client (contractor) intends to carry out the material procurement process. The beauty of the task decomposition phase is that it visualises the intra- and inter-organisational structure of the environment (material SC). Once the decomposition is set out, it becomes clear what the interaction modules will be, that is which agent collaborates (communication and negotiation) with whom and in what circumstances. Thus, under the negotiation and delivery, the SC predecessor goes from being a Negotiation Initiator to a Consumer and at the other end a SC successor is first a Negotiation Respondent, a Producer and then a Supplier (see Figures 11.5–11.7). During negotiation, the process commences at the SC head, and finishes at the SC tail. During delivery, it starts at SC tail and ends at the SC head. Once agreement is reached during negotiation, the delivery phase is entered which also includes production. It is important to note that the decimal numbering is utilised in the figure to show SC head and previous SC participants, that their tasks cannot be completed until the subsequent chains finalise their agreement. Although the negotiation and delivery belongs to subsequent sections where the internal details are modelled, the description in this section provides the physical collaboration within the entire SC (i.e. is the intra- and inter-organisational structure).

11.4.2 Selection of suppliers and material negotiation issue

Negotiation is proposed as a means for agents to communicate and compromise to reach mutually beneficial agreements (Jennings *et al.*, 1996). The approach here examines the problem of resource allocation and task distribution among autonomous agents, which can benefit from sharing a common resource or distributing a set of common tasks. The negotiation strategy adopted in this model is based on the multi-round contract-net (Davis and Smith, 1988), which involves one or more initiators that issue a request for quotation (RFQ), and one or more respondents that reply (see Figure 11.7). The figure depicts the relationship between protocols and interaction strategies. At each state, the agent may need to make decisions about how to behave or respond to its current circumstances, these decisions are made using strategies. To understand the behaviour of the

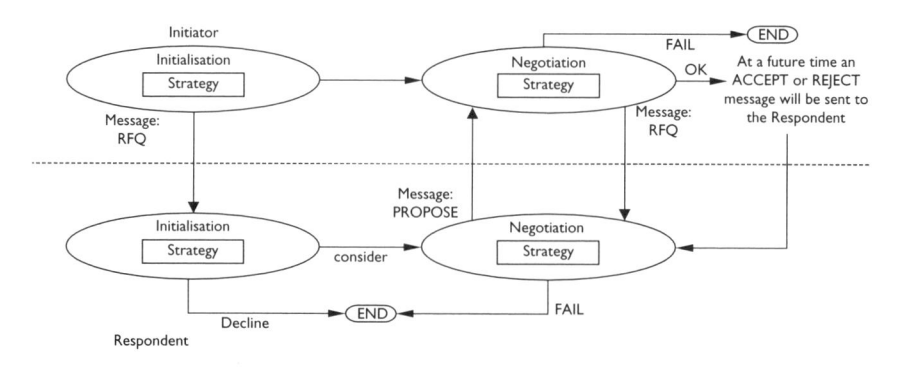

Figure 11.7 Transition diagram of a typical negotiation.
Source: Adapted from URL2.

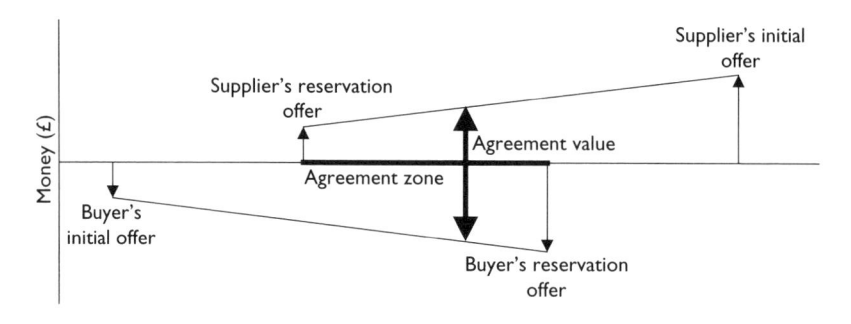

Figure 11.8 Negotiation growth and decay functions of buyer/supplier.

negotiation strategy here, this work considers a contractor who needs a material or component, but cannot provide it locally and must therefore contact a sub-contractor/supplier to supply it. Hence, the Initiator illustrated in Figure 11.7 starts the process by analysing its requirements and determining how much it is willing to pay for the material and how quickly it needs it. Using the expertise encoded into its tendering strategy it formulates a RFQ message containing its requirement, this is then broadcast to all potentially interested parties and the agent moves into the negotiation state to await responses (see Figure 11.7).

The arrival of a RFQ message on the Respondent side causes it to move into initialisation state. If the Respondent decides to respond, it moves into negotiation state. The negotiation process involves an interactive process of offers and counter-offers in which each agent chooses a deal which maximises its expected utility value. The scheme for contractor (buyers) and suppliers involves growth-function and decay-function respectively (see Figure 11.8). The figure depicts that the supplier offers the material at the highest desired price, and then decreases this price according to the decay function (which is specified as being linear, quadratic or cubic). However, when the desired date to sell the material arrives, the asking price should be about the lowest acceptable price. The converse is true with the buyer and its growth-function as shown in the figure. Thus, during negotiation, the Initiator role of the SC predecessor negotiates with the respondent in the successor. It involves request message sent out to all potential suppliers, who on their part have to agree to go into negotiation or refuse to negotiate because of other commitments. The final stage of negotiation, involves the agents applying their negotiation strategies to reach an agreement. This process commences at the supply chain head, and finishes at the SC tail (see Figure 11.7).

The negotiation mechanism proposed here is a strategic approach of negotiation that takes the passage of time during the negotiation process itself into account. The distributed negotiation mechanism that was described is simple, efficient, stable and flexible in various situations. The approach considered situations characterised by complete as well as incomplete information, and ones in which some agents lose overtime while others gain overtime. Using this negotiation mechanism autonomous agents have simple and stable negotiation strategies that result in efficient agreements without delays even when there are dynamic changes in the environment.

11.4.3 Task delivery and execution issue

This issue considered here enables the settlement aspect of trading to be separated from the negotiation mechanism. From a construction perspective, this issue describes the value chains, which refers to the process by which money is exchanged through the SC in response to an initial supply offering (see Figure 11.9). The figure shows the exchange forms part of the contractual agreement. It flows in the opposite direction of the SC. Although it shares the same corporate structures as the SC, it differs in the tasks allocated and the methods-tasks are carried out among the actors within the chains. Thus, unlike the negotiation approach that starts from the client to the last supply tier, the delivery/execution begins at the tail end and terminates when the initial offering has been satisfied and as this is going on, money is being exchanged. Therefore in an agent modelling, the Initiator role in Figure 11.7 becomes a consumer and at the other end the respondent becomes a producer, and then a supplier. The approach is modelled in such a way that if negotiation agreement is reached, the production and delivery of the task follows immediately. This pattern is repeated for the length of the SC – agent environment, as it is apparent in a construction context. During delivery, it starts at SC tail and ends at the SC head. Once agreement is reached during negotiation, the delivery phase is entered which also includes production.

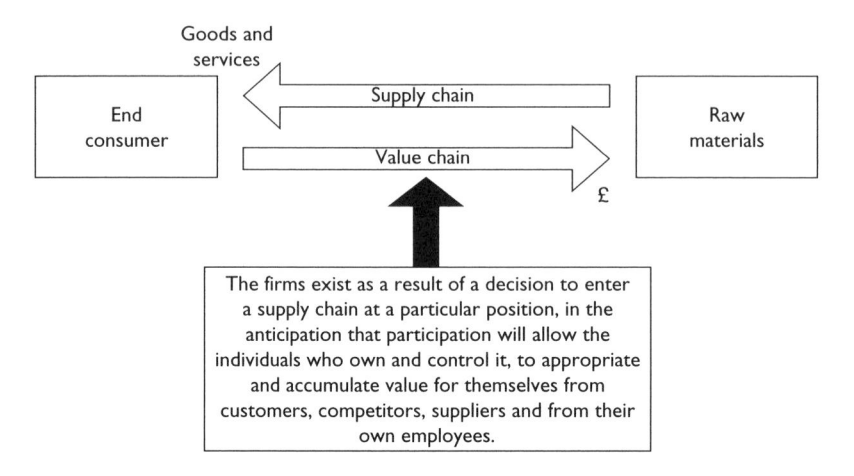

```
                    Goods and
                    services
    ┌──────────┐ ⇐═══════════════════════  ┌──────────┐
    │   End    │      Supply chain         │   Raw    │
    │ consumer │ ─────────────────────────▶│materials │
    └──────────┘      Value chain          └──────────┘
                          ▲            £
                          │
         ┌────────────────┴────────────────────┐
         │ The firms exist as a result of a decision to enter │
         │ a supply chain at a particular position, in the    │
         │ anticipation that participation will allow the     │
         │ individuals who own and control it, to appropriate │
         │ and accumulate value for themselves from           │
         │ customers, competitors, suppliers and from their   │
         │                own employees.                      │
         └────────────────────────────────────────────────────┘
```

Figure 11.9 Supply and value chains.
Source: Adapted from Cox and Thompson, 1998.

11.4.4 Ontological issues

The previous sections have described the key issues relating to the material procurement process. Although, the issues discussed are able to handle agent problems in construction material procurement to some extent, there is an endemic problem as a result of differences in representation of product/ knowledge data. In a construction material SC environment, various participants from different functional areas and disciplines work together to procure materials. A critical problem is to generate a common representation of the problem-solving domain from the various participants (Owen *et al*., 1999). Although the suppliers often share the same objectives such as preparing a material specification that meets the client's demand, they do not necessarily use the same terminology to communicate in the material SC process. Thus, even in a human-centred material SC environment, communication is often hampered because different participants may use different terms to describe different concepts. This makes it difficult to share information/knowledge or re-use knowledge associated with a given set of product/knowledge data in the domain. The problem is exacerbated when the material procurement process and product/knowledge is automated and in the form of MAS that represent the participants involved in a material procurement (Udeaja *et al*., 2003).

In a MAS environment, agents can have different terms for the same concept and identical terms for different concepts. One potential solution to this representation problem is the development of an ontology for the material SC domain concepts and terms along with their unambiguous definitions (Uschold and Gruninger, 1996). As many researchers working in this area have discovered, an ontology is required for representing the knowledge from various domains of discourse. In another instance, Odell (2002) identified that the purpose of the ontology parameter in an agent communication language (Genesereth and Ketchpel, 1994) is to define the set of terms that is used in agent communication. The existence of such ontology in a problem domain would enhance the ability of the participating actors (both human and agents) to interoperate without misunderstanding and consequently share and re-use product and knowledge data (Ugwu et al., 1999, 2001). Gruninger and Fox (1994) defined ontology as a formal description of entities and their properties; it forms a shared terminology for the objects of interest in the domain, along with definitions for the meaning of each of the terms. One can think of the ontology as an evolving theory of distinctions that are worth representing a shared product and knowledge domain (Holsapple and Joshi, 2002).

The ontology in this work is based on the principle of shared product/knowledge and process domain ontologies. Product ontologies are used by all the participants taking part in the environment to describe, match and reason about products. The approach taken in this MAS framework was to define a single, simple data format for information/knowledge exchange that is used to encode the participants' requirement specifications (see Figure 11.10). The use of the same data format for all purposes minimises

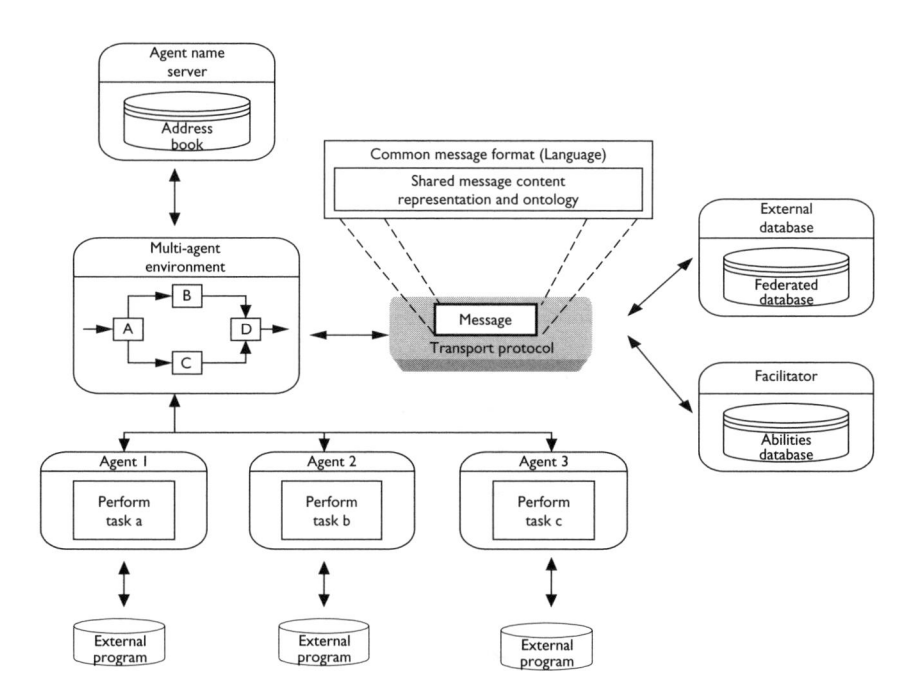

Figure 11.10 Multi-agent system ontological framework.

the complexity of matching the requirement (such as matching a set of buyer's requirements) (Odell, 2002). The figure shows that for entities in a MAS to understand and manipulate the same product and knowledge information in a material SC environment, it is necessary to define a format and protocol for information/knowledge exchange. This format essentially has two aspects (Owen *et al.*, 1999): a set of common concepts that should be understood by all software agents that participate in information/knowledge exchange – this is the vocabulary, or ontology, of the system and syntax to be used to express those concepts.

In creating a product/knowledge data representation format to describe the material supply chain environment, there are two essential characteristics that such a representation should support: the ontology creation should not be generic – that is there is no need to re-invent the wheel, but should allow for easy specification of products, services and suppliers without including domain-specific constructs within the syntax of the representation; and the representation should be sufficiently feature-rich to allow for a range of concepts to be modelled in MAS environment.

To endow the MAS framework in this work with these capabilities, the ZEUS ontology framework (Collis *et al.*, 1998) has been used as a basis to provide the necessary distinctions for describing product/knowledge data. In the ZEUS framework, the key concept is to describe the application, as shown in Figure 11.11. The scenario describes a Door application that has sub-class concepts, namely, wood, glass, key, door-handle, frame and screw. As these concepts refer to physical instances rather than abstract ones they inherit from the Entity fact, this is a child of the root ZEUSFACT concept that provides all its children with a cardinality attribute that represent the number of each concept in existence. All Entity facts also possess an

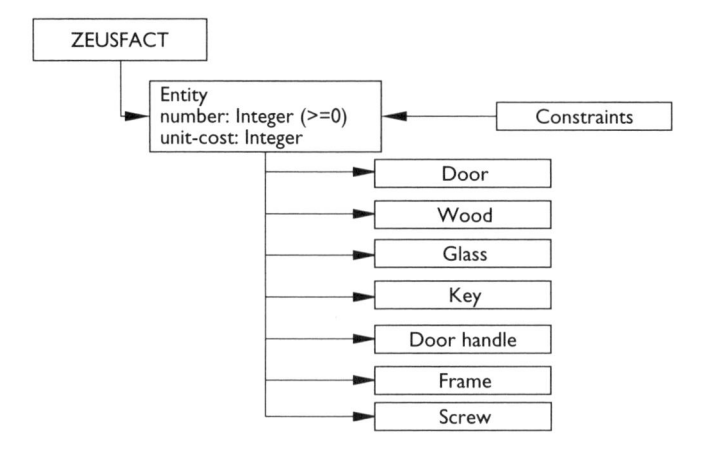

Figure 11.11 ZEUS ontology – door fact description.

attribute that refers to the concept's inherent value. In the Door scenario ontology this can be used to store information about the price of each material/component. No other attributes are explicitly mentioned in the specification of the concepts; the scenario could be easily extended with attributes like sell-by-date, colour, country of origin, environment assessment issues, value of product, etc. However, the actual developments done within this framework are used by agents to: retrieve the most appropriate collaboration model; carry out conflict management and negotiation; retrieve information; and represent them as part of the organisation.

11.5 System design of a multi-agent-based procurement for a material supply chain

The functional requirements for a MAS for material procurement have been presented in the previous section. The overall information flow in the conceptual model, the breakdown of the major functions performed within the SC processes and products have also been described in the previous section. Thus, for the solutions advocated to work properly, their implementation needs to be carried out on a system architecture that enables good collaboration between different entities that exist in the environment. However, the system architecture developed in this work provides the basic platform for integration of virtual teams in a construction material procurement. The system implemented organises the SC participants as a network of co-operating agents as shown in Figure 11.15, each performing one or more functions, and co-ordinating their actions with other agents. The focus of this architecture is to support the construction of MAS for the procurement of a material SC in a manner that guarantees that these agents will engage in an effective and efficient collaboration, communication and problem-solving approach. In doing this, the architecture has been described as a system composed of hybrid architecture each performing part of the tasks involved in the procurement of materials. These components are discussed in the ensuing sections as primary and secondary system architectures.

11.5.1 Primary system design

The primary system architecture presented here, revolve around the ZEUS building tool kit (Collis *et al.*, 1998). The tool kit provides a rapid-engineering environment for developers of collaborative agent systems (Ndumu and Tah, 1998). The framework utilises the ZEUS design philosophy (see Figure 11.12), which portrays the agent's problem-solving ability made up of the agent-level functionality and the domain-level problem solving (Collis *et al.*, 1998). The ZEUS architecture provides classes that implement generic agent-level functionality such as communication, ontology, co-ordination, planning, scheduling, task execution and monitoring and

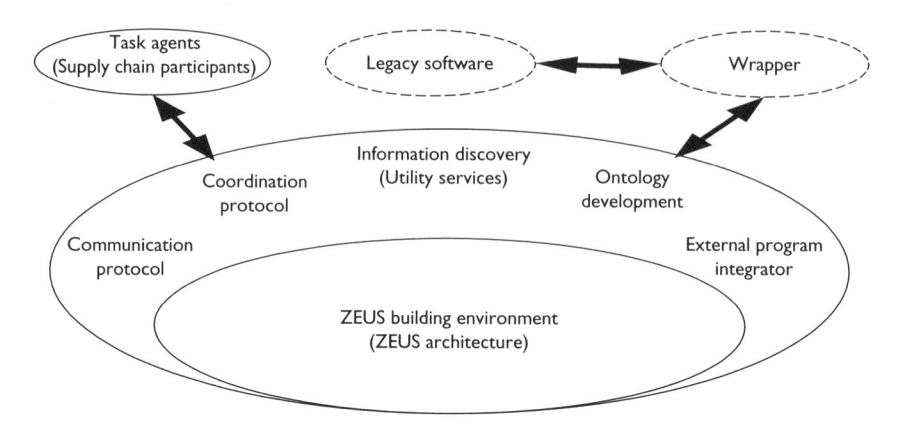

Figure 11.12 Multi-agent system framework.

exception handling. On the other hand, it expects the user to provide codes that implement the agents' domain-level (task agents as shown in Figure 11.12) problem-solving ability. That is, developing various agents with their capabilities. Each agent performs one or more functions, and co-ordinates its decisions with other relevant agents using the agent-level functionality. These task agents can collaborate as co-worker, peer and superior/subordinate agents. The co-worker representing agents of same organisational background and hierarchy, while the peer agents representing agents of same hierarchy, but of different organisations. The superior/subordinate interaction manifest itself in an organisational setting, it allows hierarchical agent system to be constructed in which the superior agent realises their task through subordinate agents, who in turn may have other subordinates.

The issue of legacy software mapping is tackled by adding codes (Wrapper) that act as a mediator between the environment and the external program (see Figure 11.12). Furthermore, the issue of agent discovery is tackled by configuring the information discovery (White and Yellow Pages) in ZEUS. The white page also known as Agent Name Server (ANS) is used to keep a register of all agents in the environment and the yellow page also known as the facilitator is used as a look-up service for agents' abilities or expertise. Figure 11.13 shows an example of a scenario interaction diagram between the task agents and the ZEUS utility agents (Yellow and White Pages). The scenario shows that when a task is defined, it knows the address of its nameserver agent. During initialisation, each task is expected to join the community by registering itself with the nameserver. The nameserver is then responsible for informing others of the presence and network address (such as TCP/IP address) of this new agent. The nameserver's

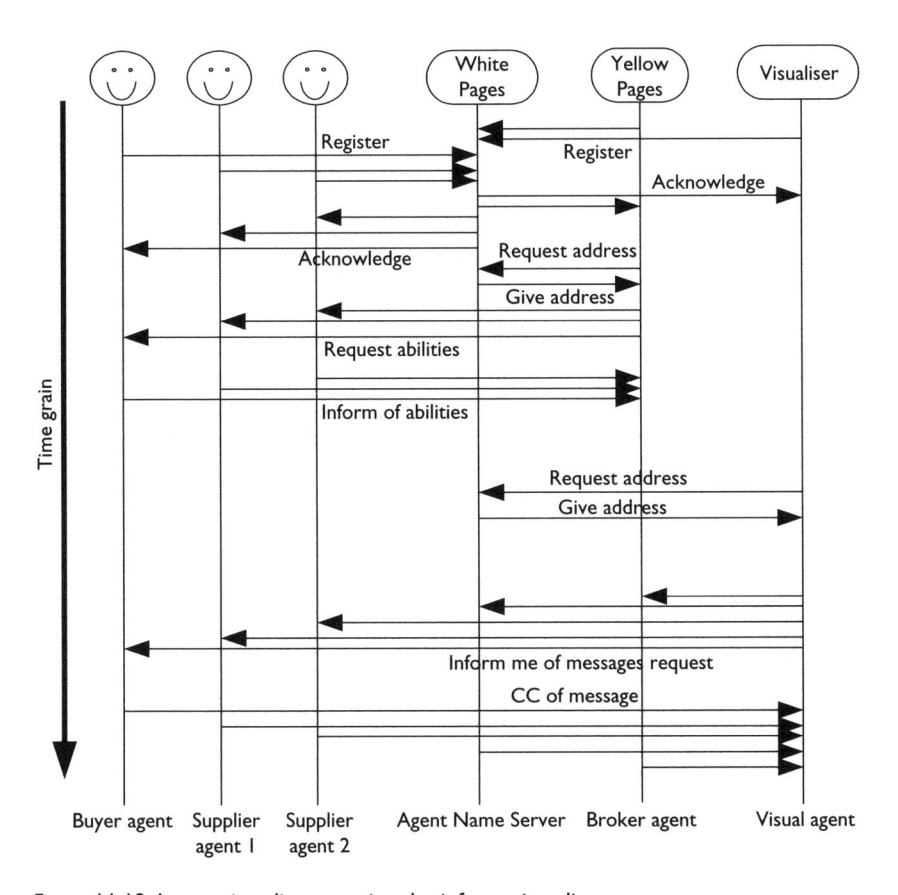

Figure 11.13 Interaction diagram using the information discovery.

announcement might be triggered either as the result of a specific request to translate an agent identifier (i.e. name) into a network protocol-specific address or the announcement might be in response to an outstanding, general subscription to the nameservice. On the other hand, ZEUS facilitator agent subscribes to the nameserver so that it receives a notification each time a new agent joins the community. Such a notification will trigger the facilitator to inform the newly created agent about its existence.

11.5.2 The secondary federated system design

The previous section has discussed how MAS architecture was conceptualised. However, building material SC agents require addressing the same basic research questions that are central to all construction procurement decision-making research, that is, how does the system manage and process information/knowledge associated within material procurement. The approach adopted in this work, is the provision of a central repository that allows common access to all participants (see Figure 11.14). The figure shows a federated system framework dedicated to an orderly and accessible repository of known facts and related data that is used as a basis for making better material procurement decisions. In a project context, it stores the different kinds of information/knowledge dealing with managerial and process activities of a construction material procurement within a given project, and it contains generic data information on construction product domain objects such as projects, materials, building elements and suppliers, etc.

The federated database system supports the sharing and exchange of information among collaborative human actors, using a centralised repository, which can also be extended to serve the multi-agent framework (see Figure 11.15). However, as shown in Figure 11.14, the secondary federated

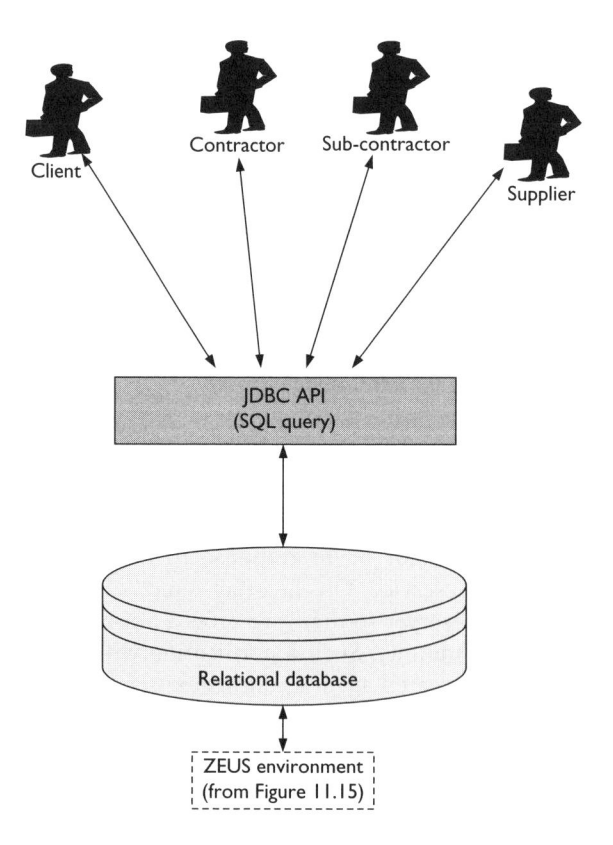

Figure 11.14 Human actor federated system framework.

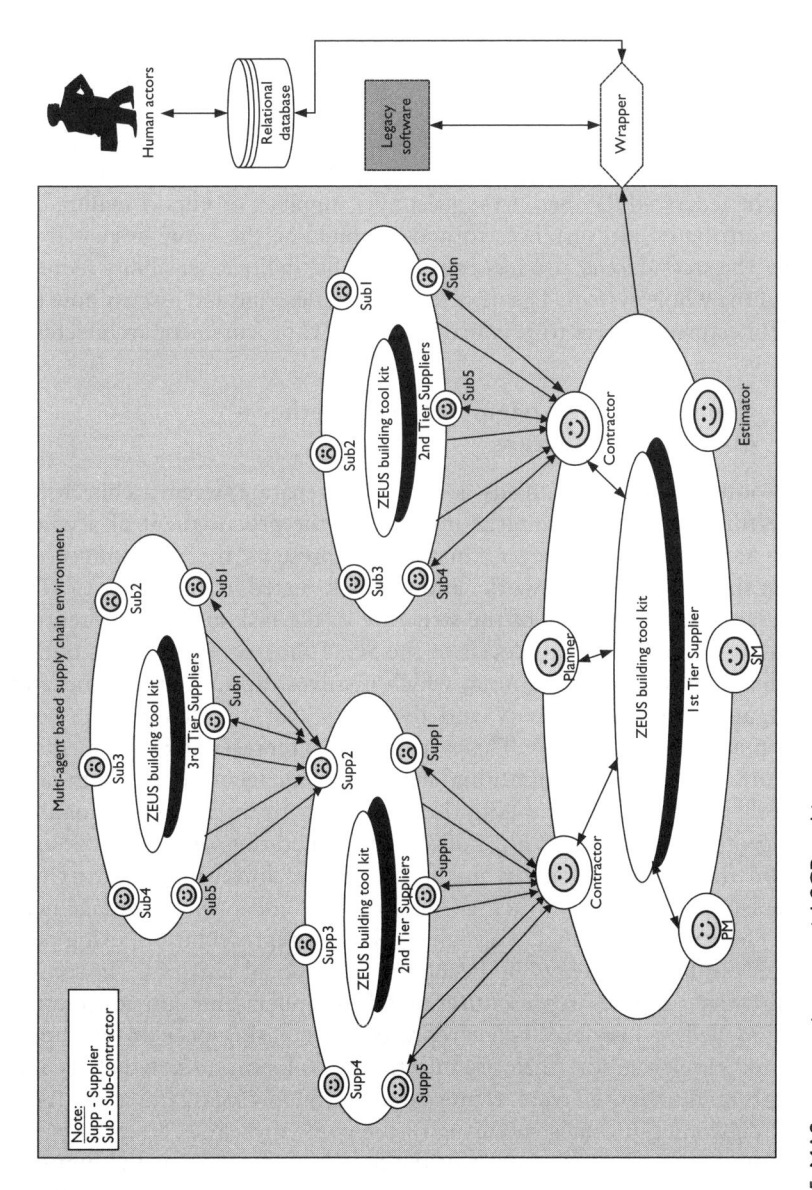

Figure 11.15 MAS-construction material SCP architecture.

system comprises of the relational database management system, query management facility, Java database connectivity Application Program Interface (API) and human actors. Applying the federated database approach for distributed decision-making has brought several new perspectives to the Multi-Agent System-Supply Chain (MAS-SC) architecture. One perspective that is addressed is the issue of selection of supplier from a list of suppliers in the database; using constraints to reduce the number of supplier that a contractor can negotiate with. The second perspective is the issue of scalability, where there is a stalemate in the negotiation, the MAS-SC environment can call on the federated database to request new suppliers or import material or element attributes and instances to avoid rebuilding the entire system from scratch. The second issue also incorporates the flexibility/extensibility requirements of the whole system. The next section will describe and discuss how the two architectures connect to produce the MAS-SC procurement architecture.

11.5.3 Connecting the system designs together – MAS-SC architecture

The previous sections have already described the separate system architectures, each performing part of the tasks involved in the procurement of a whole project or sub-parts. However, this section presents the interconnection between these system frameworks and this is depicted in Figure 11.15. The figure gives a pictorial view of the structure of the task agents representing various SC participants at all levels of the SC. The structuring solves one of the endemic problems of this area, which involves capturing the intra- and inter-organisational complexity and dynamics that characterise the construction material SC process. This is achieved by representing each member of the intra- and inter-organisational network as an agent in both high- and low-level structuring and imbuing these agents with the necessary collaboration protocols (Udeaja *et al.*, 2003). The structuring of the MAS-SC is based on the description given in the primary architecture. The first tier suppliers represent the contractor's organisation and project management teams. These were modelled at a high-level agent representation, using peer agent representation to represent the inter-organisational activities. The second and third tier suppliers representing mostly suppliers and sub-contractors were modelled at low-level agent representation, using peer group representation. However, for intra-organisational collaboration, which is not apparent in the figure, were represented as hierarchical and co-worker agency, depending on the structuring of the company's SC.

The aim here is that the output of the federated database is the input of the MAS framework and when events occur within the MAS environment, it effects a corresponding change in the federated database (e.g. when quantity of products are traded, a corresponding amount is deducted in the database). However, when an agent needs a product it first examines its local resource database, and if the necessary fact is not present it consults its federated database. The database is connected to individual agents by implementing the methods interface (wrapper). The description of the wrapper is not discussed in this section, but can be found in agent literature by Nwana *et al.* (1999).

11.6 Case study example

In the previous section the analysis of the system proposed for this work was described in detail and how the system was designed was also presented. This section presents a case study example to demonstrate the application of MAS in construction material SC. This case study was designed so as to emphasise and demonstrate that the framework is a feasible proposition, by evaluating the material SC procurement issues discussed in previous sections, which involves: tasks decomposition, planning and scheduling of tasks to achieve certain goals, material and product selection, resolving conflict (Negotiating contracts) to achieve the goals and utilising the output for other planning activities (make decisions). Hence it was deemed necessary to design the case study so that it goes through all the stages identified in the model description. It demonstrates how the computational agents that represent participants of a SC team can collaborate among themselves. In this scenario the demonstration is undertaken following the analogy made during the design stage, where the activities are modelled and designed at high level involving task decomposition and low level involving negotiation and delivery. The issue of ontology development is addressed first.

11.6.1 Knowledge modelling

At the initial stage, the various concepts (facts) are mapped into well-understood meanings using corresponding attributes and possible constraints for the problem domain as shown in Figure 11.11. The concept description leads to the ontology development and in this work it is implemented using ZEUS Agent Building Tool kit. Figure 11.16 depicts a screenshot of the development of an ontology development using ZEUS tool kit. The development is well documented by Collis *et al.* (1998) and as such the ZEUS ontology realisation will not be discussed in this section. However, once the ontology is defined, the agent realisation can follow immediately. For this case study, the connection between the ontology with database was excluded, the reason being that the case study process/product information was completely defined in the ontology development, in other words, there was no need for an external linkage with the database. Figure 11.16 is a screenshot of the ontology, its concepts refer to the physical instances rather than abstract ones and they inherit from the Entity fact, a child of the root ZEUSFACT that provides all the children with

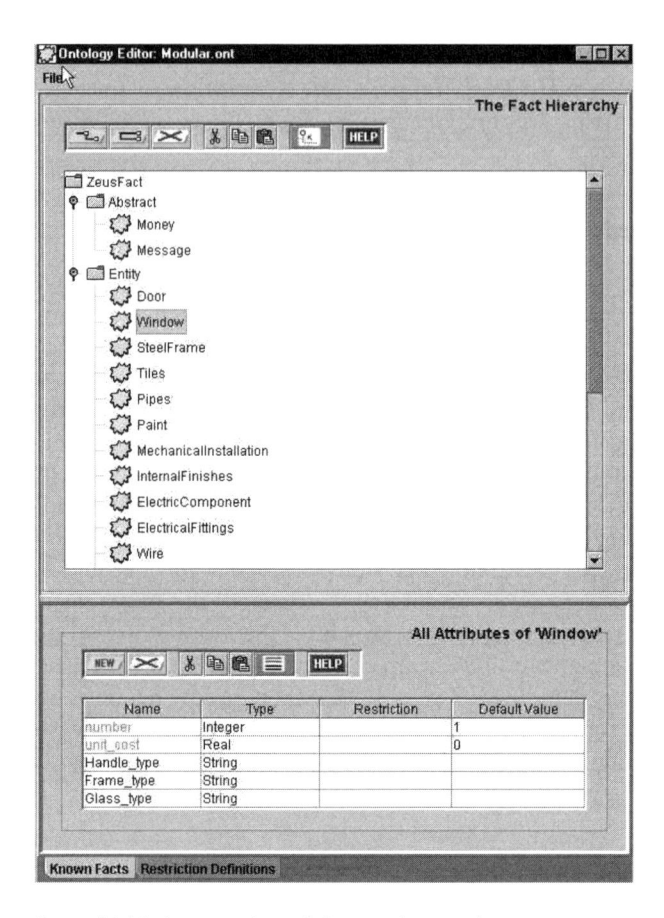

Figure 11.16 A screenshot of the ontology realisation.

cardinality attributes. Once the ontology is defined, the realisation of the design can proceed.

11.6.2 Task delegation

In the task decomposition state, ZEUS tool kit is used to construct the agents, in which a resulting visualisation tool generates a Gantt chart showing the decomposition of the task, the allocation of its constituent sub-parts to different agents in the community, and when each agent is scheduled to perform its part of the task. Other task attributes that can be shown are their costs, the priority assignment to them by the agents and the resources they require. The task decomposition/distribution graph created also shows the current status of each task, therefore either waiting, running, completed

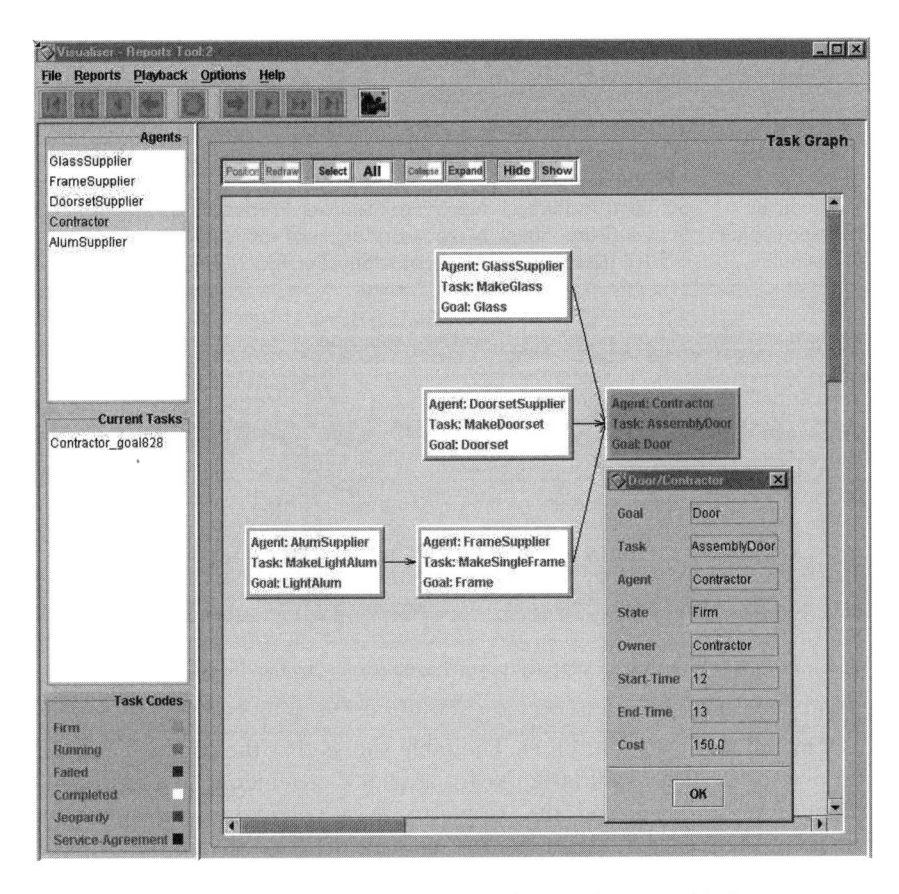

Figure 11.17 A screenshot of task decomposition for a goal to assembly doors.

or failed. Thus, from the graph a user is immediately able to determine the overall status of a goal, and if the goal fails where exactly it did so. However, the preconditions of each task specify the resources required for the task, and its post conditions specify the expected effects of performing the task. All tasks have an associated duration and cost, which could be functions of the resources used or produced by the task (see Figure 11.17). In the figure, the link between tasks in the graph specifies both precedence relation between the predecessor and successor tasks, and the resources produced by the predecessor-task which are used by the successor-task.

11.6.3 Negotiation/delivery stage

During contracting, the high-level decomposition of agents is further decomposed into low-level roles played by the same agents with different

Table 11.1 Low-level role description/decomposition table

Agent name	Role played
Contractor	SC Head (Initiator, Negotiator, Consumer)
Framesupplier	SC Participant (Respondent, Negotiator, Supplier, Producer Initiator, Consumer)
Glasssupplier	SC Tail (Respondent, Negotiator, Supplier, Producer)
Doorsetsupplier	SC Tail (Respondent, Negotiator, Supplier, Producer)
Alumsupplier	SC Tail (Respondent, Negotiator, Supplier, Producer)

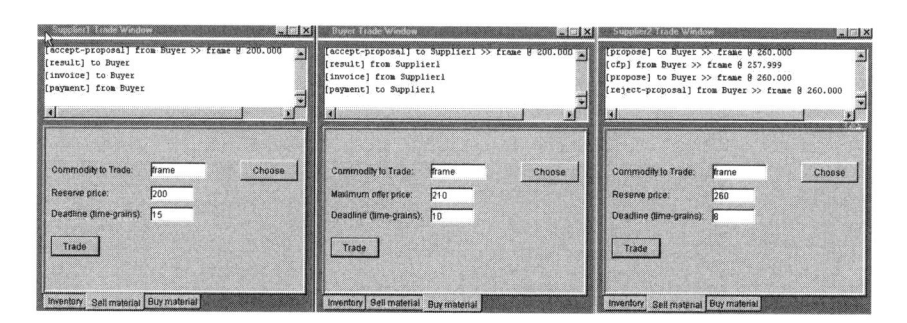

Figure 11.18 A screenshot of supplier/buyer negotiation activities (users interface).

responsibilities (see Table 11.1). The table shows that the contractor-agent can play roles such as Initiator, Negotiator or Consumer, which represents activities of a buyer and the estimator in a contractor's organisation. However, the scenario considers three task agents from the above environment, one representing the contractor's buyer and the others representing two suppliers bidding to provide the contractor with a component that make up the AlumacoDoor (example Doorset). The process is triggered off by the contractor's agent (Initiator) requesting quotes from potential suppliers (Respondent). The arrival of the announcement causes the Respondent to move into the initialisation state. If the Respondent decides to respond, it moves into the negotiation state. In which case the process of negotiation is now passed on to the Negotiator role-play. The negotiation strategy adopted in this example follows the contract-net co-ordination protocol, where the contractor's agent (Negotiator) ask potential suppliers to submit quotes for the materials in question. On receiving the quote, the contractor's agent evaluates the quotes and awards the contract to the winning agent. In this simulation, the selection of the winning agent for the supply was based on cost only, although other factors such as quality, previous performance, environmental issues, etc. could have been used in the selection criteria.

Consequently, the result of the transaction shows that the buyer agent closed the deal with supplier 1 after a series of negotiations as indicated in Figure 11.18. In the figure, three trade windows represent the interface for the various agents involved in the scenario. Each has an interface area showing what the negotiation position is and three buttons below labelled inventory that opens up a list of materials available and the amount available for transaction. The other buttons are the selling and buying buttons depending on who the buyer is and who is the seller. The interface that the buttons open has two buttons, one for choosing the material from the ontology database and the other the trade button that initiates the transaction. To further understand how the negotiation was carried out, Figures 11.19 and 11.20 show

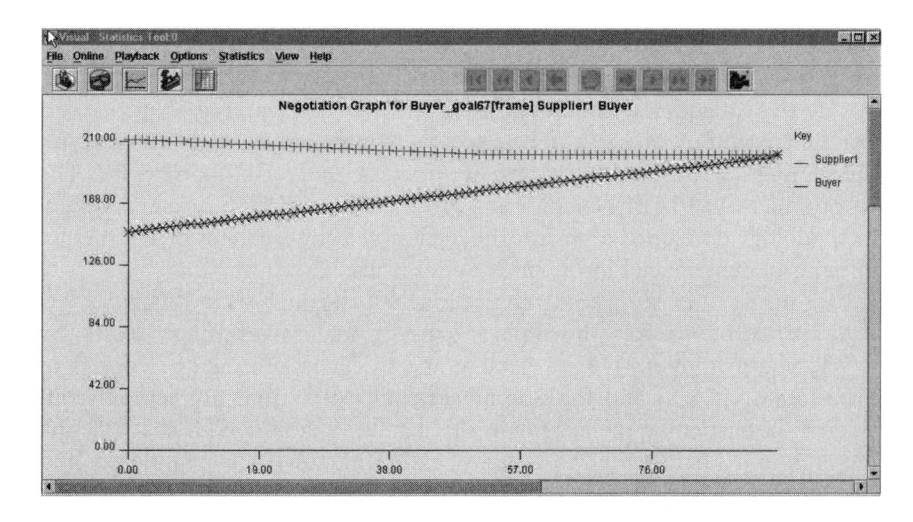

Figure 11.19 A screenshot of frame supplier 1/buyer accepted negotiation graph.

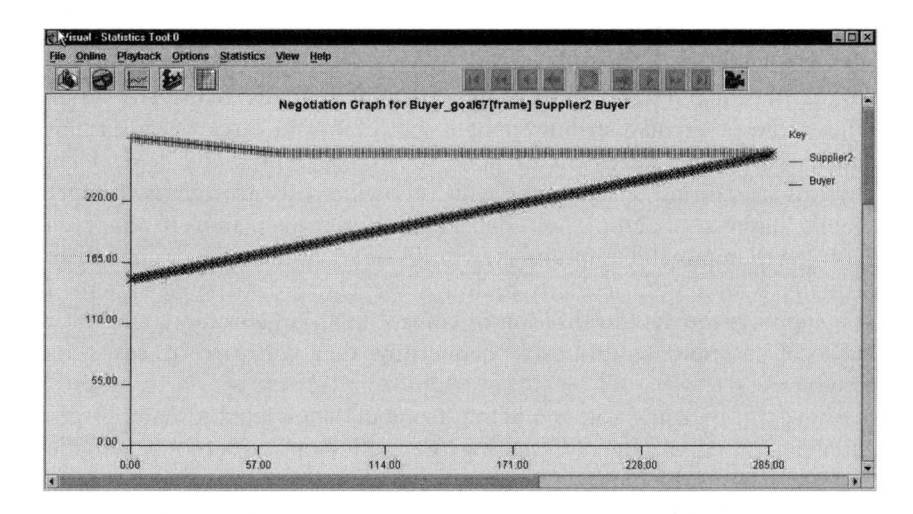

Figure 11.20 A screenshot of frame supplier 2/buyer failed negotiation graph.

graphs of supplier1/buyer and supplier2/buyer interaction respectively. The graphs show that the negotiation process between suppliers and buyer exceed the maximum price the buyer was willing to offer, in that case the proposal by supplier2 was rejected. On the other hand, the supplier1 and buyer interaction showed that the two participants reached an agreement at a price reasonable for the buyer, that is, price below the buyer's maximum value, as such the two exchanged goods and money (low-level delivery activity).

11.7 Discussion and conclusions

The efficient and effective material procurement is crucial to the success of project development, yet existing methodologies are neither comprehensive nor do they adequately reflect the complexity of the material SC processes. For a material procurement to be effective, it must reflect the processes with the entities (actors and activities) influencing the various processes. The work undertaken in this chapter is a process-co-ordination-oriented system methodology developed to mirror the inter- and intra-organisational structure difficulties experienced in previous approaches.

In contrast to existing research approaches, the analysis in this work has demonstrated some key shortfalls in current analyses of SC processes, by developing a comparative perspective and by incorporating key entities of intra-organisational and inter-organisational theory that are seldom considered. In particular, the argument that has been developed is that existing research in SC had tended to conceive rather narrowly the complexity and dynamics of SC processes and then to compound the error by focusing upon a somewhat distinctive empirical setting. Therefore, the new approach is more suitable for understanding the complex and dynamic structure of the SC procurement processes in a project development.

On the technical level, the system has several implications for adapting to the complexity and dynamic nature of large materials procurement that require the collaborative effort of more than a few individuals. The system reflects the inherent distribution of responsibility in large organisations, and makes the procurement of materials transparent to the logical and physical structuring of its components, allowing software agents to represent the interests of autonomous departments or business units to adapt and evolve with minimal disruption. New tasks or services can be defined incrementally without the need to re-design the entire distributed system. It also supports the decentralisation of control in an organisation, empowering local autonomous groups to define how they will perform tasks and activities.

Above all, by using role modelling to model construction material procurement processes, by representing the collaboration between activities and participants, by building the system on an independent platform architecture, and linking it to other planning and database packages, this research has developed a prototype for construction material SC decision-support that functions in a manner close to practice than previously implemented systems. Furthermore, the use of a ZEUS building tool kit to develop agent participants in a material SC environment guarantees that agents in the environment will use the best collaboration/co-ordination mechanisms available with minimal programming effort on the developers side. This is achieved by: developing ontologies that semantically unify agent communication and packaging the collaboration theories into agent development tool kit that ensures that agents are able to reuse standardised co-ordination and reasoning mechanisms, relieving developers from the tedious process of implementing agents from scratch.

The system is capable of solving material SC problems. They automate the provisioning of business processes as well as support business process enactment. First, since SC processes are conducted between autonomous agents through a network, the time for negotiation preparation, such as time for negotiation document presentation, waiting for negotiation meeting or gathering negotiators are reduced. Second, the negotiations between agents are more efficient than that of human negotiators since the negotiations between agents are continuous, straightforward and highly concentrated. Third, the unhealthy tricks of negotiation, such as obscure nature (e.g. a party may seek to win a negotiation by taking advantage of the other party's weakness in negotiation tactics), are avoided since agents have to reach a result within a specific time limit. Thus, the new approach addresses and solves the negotiation problem in a quick and effective manner in which all participants are satisfied (win–win scenario). Furthermore, the system is also capable of automatic generation of material plans and schedules for a particular material, thus demonstrating a degree of intelligence that provides significant added value over and above simple JIT techniques and materials requirement planning. The resulting plan and schedule can be re-planned and re-scheduled in order to accommodate changes, for example design changes can be handled by re-planning and merging the new procurement plan with the old.

This work has shown how agent technology offers a new approach to handling the complex nature of a material SC environment. The system architecture takes a radically new approach and allows us to view the domain area from a different perspective and the framework provides an extensible platform upon which a number of other components can be built. Future research is needed to extend the components on this platform to provide a richer set of functionality. These components will enable the system to support a more diverse market environment. Thus, the ability to offer a complete end-to-end process would enable a much wider range of services to be offered (i.e. complete automation). A key component of extending the capabilities would be to define a complete material ontology or link the system to an external database that describes how a high-level task is decomposed into a number of sub-task components. By providing this functionality and the ability to plan, schedule and automate the negotiation process, the system has the ability to offer more complex services in which the user can express their requirement as higher-level goals.

References

Austin, S., Baldwin, A., Hammond, J., Murray, M., Root, D., Thomson, D. and Thorpe, A. (2001) 'Design chains: a handbook for integrated collaborative design', Thomas Telford, 1st edn, UK.

Barbuceanu, M. (1997) 'The agent building shell: programming cooperative enterprise agents', http://www.ie.utoronto.ca/EIL/ABS-page/ABS-overview

Blackwell, R. D. and Blackwell, K. (1999) 'The century of the consumer: converting supply chains into demand chains', http://www.manufacturing.net/scl/scmr/fall99.htm.1999

Bond, A. H. and Gasser, L. (eds) (1988) 'An analysis of problems and research in DAI', Readings in Distributed Artificial Intelligence, Morgan Kaufmann Publishers, Los Altos, CA, pp. 3–35.

Brazier, F. M. T., Dunin-Kęplicz, B., Jennings, N. R. and Treur, J. (1996a) 'Modelling distributed industrial processes in a multi-agent framework', in G. O'Hare and S. Kirn (eds) *Towards the Intelligent Organisation – The Coordination Perspective*, Springer-Verlag, Berlin.

Brazier, F. M. T., Van Eck, P. and Treur, J. (1996b) 'Modelling cooperative behaviour for resource access in a compositional multi-agent framework', in J. L. Fiadeiro and P-Y. Schobbens (eds) in *Proceedings of Second Workshop of the ModelAge Project*, Portugal, pp. 27–40.

Brazier, F., Treur, J., Wijngaards, N. and Willems, M. (1996c) 'Temporal semantics of complex reasoning task', http://ksi.cpsc.ucalgary.ca/kaw/kaw96/wijngaards/wijngaards.html

Bresnen, M. (1996) 'An organisational perspective on changing buyer-supplier relations: a critical review of the evidence', *Organisation Articles*, 3(1): 121–46.

Briscoe, G., Danity, A. R. J. and Millett, S. (2001) 'Construction supply chain partnerships: skills, knowledge and attitudinal requirements', *European Journal of Purchasing and Supply Management*, 7: 243–55.

Case, M. P. and Lu, S. C-Y. (1995) 'Discourse model for collaborative design', *Computer-Aided Design*, 28(5): 333–45.

Christopher, M. (1998) *Logistics and supply chain management: strategies for reducing cost and improving services*, Prentice Hall, 2nd edn, UK.

Collis, J. C., Ndumu, D. T., Nwana, H. S. and Lee, L. C. (1998) 'The Zeus agent building toolkit', *BT Technology Journal*, 16(3): 60–8.

Cox, A. and Thompson, I. (1998) *Contracting for business success*, Thomas Telford, London.

Cutkosky, M. R., Englemore, R. S., Fikes, R. E., Genesereth, M. R., Gruber, T. R., Mark, W. S. and Weber, J. C. (1993) 'PACT: an experiment in integrating concurrent engineering systems', *IEEE Computer*, 26(1): 28–37.

Davis, R. and Smith, R. G. (1988) 'Negotiation as a metaphor for distributed problem solving', in A. H. Bond and L. Gasser (eds) *Readings in Distributed artificial intelligence*, Morgan Kaufmann Publishers, Inc., San Mateo, CA, pp. 333–56.

DETR (1998) CIRM Business Plans Construction Process, http://www.construction.detr.gov.uk/cirm/busplans/proc/summary.htm

Dijkstra, J. and Timmermans, H. (2002) 'Towards a multi-agent model for visualising simulated user behaviour to support the assessment of design performance', *Automation in Construction*, 11: 135–45.

Egan Report (1998) 'Rethinking construction: the report of the construction task force', Department of the Environment, Transport and the Regions, London, UK, pp. 1–39.

Elliman, T. and Orange, G. (2000) 'Electronic commerce to support construction design and supply chain management: a research note', *International Journal of Physical Distribution and Logistics Management*, 30(3/4): 345–60.

Farmer, D. H. and Ploos Von Amstel, R. (1991) 'Effective pipeline management', Gower, Aldershot, UK.

Finn, T. and Labrou, Y. (1997) 'KQML as an agent communication language', in J. M. Bradshaw (ed.) *Software Agents*, Cambridge, MIT Press, MA, pp. 291–316.

Fox, M. S., Chionglo, J. F. and Barbuceanu, M. (1991) 'The integrated supply chain management system', Department of Industrial Engineering, University of Toronto, Toronto, Ont., Canada, pp. 1–12.

Genesereth, M. R. and Ketchpel, S. P. (1994) 'Software agents', *Communication of the ACM*, 37(7): 48–53.

Green, S., Hurst, L., Nangle, B., Cunningham, P., Somers, R. and Evans, R. (1997) 'Agent Software – a review', Intelligent Agents Group (IAG), University of Dublin, Ireland, pp. 1–33.

Gruninger, M. and Fox, M. S. (1994) 'An activity ontology for enterprise modelling', *AAAI – 94*.

Halfawy, M. R., Yehia, N. A. B., Bazaraa, A. S., Dessouki, A., Hadipriono, F. C. and Duane, J. W. (1998) 'Multi-agent architecture for foundation design environments', *Computing in Civil Engineering*, pp. 200–6.

Handfield, R. B. and Nichols, E. L. Jr (1999) *Supply Chain Management*, Prentice Hall, Upper Saddle River, NJ.

Harland, C. M. (1996) 'Supply chain management: relationships, chains and networks', *British Journal of Management*, 7: S63–S80.

Holsapple, C. W. and Joshi, K. D. (2002) 'A collaborative approach to ontology design', *Communications of the ACM*, 45(2): 42–8.

Jennings, N. R., Faratin, P., Norman, T. J., O'Brien, P., Wiegand, M. E., Voudouris, C., Alty, J. L., Miah, T. and Mamdani, E. H. (1996) 'ADEPT: managing business processes using intelligent agents', in *Proceedings of the BCS Expert Systems'96 Conference*, pp. 5–23.

Kendall, E. A. (1998) 'Agent roles and role models: new abstractions for intelligent agent system analysis and design', *Intelligent Agents for Information and Process Management*, AIP'98.

Latham, M. (1994) 'Constructing the team – final report on joint review of procurement and contractual arrangement in the UK Construction industry', HMSO.

Meraghni, L., Ross, A. D. and Jaggar, D. M. (1996) 'The development of a "Requisitions and Purchase Orders Management System" Prototype', *CIB W89 Beijing International Conference*.

Ndumu, D. T. and Tah, J. M. H. (1998) 'Agents in computer-assisted collaborative design', in I. Smith (ed.) *Journal of AI in Structural Engineering*, Lecture Notes in Artificial Intelligence, No.1454, Springer-Verlag, pp. 249–70.

New, S. J. (1997) 'The scope of supply chain management research', *Supply Chain Management*, 2(1): 15–22.

Nwana, H. S. and Ndumu, D. T. (1996) 'An introduction to agents technology', *BT Technology Journal*, 14(4): 55–67.

Nwana, H. S. and Ndumu, D. T. (1999) 'A perspective on software agent research', *The Knowledge Engineering Review*, 14(2): 1–18.

Nwana, H. S., Ndumu, D. T., Lee, L. C. and Collis, J. C. (1997) 'ZEUS: a collaborative agent toolkit', in *Proceedings of second UK Workshop on Foundations of Multi-Agent Systems (FoMAS'97)*, pp. 45–52.

Nwana, H. S., Ndumu, D. T., Lee, L. C. and Collis, J. (1999) 'ZEUS: a toolkit for building distributed multi-agent systems', in *Applied Artifical Intelligence Journal*, 13(1): 129–86.

O'Brien, P. D. and Wiegand, M. E. (1996) 'Co-ordination in software agent systems', *BT Technology Journal*, 14(4): 133–40.

Odell, J. (2002) 'Key issues for agent technology', *JOOP*, http://www.Joopmag.com

Owen, M., Lee, L., Sewell, G., Steward, S. and Thomas, D. (1999) 'Multi-agent trading environment', *BT Technology Journal*, 17(3): 33–43.

Petrie, C., Goldmann, S. and Raquet, A. (1999) 'Agent-based project management', *Published in Lecture Notes in AI – 1600*, Springer-Verlag, Berlin.

Ren, Z., Anumba, C. J. and Ugwu, O. O. (2002) 'Negotiation in a multi-agent system for construction claims negotiation', *Applied Artificial Intelligence*, 16(5): 359–94.

Smith, R. G. and Davis, R. (1988) 'Negotiation as a metaphor for distributed problem solving', in A. H. Bond and L. Gasser (eds) *Reading Distributed Artificial Intelligence*, Morgan Kaufmann, Los Altos, CA, pp. 333–56.

Steven, G. C. (1991) 'Integrating the supply chain', *International Journal of Physical Distribution and logistics Management*, 19(8): 3–8.

Tavakoli, A. and Kakalia, A. (1993) 'MMS: a material management system', *Construction Management and Economics*, 11: 143–9.

Turner, E. H. and Turner, R. M. (2000) 'Selecting task decompositions for constrained heuristic search', *Workshop Notes for the 2000 AAAI Workshop on Constraints and Planning*, Austin, TX.

Udeaja, C. E. and Tah, J. H. M. (2001) 'Construction material supply chain management: towards an agent-based technology', in *The Seventeenth Annual ARCOM Conference*, University of Salford, UK.

Udeaja, C. E., Tah, J. H. M. and Ndumu, D. T. (2003) 'Construction material supply chain integration using multi-agent system technology: a conceptual framework', *International Journal of IT in Architecture, Engineering and Construction*, 1: 67–84.

Ugwu, O. O., Anumba, C. J., Newnham, L. and Thorpe, A. (1999) 'Agent-based decision support for collaborative design and project management', *The International Journal of Construction Information Technology*, 7(2): 1–16. (Special Issue: Information technology for Effective Project management and Integration).

Ugwu, O. O., Anumba, C. J. and Thorpe, A. (2001) 'Ontology development for agent-based collaborative design', *Engineering, Construction and Architecture Management*, 8(3): 211–24.

Ugwu, O. O., Kumaraswamy, M. M., Ng, T. S. and Lee, P. K. K. (2003) 'Agent-based collaborative working in construction: understanding and modelling design knowledge, construction management practice and activities for process automation', *Hong Kong Institution of Engineers (HKIE) Transactions, in the tenth Anniversary Edition*, 10 (4): 81–7. (Special Issue on Emerging Technology in the 21st Century).

URL1: The Zeus Technical Manual, http://www.labs.bt.com/projects/agents/index.htm

Uschold, M. and Gruninger, M. (1996) 'Ontologies: principles, methods and applications', *The Knowledge Engineering Review*, 11(2): 93–113.

Vrijhoef, R. and Koskela, L. (2000) 'The four roles of supply chain management in construction', *European Journal of Purchasing and Supply Management*, 6(3/4): 169–78.

Wooldridge, M. and Jennings, N. R. (1995) 'Intelligent agents: theory and practice', *The Knowledge Engineering Review*, 10(2): 115–52.

Chapter 12

Conclusions

O. O. Ugwu, C. J. Anumba and Z. Ren

This book has presented a wide range of applications of agent-based systems in construction. It reflects the current trend and focus of agent-based research in the industry. This chapter summarises the key benefits and issues associated with agent-based systems in construction.

12.1 Key benefits and issues

There are a number of benefits associated with using agent-based systems in industrial applications. In a wider context, these benefits are outlined below (Jennings and Wittig, 1991):

- *Modularity* – Multi-agent systems (MAS) offers advantages over conventional software since a complex system can be broken into sub-systems to reduce complexity and improve clarity;
- *Speed* – the resulting sub-systems can frequently operate in parallel;
- *Reliability* – there is greater potential for a system to continue even if parts of it fail;
- *Knowledge acquisition* – it is easier to find experts in a narrow domain than one expert that has general problem-solving knowledge of a wider domain;
- *Reusability* – a small independent system could be part of many co-operating software entities.

In addition to the above, the additional benefits of agent applications in construction are as follows (Anumba and Newnham, 1998):

- *Collaborative and concurrent working* – MAS offer major scope for facilitating collaborative and concurrent engineering in construction. It directly addresses the integration of multi-disciplinary perspectives and provides a framework for resolving design conflicts between members of a construction project team (see Chapter 5).

- *Reduction in information overload* – The use of intelligent agents can also help to reduce information overload (see Chapter 9), and to facilitate interoperability between the many diverse and heterogeneous (legacy) IT systems in the construction industry.
- *Integration of construction expertise* – The distributed problem solving approach allows individual areas of expertise to be encoded into particular agents, thus modelling the real-world problem of collaborative and concurrent design development in an intuitive, modular and hence expandable manner. In such an agent-based system as compared with a centralised knowledge-based (k-b) system, decisions can be taken locally according to local knowledge allowing greater flexibility as this changes. Moreover, having agents communicate with each other across the Internet brings greater increases in the speed of convergence to a satisfactory design, compared with the traditional inter-disciplinary interactions.

Some of the key aspects of agent-based systems presented in this book are summarised below:

- Agent-based systems paradigm investigate a broad range of issues that relate to the distribution and co-ordination of knowledge, plans and actions that are required by multiple software entities (agents) to solve problems in a given domain. In a classical MAS environment, the agents often act collectively as a society and they collaborate to achieve their own individual goals as well as the common goal of the community to which they belong. This collaboration demands effective negotiation if the agents are to resolve conflicting decisions.
- MAS techniques can facilitate effective collaborative working between the parties in a construction project team. This is essential in today's increasingly competitive environment in which timely delivery of data and information gives a construction firm unique advantages in decision-making.
- Collaborative agents that interact using available agent technologies have several potential applications and advantages. In the context of the Architecture, Engineering and Construction (AEC) sector such agents can:
    - Support information filtering and reductions in information overload on users, so that a user receives only the information that is essential for decision-making.
    - Support effective negotiation between different members of a project team. Such negotiations are often complex and project participants at the downstream end of the construction process need to be involved at appropriate decision points. This will reduce the incidence of

change orders and associated downtime costs during the course of a project.
- Support the integration of various expert and (k-b) system technologies, thereby facilitating the development of more robust decision-support systems (through efficient knowledge exchange). Such systems have the potential to enhance training within the academia and industry.

12.2 Implications for industry

The application of agent-based systems has important implications for the construction industry. This requires a major rethink of the traditional relationships, as well as existing fragmented and adversarial approaches to delivering constructed facilities. Construction organisations and indeed any sector that holds on to traditional modes of working cannot survive in the new business environment engineered by e-business/e-commerce and the new business models that they generate. This requires more collaborative working practices to which MAS can make significant contributions.

Jennings and Wittig (1991) identified some characteristics of domain problems for which MAS could be considered suitable. These characteristics are outlined below:

- there is a substantial number of pre-existing software;
- industrial systems are complex and require many diverse types of activities to be performed; and
- the operator (project team member) is an integral member of the problem-solving community.

There are clearly many areas within the construction industry that meet the above criteria and are therefore suitable for MAS deployment. In particular, intelligent agents implemented in the form of MAS provide a natural metaphor for collaborative and concurrent working in construction. The following sub-sections summarise the main application areas some of which have been addressed in different chapters of this book.

12.2.1 Collaborative design

Agent-based systems have considerable potential to facilitate collaborative design as demonstrated in Chapter 5. Specific ways in which this can be done include:

- Information filtering and retrieval of design data.
- Information customisation to meet the needs of various users in distributed decision-making environments including just-in-time (JIT) information delivery.

- Automation of basic design tasks including co-ordination and negotiation to achieve an optimum feasible design. This relieves users from basic mundane tasks so that they can focus their energy on more complex design activities.

12.2.2 E-business and supply chain management

Another application area for intelligent agents is in e-business, particularly supply chain management. In this context the construction process may be viewed as an integrated business for which delivering the right product to the right place, at both the right time and price, is the desired goal (CACM, 1994, 1999; Grosof, 1997; Petrie, 1998, 1999; Petrie *et al.*, 1995). E-commerce already promises to revolutionise supply chain management by providing networked customers, suppliers and manufacturers with instant access to the respective data they require for rapid and efficient decision-making. From the manufacturer's perspective, rapid and accurate delivery of data/information on construction products will shorten the demand/supply planning cycle and improve productivity (see Chapters 7 and 11).

The main attraction here is the currency and reliability associated with such data and the ability of the data/information retrieval mechanisms to cut across functional, geographical and enterprise boundaries. The Ovum Report in 1997 (Knapik and Johnson, 1998) predicted that the market share of agent-based products will grow to £4.6 billion by 2006. Hence there is tremendous opportunity for the construction sector to deploy intelligent agents for collaborative supply/demand management between the construction/engineering design team and construction product manufacturers over the Internet. Such collaboration ranges from product specification to design, development, manufacture and delivery. The ongoing research projects in the USA and Europe dedicated to distributed trading and project management in construction demonstrate the opportunities of e-commerce within the construction sector (URL1; URL2). Agent-based systems have a major role to play in this regard.

12.2.3 Knowledge management

Knowledge management at the organisational level is now an area of growing interest. This often involves identifying the knowledge associated with a given business process, documenting/storing the knowledge and/or its sources and then making the documented/stored knowledge available to the workforce. Electronic communication is sometimes the most cost-effective means of distributing such knowledge and organisations in the financial and business sectors are increasingly employing various techniques to manage and distribute knowledge among the workforce and improve overall efficiency/productivity. There is considerable scope for MAS in facilitating knowledge management in this industrial context (Malone, 1994; Gilman et al., 1997; Knapik and Johnson, 1998). This may involve user profiling, context-specific knowledge delivery, more efficient knowledge retrieval and more effective and ambient knowledge capture.

12.2.4 Information search/retrieval

There is growing application of intelligent agents in order to manage the problem of information overload on the WWW (see Chapter 9). The specific tasks undertaken by agents include:

- integrated searching over different search engines on the WWW;
- relevance ranking of documents which can be adapted for user feedback;
- document analysis;
- automatic document profile and summary;
- adaptive user profiling;
- query expansion and search contexts.

The construction industry could benefit from deploying agents in such a wider context. The application of agents in the above context cuts across the entire spectrum of project life cycles – briefing, design, construction, maintenance and disposal.

12.2.5 Project co-ordination and management

The distributed nature of functions and activities associated with project co-ordination management makes it a suitable candidate for the application of MAS. Some chapters in this book have reported the applications of MAS to contract bidding, project design and co-ordination (including supply chain management) using the contract net protocol (Ndumu and Tah, 1998). However, further research work is required to investigate the issues associated with complex co-ordination and negotiation necessary in such wide industrial applications.

12.3 Barriers to the application of agent-based systems in construction

The barriers to the application of agents and MAS in the construction industry can be discussed under the following headings: security, safety, user acceptance, level of investment in IT infrastructure, and the development of MAS architecture and framework for particular construction problems.

12.3.1 Security

Based on assessments of the current state of computing technology, there are a number of security concerns in the automation of construction processes. One of these relates to *wayward agents* that could install viruses, compromise the host or pilfer through databases. The second security consideration relates to organisational information safety since an unauthorised agent can modify or delete data from a system leading to business failure. Mobile agents (i.e. those that execute on remote machines) represent the biggest security risk and consequently most organisations may be reluctant to deploy *'mobile agents'* over their networks without adequate security guarantees (Chess *et al.*, 1994; Grosof, 1997; Knapik and Johnson, 1998; CACM, 1994, 1999). Although non-mobile agents present far less of a risk as they do not execute on remote machines, they still need access to data, information and knowledge from remote machines. This represents a security risk that needs to be managed on the local machine.

12.3.2 Safety

There are potential legal implications in abdicating responsibility for critical design decisions to autonomous agents that have independent execution autonomy. Although, the company that designs or builds the building is clearly responsible for structural safety, people do not like to hand over too much responsibility to machines that may be less than 100% reliable, especially if the results are too complex to be adequately checked by a human. Machines are still dumb in the sense that they do not have the robust intelligence that humans have and may thus make quite catastrophic decisions in the same unnoticed way a human might make a minor design flaw. Systems such as those presented in this book only provide intelligent advice and guidance to designers and other practitioners in the same way as consultants do. Therefore, from practical considerations, *reactive agents* with limited execution autonomy may be more acceptable for applications in the construction industry. This will achieve the twin objectives of giving practitioners ultimate control over critical decisions, whilst facilitating distributed collaboration.

12.3.3 User acceptance

The construction industry is still generally conservative in its uptake of new technologies. The perceived failure of AI to deliver its previous promise in the 1980s has also contributed to users' reluctance in accepting new AI technologies. This is principally due to the overselling of AI systems (in the early years of (k-b) and expert systems), with regard to their capabilities to mimic human intelligence in solving complex industrial problems. This experience at industry level has resulted in most AI research projects being confined to academic research labs. The same hype now characterises the promises of vendors that market themselves as specialists in 'Intelligent Agent Systems'. Agent researchers/developers need to take this on board and ensure that the specific concerns and needs of industry are addressed in developing MAS. An evolutionary approach to technology uptake could be beneficial.

There is also another important social dimension to the user acceptance of MAS at the organisational (job function) level. This relates to the potential threat to job security that is often associated with the introduction of a new technology for process innovation. This could have negative impacts with respect to adopting and diffusing the technology to enhance productivity improvement. A solution would involve appropriate management intervention that is underpinned by appropriate human resource management policies.

12.3.4 Possible architectures and agent frameworks

In the light of the above requirements for MAS within the construction industry, it is necessary to consider possible architectures. There are three basic types of architecture for agent-based systems:

- *Deliberative* – in which an agent contains an explicit symbolic world model, which develops plans and makes decisions in the way proposed by symbolic AI;
- *Reactive* – in which an agent is capable of reacting to events without complex reasoning;
- *Hybrid architectures* – these integrate deliberative and reactive agents to develop more robust MAS.

The distributed nature of construction processes is inherently complex and difficult for an explicit symbolic representation. For example, reactive architectures implemented in the form of expert systems, are generally rigid, limited and inflexible for application in solving construction problems. Hybrid architectures, on the other hand, offer great potential to integrate existing (legacy) IT systems and develop MAS for decision-making in construction. System developers always need to choose/design an appropriate agent architecture which is suitable to model the practical engineering problem at hand.

12.3.5 Agent collaboration mechanism

The collaboration mechanism is another vital issue for agent-based applications. Irrespective of the application, agents need to collaborate and compete to achieve their own and/or group objectives. Negotiation is at the core of agent collaboration and competition. Thus, the development and implementation of an appropriate agent negotiation mechanism is always the top concern for the development of MAS. However, the development of an agent negotiation mechanism (negotiation protocol and strategies) is indeed a very difficult task, considering the complex and dynamic nature of construction problems. As a result, many MAS systems in construction adopt simple and similar negotiation mechanisms which may not be the best for the particular applications. This greatly limits the potential benefits which MAS could bring to the industry.

The discussion in Chapter 4 with regard to various negotiation theories provides a theoretical background for the solution of such problems. It also summarises a number of important issues to be considered, which could be regarded as a general methodology for the design of MAS negotiation mechanism:

- analysing the particular industrial problem;
- identifying the key negotiation elements (objective, strategy and reasoning approach);
- addressing human and agent roles in negotiation;
- selecting an appropriate theoretical approach and tailoring it for the particular application;
- developing a negotiation protocol;
- designing negotiation strategies for each party;
- considering the implementation of the negotiation mechanism; and
- addressing the evaluation criteria.

Besides these factors, learning is often an important aspect to consider, either for negotiation or other collaboration purposes. Many mature learning models exist and could be usefully deployed in agent-based systems for construction.

12.3.6 Low level of IT investment and economic considerations

The difficulty in quantifying the immediate return on IT investments (economic, productivity and efficiency gains) often makes it difficult to embrace new IT solutions including agent-based systems. In particular, the headlines and statistics of failed software projects have contributed to the construction industry being very risk-averse with respect to IT investments. On the other hand, the dynamic nature of the IT sector generally demands continuous investment to acquire the necessary IT infrastructure, and improve the skills of the workforce. The construction industry needs to address this problem both in the medium- and long-term.

Future research and development into agent-based systems for construction needs to address the following research questions/issues amongst others:

- What is the most suitable agent platform, framework and architecture to adopt in solving functionally distributed problems? This includes platform standardisation issues.
- How to replicate the various complex patterns and levels of interaction, co-ordination, communication and messaging in distributed construction organisations, while maintaining appropriate level(s) of security in an open networked system and also ensuring that the resulting '*multiple agents*' do not degenerate into a state of chaos?
- How to develop a domain ontology to enhance communication, information and knowledge sharing between collaborating agents and also maintain the syntax and semantics of any communicated messages. In addition, the ontology must be re-usable across similar problem domains, to avoid duplication of the ontology creation effort. This requires compliance with the Open Knowledge Base Connectivity (OKBC) protocols.
- Another research issue is agent–legacy software integration across construction enterprises. This is vital for ensuring inter-operability with existing systems. Also, standardisation including XML-driven solutions, remain an important focus of action for successful agent-based enterprise integration.

Examples of specific areas for further investigation include; e-business (and the new business models required for agent applications), education (teaching and learning), ontologies and knowledge management, agent-system (software) integration and user requirements capture. Other specific research areas include; standardisation (e.g. using eXtensible Markup Language (XML) solutions), materials management, workflow modelling and management, information and communication management in design and project management (i.e. automated and instant messaging and communication), agent-based structural health monitoring and others.

References

Anumba, C. J. and Newnham, L. (1998) 'Towards the use of distributed artificial intelligence in collaborative building design', in E. T. Miresco (ed.) *Proceedings of the first International Conference on New Information Technologies for Decision-making in Civil Engineering*, Sheraton Hotel, Montreal, Canada, 11–13 October 1998, pp. 413–24.

Chess, D., Harrison, C. and Kershenbaum, A. (1994) 'Mobile agents: are they a good idea?', *IBM Research Report RC, Volume 19887*, 16 March 1995, T. J. Watson Research Center, Yorktown Heights, New York.

Communications of the ACM 34(12), Special Issue on Computer Supported Cooperative Working, 1991.

Communications of the ACM 37 (7), Special Issue on Intelligent Agents, 1994.

Communications of the ACM 42 (3), Special Issue on Multi-Agent Systems on the Net and Agents in E-Commerce, 1999.

Gilman, C., Aparicio, B., Barry, J., Durnaik, T. and Ramnath, R. (1997) 'Integration of design and manufacturing in a virtual enterprise using enterprise rules, intelligent agents, STEP, and Workflow', in B. Gopalakrishnan, S. Murugesan, O. Struger and G. Zeichen (eds) *Architectures, Networks, and Intelligent Systems for Manufacturing Integration* in the *Proceedings of SPIE Volume 3203*, pp. 160–71.

Grosof, B. N. (1997) 'Building commercial agents: an IBM research perspective', *IBM Research Report RC 20835*.

Jennings, N. R. and Wittig, T. (1991) 'ARCHON: theory and practice', in N. M. Avouris and L. Gasser (eds) *Distributed Artificial Intelligence Theory and Parxis*, Kluwer Academic Publications, London, pp. 178–95.

Jennings, N. R. and Wooldrige, M. (1995) 'Applying agent technology', *Applied Artificial Intelligence*, 9: 357–69.

Knapik, M. and Johnson, J. (1998) *Developing Intelligent Agents for Distributed Systems – Exploring Architectures, Technologies, and Applications*, McGraw-Hill, New York.

Malone, T. W. (1994) 'The interdisciplinary study of coordination', *ACM Computing Surveys*, 26(1): 87–119.

Ndumu, D. T. and Tah, J. H. M. (1998) 'Agents in computer assisted collaborative design; AI in Structural Engineering', in I. Smith (ed.) *Lecture Notes in Artificial Intelligence (1454)*, pp. 249–70.

Petrie, C. (1998) 'The Redux Server, in readings', in M. N. Huns and M. P. Singh (eds) *Agents*, Morgan Kaufmann Publishers Inc., San Francisco, CA, pp. 56–65.

Petrie, C. (1999) 'Agent-based project management', *Lecture Notes in AI – 1500*, Springer-Verlag, Berlin.

Petrie, C. J., Webster, T. A. and Cutosky, M. R. (1995) 'Using pareto optimality to coordinate distributed agents', in *AIEDAM Special Issue on Conflict Management*, (9): 269–81.

URL1: http://cic.vtt.fi/links/europroj.html

URL2: DESSYS Project http://www.ds.arch.tue.nl/Research/Agents/DessysIntro.stm

Index

Note: Page numbers in italics indicate illustrations.

ACC *see* Agent Communication Channel
ACE *see* Agent Collaboration Environment
ACL *see* agent communication language
activity diagrams 124, *126–7*
ADEPT system 12, 41, 281
ADL *see* Agent Description Language
Adler, M. R. 48, 52
ADLIB project *see* Agent-based Collaborative Design of Light Industrial Buildings project
Agent-based Collaborative Design of Light Industrial Buildings (ADLIB) project 14, 104, 109, 157–8, 191; agent registration and identification 137, 139; agent-system integration 143; code fragments for CCA 144; domain agents 117–19, 121, 122–3, 124; flow chart *150*; knowledge fragment 137, *139, 140, 141–2*; knowledge modelling 107, 134–6; methodology 113–16; negotiation protocols 137, 139, *141–2*; negotiation strategies 139, 142; objectives 109; ontology 114–15, 135, *136*, 137, *138*; outputs 150–6; package level resource definition 145–9; specifications 116–17, 120; validation 156–7
agent-based systems 36–7, 310
agent behaviours: multiple-agent 27–8; representation 26–7; single-agent 27
Agentcities 188, 206
Agent Collaboration Environment (ACE) 73–4
Agent Communication Channel (ACC) 108
agent communication language (ACL) 228, 230
agent community 253–4
agent co-ordination 42–3
Agent Description Language (ADL) 255
agentification 11, 189
AgentLink 188, 206
Agent Management System (AMS) 108
Agent Name Server (ANS) 150, 153, 217, 225–6, 294
agent negotiation *see* negotiation
agent platform 107–8
agent research and development 3; future studies 318
agents 10–12, 187–8; applications 12–14; definition and categorisation 11, 15–16, 187–8; social models 13–14; taxonomy 37–9
agent technology 188, 240; commercial applications 205–6, 305; phases *188–9*
Akkermans, H. 56
Albert, A. 249
altruistic agents 41
Amor, R. 235
AMS *see* Agent Management System
ANS *see* Agent Name Server
Anson, R. G. 162, 163
Anumba, C. J. 2, 14, 72, 106, 113, 143, 158, 163, 189, 310
APRON project 3, 187, 188–9, 204–5; accessing structured data 198–9; deployment 193–4; formative evaluation 200–1; functional components 191–3; functionalities 194–8; related work 190–1; for specification and procurement of light bulbs 193–4, 200–1; summative evaluation 201–4

ARCHON 45
Arciszewski, T. 6, 11, 16, 17, 21, 23, 34
arithmometers 8
ARPA Knowledge Sharing Effort 254
artificial intelligence (AI)-based
  negotiation 59–60
ASPEN *see* A Standards Processing
  Engine
A Standards Processing Engine
  (ASPEN) 256, 267; external
  functions 257–8; planning system
  258–60, 261–5; representation of
  operators 260–1
auctions 13; types 213–14
Austin, S. 272, 273
AutoCAD 187, 199, 203
Autonomy's AgentWare 238
Aylett, R. 188

Bacharach, S. B. 90, 91, 93, 164
Badrah, K. 249
Baeza-Yates, R. A. 186
Balasubramanian, S. 191
Barbuceanu, M. 279
Bartos, O. J. 88
Bayardo, R. 191
Bayesian learning mechanism 97, 168,
  170–3, 183
Beck-Texte 214
Beer, M. 48
behaviour negotiation theory 60, 95,
  99; dual responsiveness model 97–8;
  joint decision-making model 98;
  learning model 96–7; psychological
  model 60–1, 96
Benjamins, V. R. 205
Berman, D. 257
Berners-Lee, T. 186, 187
Billsus, D. 238
Bitzan, S. 220
blackboard architecture 254
Blackwell, K. 273
Blackwell, R. D. 273
Blake, M. B. 187
BOCA *see* Building Officials and Code
  Administrators
Bond, A. H. 285
Boone, G. 238
bots *see* agents
Böttcher, S. 200
Bradshaw, J. 10, 11
Brams, S. J. 90

Brazier, F. M. T. 20, 284, 285
Bresnen, M. 272
Briscoe, G. 276
British Telecommunications 281
Brown, B. R. 89
Brown, D. C. 63, 64, 65, 66, 68
Brümmer, A. 235
Brustoloni, J. C. 32, 37, 187
Bui, H. H. 171
Building Information Warehouse
  (BIW) Technologies Plc (UK) 187
Building Officials and Code
  Administrators (BOCA) 249, 250
Bussmann, S. 49, 60

Cammarata, S. 43, 45, 48, 54
Campos, J. R. 36
Carbonell, J. M. 67
Carnevale, P. J. 88
Casasola, E. 15
Case, M. P. 281
CCIN *see* Collaborative Construction
  Information Network
Chavez, A. 13, 57
Chen, Y. 47
Chess, D. 315
Chiou, J. D. 73
Christopher, M. 273
CIRC 103
Clarke, D. W. 6
class diagrams 124, 126–8, *129–33*;
  public key infrastructure *230*
Clay, G. A. 87
CNP *see* Contract Net Protocol
Cockburn, D. 46
Coen, M. H. 32, 47
Cohen, R. 89
Cohen, W. W. 238
Cohen's Ripper system 238
collaborative browsing 238–40
Collaborative Construction Information
  Network (CCIN) 241–6
collaborative design in construction
  103, 105; business processes
  110–12; knowledge representation
  109; negotiation process 128, 134,
  142–3, *144*; practices 105–6
Collis, J. C. 292, 293, 299
competitive negotiation 49, 62; *see also*
  co-operative negotiation
computers 8
Conry, S. E. 48, 55

construction claims negotiation: characteristics 164–5, 183; *see also* MASCOT model
construction industry: fragmentation problems 2
Construction Industry Gateway (UK) 235
construction information technology 3–4, 207; socio-technical dimensions 4
construction material supply chain 274, 276; characteristics 278–9, *286*
Construction Task Force (UK) 103
content-based information filtering systems 237, 239; *see also* collaborative browsing
contract-based negotiation 54–5
contracting 43
Contract Net Protocol (CNP) 43, 54–5, 61
CONVINCER model of conflict resolution 75, 76
co-operative negotiation 61, 88; *see also* competitive negotiation
Copernic (search engine) 236
Corkill, D. D. 12
Cottam, R. 24
Cox, A. 273
crawlers *see* agents
Crites, R. 63
critic agents 73
Cross, J. G. 65, 94, 164
Cross's negotiation model 94, 164
cryptography 216–17
Cutkosky, M. R. 281
cyclic negotiation model 60
Cycorp Co. 14
CYC project 14

DAI *see* distributed artificial intelligence
Daley, R. 69
Davidsson, P. 13
Davies, J. 240, 243
Davis, R. 48, 54, 61, 287
Decker, K. 47
De Jong, K. A. 26
deliberative agents 39–40, 316
Deloitte Research 186
Denk, H. 221, 222, 226
DESSYS project 74
DETR 274
dialogues 53

digital library service: market-based negotiation 56
Digital Signature Act (Germany) 211–12, 214, 215, 217, 222, 230
Dijkstra, J. 280
Directed Evolution 7–10
Directory Facilitator 108
Disciple (intelligent agent) 15
distributed artificial intelligence (DAI) 31, 44, 88, 103, 240
Distributed Problem Solving (DPS) 44–5, 78; applications 45–6; negotiation 61
Distributed Scheduling Protocol (DSP) 55, 61
distributed sensing agents 12
distributed systems 12
distributed vehicle monitoring 45
document editor 11
Dogpile (search engine) 236
domain agents 117–19, 121, 122–3, 124, 127
Doorenbos, R. 13
DPS *see* Distributed Problem Solving
DSP *see* Distributed Scheduling Protocol
Durfee, D. H. 42
Durfee, E. H. 31, 45, 46, 48, 54, 55
Dutch auctions 213

EA *see* Evolutionary Algorithms
e-business 186; agent-based systems 13, 187, 206–7, 313
EC *see* evolutionary computation
ECI *see* European Construction Institute
economic-theoretical based negotiation 93, 164; vs. game theory-based negotiation 95; models 93–5
eConstruct project 190
Egan, J. 272, 279
Egan Report 103
Ekenberg, L. 57
Elam, S. 249
Elliman, T. 274
email filtering 13, 37, 237
Engineering and Physical Sciences Research Council (EPSRC) 109
English auctions 213
EPSRC *see* Engineering and Physical Sciences Research Council
European Construction Institute (ECI) 165

European Parliament 215
Evolutionary Algorithms (EA) 25–6
evolutionary computation (EC) 2–3, 12; principles 25–6
external functions 251, 256–8, 260, 262, 267, 268; remote execution 263

Farmer, D. H. 273
FASTRAK 109, 111, 143, 150, 156
federation architecture 252–3
Fédération Internationale des Ingénieurs-Conseils (FIDIC) 165
Fenves, S. J. 73, 249, 259
Ferber, J. 163, 187
Ferguson, I. A. 40, 41
FIDIC *see Fédération Internationale des Ingénieurs-Conseils*
Fikes, R. E. 259, 260
Fingar, P. 188
Finin, T. 254, 255
Finn, T. 280
FIPA *see* Foundation on Intelligent Physical Agents
first price sealed bid auction 213–14
Fischer, K. 42
Fisher, R. 89
Foundation on Intelligent Physical Agents (FIPA) 108, 137, 138, 153
Foundation on Intelligent Physical Agents (FIPA) Architecture Board 187
Fowler, A. 89, 234
Fox, M. S. 48, 280, 291
Franklin, S. 11, 15, 17, 188
Froese, T. 195

game theory-based negotiation 57–9, 89–93, 164; vs. economic-theoretical based negotiation 95
Garrett, J. H. 249, 259
Gasser, L. 76, 252, 285
Gauch, S. 15
generator agents 73
Genesereth, M. R. 54, 250, 251, 252, 254, 291
GENIAL project *see* Global Engineering Network Intelligent Access Libraries project
Georgeff, M. P. 46
Gero, J. S. 20
Gilman, C. 314
Gilster, P. 235
Gini, M. 42, 187

Global Engineering Network Intelligent Access Libraries (GENIAL) project 74, 190, 199
Goldberg, D. 239
gopher 15
Gordon, D. 69
Graesser, A. 11, 15, 17, 188
Grecu, D. L. 63, 64, 65, 66, 68
Green, S. 280
Grefenstette, J. 69
Grosof, B. N. 313, 315
Grosz, B. 46
GroupLens (collaborative browsing system) 239
Gruninger, M. 290, 291
Guha, R. V. 249, 255
Gulliver, P. H. 60, 87, 88, 89, 91, 93, 96

Hafner, C. 257
Hakim, M. M. 249
Halfawy, M. R. 281
Hammer, J. 190
Hammer, W. C. 87
Handfield, R. B. 273
Harland, C. M. 273, 274
Harsanyi, J. C. 92, 164, 167, 169
Hayes-Roth, B. 32, 254
Heckel, J. 73
Hendler, J. 187
Hermans, B. 236, 240
Hewitt, C. E. 46
hierarchical directories 234, 235
Hill, W. 239
Holsapple, C. W. 291
Hu, J. 66
Hu, J. X. 162
Huhns, M. N. 46, 76
hybrid agents 40–1, 316

IA *see* intelligent agents
IBDE project *see* Integrated Building Design Environment project
industrial process control 45–6
information 102; filtering 237–8, 312; lifting 255; retrieval agents 4, 12–13; search techniques 237–40; sharing 102–3, 106; shifting 255
Infosphere 41
Institute of Concrete Structures and Materials 211
Integrated Building Design Environment (IBDE) project 73

integrated project management (IPM) 47
intelligent agent-based information search and retrieval 240; advantages 240–1; *see also* Collaborative Construction Information Network
intelligent agents (IA) 1, 2, 6–7, 14–16, 32–3, 78, 103; applications in construction industry 2–3, 28; architecture 35; attributes 16–17, 19, 33–4, 37–8; classification 19–23; development 9–10, 34–7; interaction range and depth 17–18; learning and knowledge 18–19
interacting agents 39, 40–1, 44
interaction diagrams *see* sequence diagrams
Internal Platform Message Transport (IPMT) 108
International Workshop on Agents in Design (1st: Sydney) 19–20
Internet 71; search and retrieval 234–5, 246, 314; usage figures 233; use in construction industry 190
Internet search engines 235–6
inverse auctions 213, 214
IPM *see* integrated project management
IPMT *see* Internal Platform Message Transport
ISIS-MTT Standard 215–16
Iversen, G. R. 170

Jain, D. 249
Janca, P. 240
Jelassi, M. T. 162, 163
Jennings, N. R. 11, 12, 15, 16, 17, 31, 33, 36, 39, 41, 44, 46, 47, 48, 49, 51, 52, 54, 68, 76, 77, 187, 188, 280, 281, 287, 310, 312
Johnson, J. 313, 314, 315
Johnson, P. 249
Johnson Space Center Software Technology Branch 256
Jordan, J. S. 97, 170
Joshi, K. D. 291

Kaelbling, L. P. 71
Kakalia, A. 277
KASBAH project 57
Kashyap, N. 188
k-b systems *see* knowledge-based systems

Kendall, E. A. 284
Ketchpel, S. P. 291
Kiliccote, H. 249, 255, 257, 259, 267
Kim, K. 75
Klein, M. 48, 65, 250
Knapik, M. 313, 314, 315
Knight, K. 259, 261
knowledge-based (k-b) systems 2, 6; evolution 7–10
knowledge management 313; and multi-agent systems 313–14
Knowledge Query and Manipulation Language (KQML) 254–5
Konstan, J. A. 237, 239
Koskela, L. 276
KPMG Consulting 186
KQML *see* Knowledge Query and Manipulation Language
Kraus, S. 46, 47, 54, 57, 59, 61, 89, 90, 163, 166
Kreifelt, T. 56
Kreps, D. M. 94, 97
Kumar, B. 249

Labrou, Y. 280
Laird, J. E. 259
Lakmazaheri, S. 249
Lander, S. E. 59, 60
Latham, M. 103, 272, 279
Latham Report 103
Law, K. 249
Lawler, E. J. 90, 91, 93, 164
Lee, L. C. 57
Lehmann, D. 61
Leitao, P. 16, 17
Lenat, D. B. 14, 249
Lesser, V. R. 12, 42, 46, 47, 55, 57, 59, 60, 62, 63, 87, 88
Levesque, H. J. 46
Lieberman, H. 187, 237
Lieberman's Letizia 237
Lin, F. 47
Lipp, P. 217
Lockley, S. R. 235
Logcher, R. D. 73
Lopez, L. A. 249
Loughborough University 109, 187, 190, 201
Lu, S. C.-Y. 281
Luce, R. D. 89
Luck, M. 188, 189, 205, 206
Lynch, C. 234

M4I *see* Movement for Innovation
McCarthy, J. 10
machine learning 67–70, 103–4; research issues 70–1
MACIV model 75
Mckersie, R. B. 89
Madhusudan, T. 186
Maes, P. 13, 32, 187, 237, 239
Maes' agents 237
Mahapatra, T. 188
Malone, T. W. 55, 314
Manning, G. 235
market-based negotiation 56–7
MAS *see* multi-agent systems
MASCOT model 3, 76, 77, 79, 162–3, 183–4; applications in water supply project 175–81; design 165–6; implementation 181–3; negotiation mechanism 183; negotiation protocol 167–8; negotiation strategies 168–70, 173–4
materials procurement 4, 276–8, 304
Matos, N. 57
Matrix Information and Directory Services 233
MCP *see* Monotonic Concession Protocol
Mead, D. 249
mechanical theory of negotiation 89–90, 99; models 57–9, 90–5
Meraghni, L. 277
MetaCrawler (search engine) 236
meta search engines 235, 236–7, 245
Meyer, J. J. 12
Microsoft Internet Explorer (browser) 241
Mishra, S. 188
mobile agents 240, 315
Monotonic Concession Protocol (MCP) 75, 137, 167–8
Montgomery, T. A. 48, 54
Morales-Morell, A. 15
Morgenstern, O. 91
Morley, I. E. 89
Moss, S. 13
Movement for Innovation (M4I) 103
Mullen, T. 56
Muller, H. J. 49, 50, 52, 60
Muller, J. P. 39
multi-agent-based air traffic control 43, 45
multi-agent-based collaborative design in construction 104–5, 157, 312–13; requirements 106; *see also* Agent-based Collaborative Design of Light Industrial Buildings project
multi-agent-based international crisis negotiation 47
multi-agent-based procurement for the construction material supply chain 282–3, 304–6; knowledge modelling 299–300; negotiation 287–9, 301–4; ontology 290–3; system design 293–8; task decomposition 284–5, 287, 300–1; task delivery and execution 289–90
multi-agent-based standards processing 4, 251, 265–7
multi-agent-based supply chain management 46–7
multi-agent-based virtual marketplace for construction bidding 4, 211, 212–13, 230, 232, 314; *see also* SiReAM
multi-agent planning 43
multi-agent systems (MAS) 1–2, 3, 5, 31–2, 44, 46, 48, 78–9; advantages 310; applications 46–8; and knowledge management 313–14; problems and challenges 76–8; research 280–2, 318
multi-agent systems (MAS) in construction industry 71–2, 272, 311–12; advantages 310–11; applications 72–6; architectures 316; barriers 314–16; future research 318; implications 312; negotiation 74, 316–17; negotiation algorithms 75–6
multi-agent systems (MAS) learning process 62–4; elements 65–7; methods 67–70; objectives 64–5; research issues 70–1
multi-agent systems (MAS) management reference model 107–8
multi-stage negotiation 55–6

Nagayuki, Y. 63
Nash, J. F. 90, 91, 92, 164
Ndekugri, I. 165
Ndumu, D. 1, 17, 44, 47, 51, 72, 76, 272, 279, 280, 281, 282, 293, 314
negotiation 3, 48–50, 87–9, 99; argument-based 61; categories 50–7; in collaborative design 128, 134, 142–3, *144*; in Distributed Problem

Solving (DPS) 61; domains 58–9; mechanisms 51–2, 163; modelling in multi-agent systems (MAS) design space 137, 139, *141–2*; models 54, 89–90; in multi-agent systems (MAS) 62, 74–6, 163–4, 316–17; objects 52; protocols 52–3; stages 88; strategies 53, 64–5, 139, 142
negotiation support systems (NSS) 162, 163
Negroponte, N. 13
Neilson, A. 249
Netscape (browser) 241
New, S. J. 273
Newell, A. 36, 38, 71, 261
Newell, S. C. 237
Newnham, L. 163, 310
Nichols, D. M. 239
Nichols, E. L. Jr 273
Nilsson, N. J. 259, 260
non-swarm agents 22
Norvig, P. 17, 44
NSS *see* negotiation support systems
Nwana, H. 1, 17, 33, 34, 37, 44, 47, 51, 72, 76, 187, 219, 240, 279, 280, 299

object-oriented (OO) systems 36, 78
Obonyo, E. O. 47
O'Brien, P. 190, 191, 281
Odell, J. 291, 292
Oliveira, E. D. 75
on-site logistics management 14
ontology 14–15, 220
OO systems *see* object-oriented systems
Orange, G. 274
organisational structuring 42–3
Osborne, M. J. 90, 92, 94, 97
Ovum Report 313
Owen, M. 290, 292

PACT *see* Palo Alto Collaborative Testbed
Padgham, L. 16
Palm personal data management device 9
Palo Alto Collaborative Testbed (PACT) 281
Partial Global Planning (PGP) 55
Parunak, H. V. D. 35, 36, 46, 191
Pazzam, M. J. 238
peer-to-peer negotiation 43
Pena-Mora, F. 57, 75, 165

personal software assistants 12, 15
persuader negotiation model 59–60
Petrie, C. 47, 191, 281, 313
PGP *see* Partial Global Planning
Pickavance, K. 165
PKI *see* public key infrastructure
plan-based negotiation 55–6
Pleiades project 41
Ploos Von Amstel, R. 273
Polat, F. 61
Powell-Smith, V. 162
Prasad, M. V. N. 62, 63
Preist, C. 49
ProcessLink 191
Pruitt, D. G. 88
psychology-based negotiation 60–1, 96
public key infrastructure (PKI) 4, 221–6; class diagram *230*
public tendering procedures 214, 230

Radeke, E. 74, 190
Radjou, N. 207
Raiffa, H. 89
Rapoport, A. 92
RAPPID *see* Responsible Agents for Product-Process Integrated Design
Rasdorf, W. J. 249
Rasmusen, E. 94
reactive agents 39, 315, 316
recommendation systems *see* collaborative browsing
Redux model 47, 282
Reeds, C. 53
Ren, Z. 51, 76, 182, 282
Responsible Agents for Product-Process Integrated Design (RAPPID) 191
Restivo, F. 16, 17
Rich, E. 259, 261
Ringo (collaborative browsing system) 239
Rojot, J. 89, 90
Rosenschein, J. S. 42, 47, 48, 54, 57, 58, 59, 62, 87, 89, 90, 137, 163, 167, 170
Rossnagel, A. 232
Roth, A. E. 57
Rubin, J. Z. 89
Rubinstein, A. 90, 92, 94, 97
Russel, S. J. 17, 44

Sahin, F. 171
Samtani, G. 186, 187

Sandholm, T. 42, 47, 55, 57, 63, 89, 90, 163
sapient agents 3, 23–4
Sathi, A. 48
Schatz, B. 233
Schneier, B. 216
Schnellenbach-Held, M. 47, 221, 222, 226, 249
Searle, J. R. 117, 139
self-interested agents 42, 48
self-motivated agents 59
Selfridge, Oliver G. 10
semantic networks 254, 260
Semantic Web 187
Sen, S. 34, 42, 48, 69, 72
sequence diagrams 124, *225*, 226, *229*
Shapiro, S. C. 254
Shardanand, U. 239
Shaw, M. 68
Shehory, O. 205
Shell, G. R. 163
Shen, W. 47, 49, 51, 54, 56
Shubik, M. 91, 92
Sidler, Gabriel 239
Sierra, C. 61
Singh, N. A. 254
SiReAM 230; architecture 218, *219*; development 217–18; ontology 220–1; public-key infrastructure (PKI) 221–6, 230–1; public tendering procedure 226–8, *229*
Skolicki, Z. 6, 11, 16, 17, 21, 23
Smith, D. C. 32
Smith, G. J. 13
Smith, R. G. 4, 43, 48, 49, 54, 55, 61, 287
SML *see* Standards Modeling Language
social resource discovery *see* collaborative browsing
software agents *see* agents
Sowa, F. 249
Spector, B. I. 88, 95
SPF *see* standards processing framework
SPF-CL *see* standards processing framework Communication Language
spiders *see* agents
Standards Modeling Language (SML) 255, 266
standards processing 249–50; models 249
standards processing framework (SPF) 251, 265, 267–8; architecture 251–4

standards processing framework (SPF) agents 251, 255–6, 265, 267
standards processing framework Communication Language (SPF-CL) 254–5, 256, 258, 267
Standards Usage Language (SUL) 255
state chart diagrams 124
Stephens, J. 190
Stephenson, D. 89, 162
Steven, G. C. 273
strong agents 16
structural systems 9
structured programming 35
Subramanian, D. 69
Suhl, L. 200
SUL *see* Standards Usage Language
Sun, L. 13, 241
supply chain management 46–7, 273–4, *275*, 313
swarm agents 9, 19, 20; characteristics 22–3
Sycara, K. P. 41, 48, 54, 59, 63, 171, 172, 182
Syskill & Webert (Web browsing assistant) 238

Tah, J. H. M. 47, 107, 272, 280, 281, 282, 293, 314
Tambe, M. 46
Tapestry (collaborative browsing system) 239
Tavakoli, A. 277
Taylor, G. 16
TEAM negotiation system 60
Tecuci, G. 15, 17
Terveen, L. 239
Thiels, L. 190
Thomason, R. H. 254
Thompson, I. 273
Timmermans, H. 280
Topping, B. H. V. 249
Touretzky, D. S. 254
Touring Machines hybrid 41
traditional auctions 213–14
Tsvetovatyy, M. B. 42
Turing, A. M. 14
Turner, E. H. 284, 285
Turner, R. M. 284, 285
Twidale, M. B. 239

Udeaja, C. E. 47, 107, 272, 290, 298
Ugwu, O. O. 14, 31, 33, 38, 46, 57, 72, 75, 103, 108, 112, 113, 114,

115, 120, 133, 143, 156, 158, 189, 190, 191, 205, 207, 282, 291
Uschold, M. 290
Use Cases 115, 119–20, 122, 204; activity diagrams *126–7*; sequence diagrams *125*
Usenet news filtering 237
user profiles 237, 238, 239, 244, 245

VBA *see* Visual Basic Application
Velásquez, J. 16
Vickrey auctions 214
Video Recommender (collaborative browsing system) 239
Vidogah, W. 165
virtual marketplaces for construction industry 211
Visitor Hosting system 41
Visual Basic Application (VBA) 199
VOB 213, 214
Von Martial, F. 56
Von Neumann, J. 91
Vrijhoef, R. 276

Walton, R. E. 89
Wang, C. 57, 75, 165
Warham, S. M. 89
weak agents 16
Weiss, G. 62, 63, 67, 69, 71
Wellman, M. P. 53, 56, 66
Werkman, K. J. 43, 59, 60
Wiederhold, G. 254
Wiegand, M. E. 281
Wilkenfeld, J. 47, 54, 57

Wilson, P. R. 250
Winston, P. H. 67, 68
Wittig, T. 310, 312
Woodrow, Taylor 190, 203
Woods, W. A. 254
Wooldridge, M. J. 11, 12, 15, 16, 17, 33, 37, 39, 187, 188, 280
World Wide Web (WWW) 186, 233, 246, 314; development 234
World Wide Web Virtual Library 235
Wurman, P. R. 53
WWW *see* World Wide Web

Yabuki, N. 249
Yahoo 235
Yellow Pages *see* hierarchical directories
Ygge, F. 56
Young, H. P. 88
Young, O. R. 89, 91, 93, 94, 95, 99, 166

Zack, J. G. 162
Zakon, R. H. 234
Zartman, I. W. 87, 93, 95, 98
Zeng, D. 41, 63, 171, 172, 182
ZEUS agent building tool kit 181, 217, 218–19, 281, 282, 293, 299, 300, 305
Zeuthen, F. 76, 93, 164, 169, 170
Zeuthen's economic welfare model 93–4, 164, 166–7, 169–70, 182
Zlotkin, G. 42, 47, 48, 54, 57, 58, 59, 62, 87, 89, 90, 163, 167, 170